A HISTORY OF THE WARFARE OF SCIENCE WITH THEOLOGY

IN CHRISTENDOM

BY

ANDREW DICKSON WHITE

LL. D. (YALE), L. H. D. (COLUMBIA), PH. DR. (JENA)

LATE PRESIDENT AND PROFESSOR OF HISTORY AT CORNELL UNIVERSITY.

IN TWO VOLUMES

VOL. I

NEW YORK
D. APPLETON AND COMPANY
1896

THE INTERNATIONAL SCIENTIFIC SERIES.

HISTORY

OF

THE CONFLICT

BETWEEN

RELIGION AND SCIENCE.

BY

JOHN WILLIAM DRAPER, M. D., LL. D.,

PROFESSOR IN THE UNIVERSITY OF NEW YORK; AUTHOR OF A "TREATISE ON HUMAN PHYSIOLOGY;" "HISTORY OF THE INTELLECTUAL DEVELOPMENT OF EUROPE;" "HISTORY OF THE AMERICAN CIVIL WAR;", AND OF MANY EXPERIMENTAL MEMOIRS ON CHEMICAL AND OTHER SCIENTIFIC SUBJECTS.

NEW YORK:
D. APPLETON AND COMPANY,
549 & 551 BROADWAY.
1875.

Frontispiece. Title pages for *A History of the Warfare of Science with Theology in Christendom* (1896) and *History of Conflict between Religion and Science* (1874).

SCIENCE AND CULTURE IN THE NINETEENTH CENTURY

Bernard Lightman, Editor

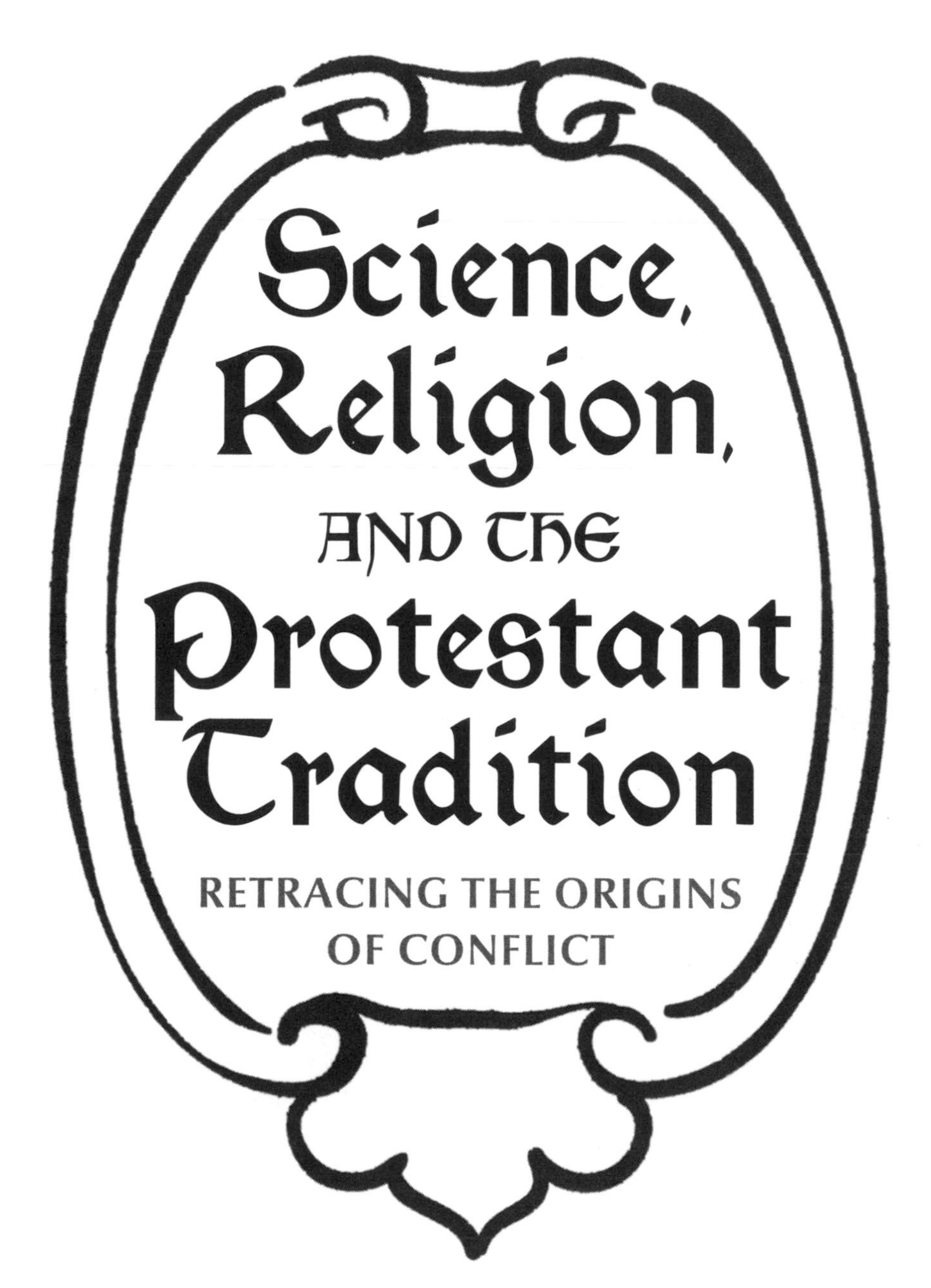

Science, Religion, and the Protestant Tradition

RETRACING THE ORIGINS OF CONFLICT

JAMES C. UNGUREANU

UNIVERSITY OF PITTSBURGH PRESS

Published by the University of Pittsburgh Press, Pittsburgh, Pa., 15260

Manufactured in the United States of America
Printed on acid-free paper
10 9 8 7 6 5 4 3 2 1

Cataloging-in-Publication data is available from the Library of Congress

ISBN 13: 978-0-8229-4581-9
ISBN 10: 0-8229-4581-9

Cover art: Joseph Ferdinand Keppler, *The universal church of the future - from the present religious outlook*, in *Puck* v. 12, no. 305, January 10, 1883 (Keppler & Schwarzmann, 1883), centerfold.
Cover design: Joel W. Coggins

CONTENTS

ACKNOWLEDGMENTS

SCORES of individuals and institutions have assisted me in completing this book, and here I can only record a special thanks to a few to whom I am especially indebted for their advice, guidance, and support. Throughout this process, I have amassed a tremendous debt to Peter Harrison. In 2012 I began writing him about doing postgraduate work in the history of science. The very next year I entered the doctoral program at the University of Queensland under his supervision. He was welcoming, accommodating, and generous. I am very grateful for his feedback on the countless drafts of chapters, paragraphs, and sentences I sent him. I would also like to thank Ian Hesketh, who went above and beyond his role as co-supervisor. With him, I shared the excitement of discovery, the frustration of writing, and the joy of discussing new research. I owe an enormous thanks to Bernie Lightman, both for his comments and for his patience in slogging through earlier versions of the manuscript. Bernie believed in this project from the beginning, and I will forever be grateful to him for including it in his series. Special thanks also go to Ted Davis for his constant support and encouragement, on this book and life in general. Thanks also to the many other scholars, from all over the world, who helped in some capacity along the way, including Simon Kennedy, James Lancaster, Leigh Penman, John Stenhouse, Ron Numbers, Steve Snobelen, Tim Larsen, Jon Roberts, Lynn Nyhart, Florence Hsia, John Brooke, David Mislin, Jon Topham, Dave Livingstone, Diarmid Finnegan, Brad Gregory, Larry Principe, James Stump, and Allison Coudert.

A word of thanks must also go to the dead. I spent many long hours and nights pouring through their writings and private letters, and to this I am especially grateful to the librarians in the Manuscript Division at the Library of Congress in Washington D.C., as well as the Library Division of Rare and Manuscript Collections at Cornell University. Of course, I also owe a special thank you to Abby Collier, Amy Sherman, Paul Schellinger, and the rest of the team at the University of Pittsburgh Press for getting this book into the hands of readers.

Since 2017 I have been in a prolonged transition. As a result, I have accumulated a great debt to the hospitality of friends and even strangers. I wanted to express my gratitude in particular to Tom Rudy and Jennifer Ondrejka, Daniel and Nicole Leatherwood, Patty Ladpli, Calvin and Ruth DeWitt, Fred and Virginia Moore, Jeremy and Lori Jensen, Greg Pechacek, and Chuck Moore. Most important, I owe so much to my wife. She has fed me, listened to me, encouraged me, and loved and supported me for better or worse in what must have seemed an interminable task. During the writing of this book we were blessed with the birth of our two children, and I am grateful for their enduring patience with Papa's work schedule. It has been a long, sometimes painful, but ultimately wonderful journey. We look forward as a family to the next chapter.

S. D. G.

SCIENCE, RELIGION, AND THE PROTESTANT TRADITION

INTRODUCTION

THE most tenacious, if not ubiquitous, view of the relationship between science and religion is that they are in perpetual "conflict" or "warfare," and in which a triumphant science ultimately displaces religion. In recent years, no group has propagated the image of warfare between science and religion more than the so-called "new atheists." These authors, which include Sam Harris, Richard Dawkins, Christopher Hitchens, Daniel C. Dennett, Victor Stenger, Jerry A. Coyne, Lawrence Krauss, and others, contend throughout their popular writings that religion has been the relentless foe of scientific progress. While most of their arguments are philosophical in character, they tacitly and explicitly appeal to the historical record. In various writings, for example, Harris often refers to the "clash of science and religion," describing the conflict as "inherent and (very nearly) zero-sum." He even postulates that if "reason" had emerged at the time of the Crusades, "we might have had modern democracy and the Internet by 1600." Provocatively, Harris proclaims that "science must destroy religion."[1] Similarly, the late journalist and social critic Christopher Hitchens referred to "the terror imposed by religion on science and scholarship throughout the early Christian centuries," and that "all attempts to reconcile faith with science and reason are consigned to failure and ridicule."[2] Physicist and former Catholic Victor Stenger also argued that "the totality of evidence indicates that, on the whole, over the millennia the Christian religion was more of a hindrance than a help to the development of science."[3] While there are many differences between each author, the new atheists all share the view that science and religion have been and still are implacable foes.

A veritable industry has emerged responding to the claims of the new atheists.[4] Crucially, these studies accuse the new atheists of misreading or even falsifying the historical record, pointing out that numerous historians, philosophers, and theologians have long disclaimed notions of an endemic conflict between science and religion. Indeed, while notions of conflict or warfare between science and religion remain surprisingly resilient—especially in popular historical writing, the media, and even science textbooks—decades of scholarship demonstrates that such essentialist tropes are wholly inadequate.

Surely, to write about how "science and religion" relate—or should relate—is a formidable challenge, and one that remains a highly contentious subject. But since the beginning of the twentieth century, scholars from various disciplines have demonstrated that no matter how visceral such conflicts appear to be, there is little if any historical basis for the "conflict thesis," the overarching view that science and religion are irrevocably at odds. The following section of this introduction, mainly historiographical, is intended to give a taste of this scholarly literature by capturing some of the vital changes in the historiography. It also illustrates the range of conclusions historians and other scholars have reached that explicitly undermine the conventional conflict historiography employed by the new atheists. Finally, and not without some irony, it will also indicate the need to completely reevaluate this revisionist historiography.

A SHIFTING HISTORIOGRAPHY

Starting in the 1920s, when the nascent discipline of the history of science was first emerging, a number of scholars were already arguing that the historical relationship between science and religion was far too complicated to categorize as one of "conflict." English mathematician and philosopher Alfred North Whitehead (1861–1947), for example, warned readers of the difficulty in approaching the subject. Although "*conflict* between religion and science is what naturally occurs to our minds when we think of its subject," he wrote, "the true facts of the case are very much more complex, and refuse to be summarised in these simple terms." The terms in question, "religion" and "science," according to Whitehead, "have always been in a state of continual development." While theology exhibits "gradual development," science "is even more changeable."[5]

Whitehead also observed that the very foundations of modern science were laid in the soil of medieval religious thought, which had insisted "on the rationality of God," and concomitantly on a rational and orderly creation. Science arose in Europe, according to Whitehead, because of the "faith in the possibility of science." Modern scientific theory, in short, was "an unconscious derivative from medieval theology."[6] Indeed, in the early decades of the twentieth century, a number of scholars were beginning to question one of the most salient features of conflict historiography: that medieval Christianity suppressed the growth and progress of science. For example, French physicist and philosopher Pierre Duhem (1861–1916) discovered massive medieval precedents to modern physics, particularly the work of medieval mathematician Jordanus de Nemore. As he continued his research, he discovered the scientific achievements of other fourteenth-century thinkers, such as Albert of Saxony, Jean Buridan, and Nicole Oresme. Duhem became convinced that, far from being a period of scientific stagnation, the medieval period actively laid the foundations of modern science and that consequently the entire concept of the "dark ages" had to be reassessed.[7]

Duhem's revolutionary discoveries had "rehabilitated" medieval science and was soon joined by a number of other prominent scholars, including Charles H. Haskins (1870–1937), Lynn Thorndike (1882–1965), Alexandre Koyré (1892–1964), and Marshall Clagett (1916–2005). This continues to be a fertile area of investigation for historians of science.[8] At roughly the same time, other scholars were beginning to reexamine another aspect of standard conflict historiography: that the scientific advances in the early seventeenth century stood independently from theological views of man, God, and the cosmos. American philosopher of religion Edwin Arthur Burtt (1892–1989), for instance, found the foundations of modern physical science in the philosophical or "metaphysical" assumptions of a number of seventeenth-century thinkers.[9] Central to Burtt's argument was that the seventeenth-century conception of the absolute "uniformity of nature," a basic premise of modern science, was grounded in the constancy and fidelity of God. For example, Nicholas Copernicus and Johannes Kepler were thoroughly convinced, for religious reasons, of the uniformity of motion. Galileo believed God was a "geometrician" and that he made the world through a "mathematical system." Similarly, René Descartes's geometrical conception of the physical universe saw God as extending and maintaining things in motion by his "general concourse." Ralph Cudworth was also confident that the new mechanical philosophy would reveal "incorporeal beings, especially one supreme spiritual Deity." Robert Boyle believed that some divine ends were readable to all, and that God's admirable "workmanship" was displayed throughout the universe. Finally, Isaac Newton's absolute space and time are God's "sensorium."[10] According to Burtt, "human nature demands metaphysics for its full intellectual satisfaction [since] no great mind can wholly avoid playing with ultimate questions."[11]

Rather than oppressive or obstructionist, scholars were beginning to view religious values and beliefs as important, if not essential, to the growth of modern science. But by the 1930s, scholars had shifted from the medieval world to Protestant forerunners of experimental science. Works by Dorothy Stimson (1890–1988), Michael B. Foster (1903–1959), and Robert K. Merton (1910–2003) all argued that English Protestantism in particular had shaped, nurtured, and encouraged science as a noble pursuit in the early modern period. The American sociologist Robert Merton, for instance, adopting the theoretical work of R. H. Tawney and Max Weber, famously argued that the "Puritan ethic" was an important element in increasing the "cultivation of science."[12]

While Merton wrote on how the Puritan *ethos* cultivated modern science, later studies by Paul H. Kocher (1907–1998), Richard S. Westfall (1924–1996), John Dillenberger (1918–2008), Reijer Hooykaas (1906–1994), and others defended the claim that Protestant *theology* in seventeenth- and eighteenth-century England inspired a new empirical and experimental approach to understanding nature. Paul Kocher's study on science and religion in Elizabethan England found that most theological writers during Queen Elizabeth's reign believed scientific studies were amicable to theological ones, and often proclaimed that "natural science is a gift of God to man." Natural philosophy had revealed to mankind the wonders of God's handiwork, glorifying God and promoting religious faith. Richard Westfall, examining the late seventeenth-century figures known as the "English virtuosi," many of whom were founders of the Royal Society of London, similarly concluded that not only were they enthusiastic promoters of natural philosophy, but deeply religious men who regarded natural phenomena as revelatory of God's glory and power.[13]

If the work of early twentieth-century scholars rejected the conflict thesis, by mid-century many more apologetically inclined scholars were arguing that Christianity, in fact, made science possible. These scholars went so far as to argue that, historically, science and Christianity were essentially harmonious. Dutch historian Reijer Hooykaas, for instance, exemplified this harmonist position. He argued that Reformed theology inspired a new empirical and experimental approach to natural philosophy and maintained that the rise of modern science "is more a consequence than a cause of certain religious views," particularly a "biblical world view." Indeed, Christianity, and especially its Protestant variety, taught that, "in total contradiction to pagan religion, nature is not a deity to be feared and worshipped, but a work of God to be admired, studied and managed." The recovery of the biblical worldview by Protestants led to the "'de-deification' of nature, a more modest estimation of human reason, and a higher respect for manual labour," and thence to the rise of modern science.[14]

Other scholars sought a more fruitful balance that avoided the triumphalist narratives of both "conflict" and "concord" in the history of science and religion. A young and relatively unknown scholar, Herbert Butterfield (1900–1979), published his essay *The Whig Interpretation of History* (1931), which objected to presentism in historical writing.[15] More precisely, what Butterfield disapproved of was the habit of taking the present as the standard of what is "good," the test of "progress," and then tracing the line of that progress toward the good, by way of, for instance, Luther and the Protestant Reformation, the French Revolution, or whatever other men and events appear to have "contributed" to the present good. In short, "Whigs" studied the "past with reference to the present," treating history as a series of stepping-stones to human progress—progress, that is, toward their own point of view. Although not a professional historian of science himself, Butterfield nevertheless called upon historians of any subject to forget the present and study the past "for its own sake." Butterfield's rejection of Whig history is directly relevant to the historiographical question, for as historian of religion Thomas McIntire recently put it, the historiography of science is "riddled with Whiggish history."[16]

In point of fact, Butterfield did indeed contribute his own substantive treatment to the history of science in his *The Origins of Modern Science* (1949).[17] While not immune to charges of "whiggism" of his own, Butterfield nevertheless carried over a new and important historiographical approach to studying the relationship between science and religion. If we avoid using present constructions and definitions of "science," he argued, we discover that scientific change and revolution occurred not by new facts or observations, "but by the transpositions that were taking place inside the minds of the scientists themselves." Examining the canonical figures of Copernicus, Kepler, Galileo, and others, he contended that their achievements were not the discovery of new data, but the placement of old or current data "in a new system of relations with one another by giving them a different framework, all of which virtually means putting on a different kind of thinking-cap for the moment."[18] To understand the development of modern science, then, one must not merely report discoveries and ideas, but trace the development and succession of philosophies and worldviews.

By the mid-1970s, another major historiographical shift occurred within the scholarship. It became clear that the notion of "conflict" between science and religion had been mostly confined to the nineteenth century, to the controversies that broke out in the fields of geology and biology, and specifically to religious reactions to the publication of Charles Darwin's *On the Origin of Species by Means of Natural Selection* (1859). But even here things were not so clear-cut. In an early statement on the topic, for instance, theologian Charles E. Raven (1885–1964) contended that the controversies amounted to little more than a "storm in a Victorian tea-cup." The conflict was a clash of personalities, between old and new worldviews, not between some reified "science and religion."[19] Historian of science Charles C. Gillispie (1918–2015) similarly argued that controversy arose not between science *and* religion but from religion *in* science—that is, from religious attitudes within the new science of geology.[20] Church historian Owen Chadwick (1916–2015) concurred, arguing that one of the most conspicuous features of Victorian England was, in fact, its religiosity. If there was religious doubt (and undoubtedly there was), it had existed before the *Origin of Species*. "Darwin was only a sign of a movement bigger than Darwin," he wrote, "bigger than biological science, bigger than intellectual enquiry." He insisted that scholars need to make the important distinction "between science when it was against religion and the scientists when they were against religion."[21]

By the end of the decade, scholars were making just such distinctions, investigating specific themes, concepts, events, and people of the nineteenth century. Some of the most important studies were produced by Walter F. Cannon (1925–1981), Robert M. Young (1935–), and Frank M. Turner (1944–2010). Cannon, for example, argued that "science and religion had developed a firm alliance in England." This unity was shattered, however, when theologians began attacking Darwin. It was not until the collapse of what Cannon called the "truth complex" in the second half of the nineteenth century that science and religion really first came into conflict.[22] Similarly, Young urged the importance of what he termed the "common intellectual context," which viewed science as integrated within the ideology of the Victorian social, political, and religious middle class. He argued that this shared context was largely defined by the enterprise of natural theology. Each new discovery of science, he contended, was to the early Victorians "a separate additional proof of the wisdom, power, and goodness of the Deity."[23] Turner transferred the apparent hostility between science and religion to a "shift of authority and prestige . . . from one part of the intellectual nation to another." He maintained that the conflict in the nineteenth century was the result of the rising power of a new professional scientific class vying for cultural hegemony. "The primary motivating force behind this shift in social and intellectual authority," he argued, "was activity within the scientific community that displayed most of the features associated with nascent professionalism." This "young guard of science," which consisted of figures such as Thomas H. Huxley, John Tyndall, Joseph D. Hooker, George Busk, Edward Frankland, Thomas A. Hirst, John Lubbock, William Spottiswoode, Herbert Spencer, and others, "had established themselves as a major segment of the elite of the Victorian scientific world." They advocated a new and exclusive epistemology that came to "discredit the wider cultural influences of organized religion." This exclusivity eventually came to serve as a weapon against the cultural influence of religion in general. While in previous generations science and religion were both compatible and even complementary, by the 1840s a "naturalistic bent of theories in geology, biology, and physiological psychology drove deep wedges into existing reconciliation of scientific theory with revelation or theology."[24]

Working within this revised historiographical framework, at the turn of the decade James R. Moore (1947–) published his seminal book on the *Post-Darwinian Controversies* (1979), perhaps the most exhaustive treatment of Protestant reactions to Darwin's theory of evolution. Moore contended that the "baneful" effects of such notions as "conflict" had made historians "prisoners of war," blinding them from seeing how easily leading Christian thinkers accepted, absorbed, and accommodated Darwin's theory of evolution. Indeed, he provocatively argued that Calvinists in particular were more willing to accept the outcome of Darwinism than liberal Protestants.[25] Moore's work made a significant impact on later historians, and subsequent studies by David N. Livingstone, Jon H. Roberts, and Ronald L. Numbers, for instance, continued to show the complexity of religious responses to Darwin.[26]

The notion of a "complex" relationship between science and religion became the clarion call of most historians of science in the later part of the twentieth century. In 1981, an international conference of historians met at the University of Wisconsin-Madison. The essays presented at the meeting were collected, edited, and published by historians of science David C. Lindberg and Ronald L. Numbers with the purpose of providing "the best available scholarship" on the historical relations between Christianity and science. The volume collectively dismantled a web of narratives developed by a conflict historiography, from the patristic period to twentieth-century Protestant theology. As Lindberg and Numbers observed in their introduction, "almost every chapter portrays a complex and diverse interaction that defies reduction to simple 'conflict' or 'harmony.'"[27]

Eschewing triumphalist narratives of either conflict or concord, historian of science John Hedley Brooke consolidated almost a century of scholarship on historical perspectives on science and religion in his magisterial *Science and Religion: Some Historical Perspectives* (1991). Brooke aimed to "reveal something of the complexity of the relationship between science and religion as they have interacted in the past." He emphatically rejected any generalizations about the relationship between science and religion. Following Whitehead's insight, Brooke similarly maintained that the shifting nature of the boundaries between science and religion over time makes it impossible to analyze their relationship according to any single conceptual model, be it "conflict" or "harmony." At the same time, he demonstrated that religious and metaphysical beliefs had provided a number of important presuppositions, sanctions, and motives for studying nature. Despite this challenging typology, Brooke concluded that "serious scholarship in the history of science has revealed so extraordinarily rich and complex a relationship between science and religion in the past that general theses are difficult to sustain. The real lesson turns out to be the complexity."[28]

Brooke's revisionist historiography and his call to "relish in the differentiation" has issued numerous dividends.[29] Complexity is now the central theme of most historical scholarship on science–religion relations. Indeed, since the turn of the millennium, there has been a seemingly endless stream of articles, books, and surveys published almost every year emphasizing the complex historical relationship between science and religion.[30] Celebrated episodes of conflict, such as the so-called Galileo affair, or the religious response to Darwin's theory of evolution, have been reinterpreted and reappraised. What historians of science have demonstrated is that for every particular episode where religious faith seemed to obstruct scientific progress, there are a host of other variables to consider, including political, philosophical, theological, and even scientific.

Brooke held the first Andreas Idreos Professor of Science and Religion at Oxford University from 1999 to 2006, where he also served as Director of the Ian Ramsey Centre for Science and Religion. Upon retirement, he was replaced by Peter Harrison, a leading intellectual historian of the early modern period whose unequivocal affiliation with Brooke's revisionist historiography is obvious. In his *The Bible, Protestantism, and the Rise of Natural Science* (1998), for example, Harrison reasserted a revised version of Merton's thesis, paradoxically locating the hermeneutical preconditions of modern science in the Protestant, literal understanding of Scripture. When Protestants stripped the Book of Scripture from its symbolic or emblematic meaning, all texts, including the Book of Nature, became open to new interpretation. Whereas we may view biblical literalism as an obstacle to science, according to Harrison, in the seventeenth century it "brought with it an alternative conception of the natural order," and this new conception was "the precondition for the emergence of natural science."[31]

Harrison has also spoken of the significance of the Augustinian doctrine of the Fall and how it influenced methodological developments in the natural sciences. In his *The Fall of Man and the Foundations of Science* (2007), Harrison argued that for many seventeenth-century natural philosophers, the accumulation of knowledge of the natural world was seen as ushering in the prelapsarian world of Adam before the Fall. This was once again a particular Protestant emphasis. According to Harrison, "contrary to first impressions, the anthropology of the reformers, informed as it was by the biblical account of Adam's Fall, had the potential to promote a new, more critical, appraisal of human intellectual capacities." The Protestant doctrine of the priesthood of all believers, moreover, led seventeenth- and eighteenth-century natural philosophers to see themselves as priests of the book of nature tasked with interpreting God's creation.[32]

More recently, building on the work of Whitehead, Brooke, and many others, including history of religion scholarship, Harrison offers a philological critique of conflict historiography in his *The Territories of Science and Religion* (2015). Harrison writes that "much contemporary discussion about science and religion assumes that there are discrete human activities, 'science' and 'religion,' which has had some unitary and enduring essence that persists over time." But according to Harrison, both *scientia* ("science") and *religio* ("religion") are historically unstable concepts. That is, the contours or "territories" of science and religion are themselves historically contingent. Traditionally, "science" and "religion" began as "inner qualities of the individual," or "virtues," before becoming concrete and abstract entities in the sense of doctrines and practices. Thus modern conceptions of the relationship between science and religion, whether in terms of conflict or concord, are in fact a "distorting projection of our present conceptual maps back onto the intellectual territories of the past." In this sense, Harrison views contemporary debates between science and religion as "proxies for more deep-seated ideological or, in the broadest sense, 'theological' battles." In short, the conflict between science and religion seems irresolvable only "because the underlying value systems—which are 'natural theologies' of a kind—are ultimately irreconcilable."[33]

THE "FOUNDERS" OF CONFLICT

In criticizing conflict historiography, most historians of science have traced the origins of these narratives to two late nineteenth-century works: John William Draper's *History of the Conflict between Religion and Science* (1874) and Andrew Dickson White's *A History of the Warfare of Science with Theology in Christendom* (1896). Indeed, as early as 1970, Robert Merton accused Draper and White of propagating the belief that the "prime historical relation between religion and science is bound to be one of conflict."[34] Four years later the first professor of history of science at the Open University Colin Russell more directly associated the conflict thesis with the work of Draper and White.[35] But it was James Moore's monumental study of the post-Darwinian controversies that led almost every other subsequent historian of science to designate Draper and White as the "co-founders" of the conflict thesis. According to Moore, Draper and White set "the terms of the debate," and therefore must be "regarded as the principal *casus belli*" of the conflict narrative.[36]

By the 1980s, it had become abundantly clear that Draper and White were now the official whipping boys of the new revisionist historiography. A consensus emerged among historians of science that Draper and White developed, defined, and defended the conflict thesis. Numbers, Lindberg, and Russell in particular have placed a good deal of blame on these two historical figures. In an early statement on the subject, Numbers wrote that "Military metaphors have dominated the historical literature on science and religion since the last third of the nineteenth century, when the Americans Andrew Dickson White and John William Draper published their popular surveys of the supposed conflict between religion and science."[37] In their joint projects, Numbers and Lindberg declared that no work has done more to "instill in the public mind a sense of the adversarial relationship between science and religion" than that of Draper and White.[38] Russell agreed, tracing the "social origins" of the conflict "myth" to the "whiggish" historiography of Draper and White.[39]

In the following decades and up to the present, numerous revisionist historians continued to cite Draper and White as instigators of the conflict thesis. Brooke thus adhered to the precedent when he argued that Draper and White put "forward a principle of interpretation that still enjoys popular support."[40] In several places, Harrison also designates Draper and White as the "chief architects" of the "conflict myth," arguing that the "general tenor" of their positions can be "gleaned" from the titles of their respective works.[41] At a more popular level, revisionist historiographical surveys continued to beat the same drum, citing the *bêtes noires* Draper and White as the principal exponents of the conflict thesis. Indeed, in introducing the subject, it has almost become obligatory to begin discussion by citing Draper and White as its cofounders. Numbers has perhaps put it most succinctly, writing that "no one bears more responsibility for promoting this notion [of conflict] than two nineteenth-century American polemicists: Andrew Dickson White and John William Draper."[42]

Scholars have also attempted to make distinctions between Draper and White. In most of the scholarly literature, the pervading assessment is that while Draper regarded "the struggle as one between Science and Religion," White saw it as "a struggle between Science and Dogmatic Theology." More precisely, Draper's work has been characterized almost exclusively as a diatribe against Roman Catholicism, prompted by the encyclical *Quanta Cura* (1864) and the assertion of papal infallibility at the first Vatican Council (1869–70). White, scholars maintain, recognized Draper's rhetoric as exaggerated, and therefore argued instead that the conflict resided in religious dogmatism, not religion, and that his position was a reaction against the sectarian opposition he encountered as cofounder of Cornell University.[43]

DEMYTHOLOGIZING THE "CONFLICT THESIS"

As the above historiographical survey demonstrates, Draper and White remain consequential figures today. While this volume agrees that tenuous and tendentious "myths" about science and religion need to be discredited, it challenges a number of basic assumptions about the nineteenth-century origins of the conflict thesis. My own research convinces me that Draper and White, contrary to conventional interpretations, did not in fact posit an endemic and irrevocable conflict between "science and religion." Indeed, if we examine more carefully their lives and writings, rather than perfunctorily repeating past scholarly assessments, a more nuanced interpretation emerges. A more generous reading reveals that, unlike the new atheists, who intentionally write to advance unbelief, Draper and White hoped their narratives would actually preserve religious belief. For Draper and White, science was a reforming agent of knowledge, society, and religion. Indeed, for Draper and White, science was ultimately a scapegoat for a much larger and much more important argument, one in which they pitted two theological traditions against each other—a more progressive, liberal, and diffusive Christianity against a more traditional, conservative, and orthodox Christianity. They thus conceived of conflict as occurring within a religious epistemology, between two distinct "modes" or "epochs" in human thought—one scientific or progressive and the other theological or traditional. Conflict was in this sense positively beneficial, as it would assist in the progress of religion. The titles of their most well-known works, then, were only tangentially related to their content and aim.

Their "conflict" was thus not the mere caricatures Moore, Numbers, Lindberg, Russell, and so many other revisionist historians of science have made it out to be. In fact, it is not without some irony that the actual conflict Draper and White envisioned is remarkably similar to how such historians have sought to redefine the idea of "warfare" or "conflict" between science and Christianity as one *within* religion—what Moore called, for instance, "cognitive dissonance." For his part, Moore bewailed the "zealous defenders of biblical literalism" who indulged in "monkey business" in their "campaign against evolution in education." He thus sought to "come to terms with Darwin" by redefining Christian "orthodoxy" to the total exclusion of "Biblical fundamentalism" and "literalistic" hermeneutics, and essentially concluded that Christians needed to "come to terms with Darwinism." Indeed, Moore argued that if only Christianity could be "transformed" and "rightly viewed" there would be no conflict with science.[44]

Even Whitehead, who first warned historians against using the trope "science and religion," followed Draper and White in arguing that "religion will not regain its old power until it can face change in the same spirit as does science."[45] Religion, like science, "requires continual development," and indeed "must be continually modified as scientific knowledge advances." Like Draper and White, Whitehead believed that religion "emerged into human experience mixed with the crudest fancies of barbaric imagination," and only "gradually, slowly, steadily the vision recurs in history under nobler form and with clearer expression."[46]

In more recent proposals for mapping out the historiographical way forward,

Draper and White continue to haunt the pages of scholarship. In a recent *Festschrift* honoring the scholarship of Brooke, for instance, Numbers identifies five "mid-scale patterns, whether epistemic or social, demographic or geographical, theological or scientific," where conflict remains. Similarly, in reconceptualizing the conflict narrative, historian and philosopher of science Geoffrey Cantor also highlights the tensions created within the mind of an individual when confronting "engagements with science and religion." Like Draper and White, Cantor envisions conflict as the necessary catalyst for change, for "helping sweep away a corrupt regime." As he puts it, "in the context of science and religion, conflict has been the engine of change, even perhaps of what we might call progress."[47]

Moore's cognitive dissonance theory, Numbers's small-scale patterns, Cantor's conflict as change, and even Whitehead's belief that religion continually develops are not so different from the concept of "conflict" or "warfare" that Draper and White promoted. How could such a serious error occur in otherwise excellent and exciting revisionist scholarship? This oversight raises the intriguing question of whether a scholar's religious biography plays some role in misunderstanding or obscuring the origins of the conflict thesis. For instance, Numbers and a host of other revisionist historians self-identify as either atheist or agnostic, while others come from a more liberal Protestant tradition.[48] Draper and White also considered themselves advanced "theists" of a liberal Protestant variety. Thus, while historians of science have debunked many of the "myths" found in the narratives of Draper and White, their central thesis remains, either rising again in smaller scale struggles or internal, mental dissonance. No scholarship to date, however, has explored the potential of undermining the conflict thesis by showing that Draper and White themselves did not adhere to it in its simplified form.

RELOCATING THE CONFLICT

My contention is not that historians of science are being inconsistent. Complexity, of course, allows for episodes of both concord and conflict in the history of science and religion. Rather, my point is that numerous historians have not only mischaracterized the position of Draper and White but have also mislocated the provenance of the conflict. The rift Draper and White envisioned existed long before they ever put pen to paper. Moreover, many contemporary authors discussed conflict in similar terms. Draper and White thus drew from a variety of disparate traditions, in addition to discussions and themes from contemporaries who expressed similar views. The similarities between Draper, White, and others does not necessarily demonstrate direct influence but rather convergence. As we shall see, such narratives have a long religious pedigree that can be traced back to as early as the sixteenth and seventeenth centuries, in Protestant polemics against

Roman Catholicism, which often deployed a rhetoric of history, reason, and new knowledge to undermine its cultural and religious authority.

It must be immediately emphasized that arguments directed at Catholics also appeared in disputes between Protestant sects. Similar rhetorical strategies were adopted between contending Protestant groups, particularly between the Established Church and Dissenters or Nonconformists, and even between High, Low, and Broad Church Anglicans. In defining or redefining "religion," Draper and White embodied the characteristic qualities of two distinct theological traditions. From the seventeenth to the eighteenth century, "rational" theologians attempted to solve issues surrounding faith and reason by associating or accommodating religion to reason and rationality, often reducing it to what they believed was its most essential elements. But by the late eighteenth and early nineteenth century, this way of solving the conflict proved unsatisfactory for many, for it abstracted the intellectual content of religion and ignored its more concrete reality found in experience, feelings, emotions, sentiments, intuitions, and morality. Significantly, both traditions emphasized minimal doctrinal attachments and were thus indispensable elements of a more liberal Protestant faith. As we shall see, the language Draper and White employed throughout their writings is remarkably similar to how liberal Protestants responded to the advances of the new sciences, from astronomy and zoology to biblical historical-critical scholarship.

Draper and White must be placed firmly within this Protestant heritage. The origins of the conflict narrative can thus be found in an internal, religious critique—or more precisely an intra-Protestant self-critique. By the nineteenth century, there was a long-standing tradition of theology being subject to the authority of history, reason, and science. Draper and White wished to persuade readers that history was on their side and that their minimal religious creed would ultimately not only reconcile science and religion but save religion from itself. Their narratives, therefore, were less descriptive than prescriptive.

This book also examines how the narratives of Draper and White were disseminated, popularized, and ultimately appropriated by others during a time of marked expansion in science publishing. Edward Livingston Youmans, science editor of D. Appleton and Company, one of America's most influential publishing houses in the second half of the nineteenth century, was among the chief promoters and popularizers behind this expansion. An innovator in publishing, Youmans ensured that Draper's and White's ideas would reach the widest readership. As other scholars have demonstrated, the "communication revolution" of the nineteenth century, which included new printing technologies, higher literacy rates, improved systems of transportation, and a reduction in the cost of paper, enabled publishers to communicate to a broader audience.[49] Youmans began publishing the work of Draper and White just as British and American publishing was undergoing this remarkable revolution. Indeed, he was at the forefront of this revolution, establishing new international copyright agreements and popularizing scientific knowledge in his extremely successful *Popular Science Monthly* and "International Scientific Series," both of which were ambitious projects started in the 1870s with the intent of diffusing the latest advances in science to a global audience.

In his various publishing ventures, Youmans consistently advertised, defended, and clarified the ideas of Draper and White. Like them, he too believed there was no intrinsic conflict between science and religion. It was Youmans who commissioned Draper to write the *History of the Conflict* for his International Scientific Series, which became one of the most successful books in the series. It went through fifty printings in the United States alone, twenty-four in England, and was translated into ten languages, including French, German, Italian, Dutch, Spanish, Polish, Japanese, Russian, and Portuguese. Youmans also published White's articles on the "New Chapters in the Warfare of Science" in his *Popular Science Monthly*, which ran for a decade between 1885 and 1895. Later White expanded these articles into his famous *History of the Warfare*, also published by Youmans. Thus Youmans must be seen as a central figure in the diffusion and popularization of their ideas.

But while Youmans agreed with Draper and White that science and religion were not in conflict, he did not share their hopes of a final reconciliation between modern thought and Christianity. Rather, as one of the leaders of the Free Religious Association, an organization founded in 1867 that called for the emancipation of religion from all "dogmatic traditions," and as an advocate of scientific naturalism, as defined by Huxley, Tyndall, and Spencer, Youmans appropriated the narratives of Draper and White in support of his own vision of the new religion of the future. As we shall see, the reconciliation Youmans described was the end of "orthodoxy," a term by which he meant Protestant as well as historic Christianity broadly defined. He believed that the history of science demonstrated not only the progress of knowledge but also the progress of religion. Youmans strongly believed that traditional Christianity was no longer tenable. Thus, while he appropriated the histories of Draper and White in support of his vision of the religion of the future, he believed the future of religion went beyond Protestantism. While he retained the language of Protestantism, it was bereft of any doctrinal beliefs of traditional Christianity. Youmans's new religion, in short, was a Protestantism-minus-Christianity.

Finally, this study also examines the early public reception of Draper and White. Their narratives received extensive commentary in periodicals and private letters. In examining this material, I hope to demonstrate that the positive aspects of their projects ultimately failed. How Draper and White envisioned their reformed, minimal religion was deeply contested by readers. More religiously conservative or orthodox reviewers did not accept their attempted reconciliation, for they could not accept their redefinitions of religion. They thus warned that any such attempt would only lead to a greater perception that science and religion were indeed at war. This was a prescient warning, for, by the end of the nineteenth century, freethinkers, secularists, and atheists appropriated their narratives as a weapon against all religion. As we shall see, a number of late nineteenth-century and early twentieth-century antireligious authors and publishers took up the narratives of Draper and White without any of their nuances. More remarkable still, it was this simplistic way of relating science and religion that provided the very foundations of the incipient discipline of the history of science, which was first emerging during the early decades of the twentieth century.

Draper and White, therefore, did not set "the terms of the debate." Their understanding of the conflict turns out to be remarkably more complicated than what modern historians of science lead us to believe. Indeed, the nineteenth century is in fact too late a date for the origins of that conflict. That the conflict thesis emerged out of an internal religious struggle within Protestantism should give pause to historians of science who routinely accuse Draper and White of cofounding that narrative. That accusation is yet another myth in the historiography of science.[50]

Rather than attempt to discredit their historical narratives as myths about science and religion, I aim to remedy a scholarly oversight by assessing the work of Draper and White as primary sources, explicating not only their own sources of inspiration but also examining the cultural functions they performed.[51] What Draper and White provided was a synthesis and codification of ideas about nature, man, and God—ideas, moreover, that can be traced all the way back to sixteenth-century debates between Protestants and Catholics. As such, this volume confirms a number of other studies that have maintained that the boundaries between Protestantism and secularism were remarkably porous and that certain elements of Protestant theology eased the transition from belief to unbelief. The conflict Draper and White envisioned turns out to be an incredibly sophisticated array of nested stories and myths that Protestants have told each other from the beginning of the Reformation. By the 1860s, the narrative of conflict preceded Draper and White by many years—even centuries. In one of those remarkable ironies of history, what we shall discover in this investigation is that the language of conflict was largely drawn from centuries of Protestant polemic.

This reinforces what a number of scholars have tacitly or explicitly argued in more general studies.[52] Indeed, the narratives of Draper and White are examples of what philosopher Charles Taylor (1931–) has called "social imaginaries," constructed narratives that shape how we relate and are attuned to the world. They are also related to what Taylor called "subtraction stories," which narrate the inevitable erasing of religious belief. Ironically, such stories were first articulated not in the salons or coffee houses of atheists and freethinkers but within a particularly theistic environment. Thus the separation of theology from religion can be traced to specific locations and permutations in the history and development of theological discourse.

It is often supposed that conflict historiography had existed in the anticlerical spirit of eighteenth-century French *philosophes*. The "prophets of Paris," as Frank E. Manuel once called them, presented grand narratives of progress, with religion and science in opposing roles.[53] François-Marie Arouet, or Voltaire, Julien Offray de La Mettrie, Denis Diderot, Jean le Rond d'Alembert, Anne-Robert-Jacques Turgot, Baron de Laune, Marie Jean Antoine Nicholas de Caritat, marquis de Condorcet, Claude Henri de Rouvroy, comte de Saint-Simon, François Marie Charles Fourier, Auguste Comte, and their followers, all supposedly proclaimed the old orthodoxy as stultifying moral, scientific, and material progress. While this is no doubt true, what distinguished the French *philosophes* from English intellectuals is that by the nineteenth century, Britain had already possessed a 200-year tradition of historical narratives of progress, framed within a Protestant anti-Catholic polemic. A number of studies over the last fifty years have demonstrated the ubiquitous nature of anti-Catholicism among the English. Indeed, as early as Elizabethan theologian Richard Hooker, English reformers interpreted history as the progression from the irrationality of "papism" to the light of reasoned Protestantism. Perhaps more importantly, beginning with Francis Bacon, natural philosophy and, subsequently, natural science became linked to religious themes, tacitly aligning campaigns for scientific progress with reformers' campaign for the restoration of Christian purity, and thus bestowing upon men of science a "prophetic" authority.

I hope this study encourages historians of science to take theology more seriously. Much of the narrative found in Draper and White was drawn from centuries of Protestant Christian polemic, particularly from seventeenth- and eighteenth-century English writers. To be sure, while Draper and White inherited this English tradition, there were also elements of French and German thought in the formation of their narratives. Draper, on the one hand, was educated at London University (later called University College, London), where he came under the direct influence of Unitarianism, utilitarianism, and Comtean positivism, and all this would have a lasting impact on his understanding of historical progress. White, on the other hand, was greatly influenced by German historical thought during his time at the University of Berlin, and indeed found his calling as scholar and teacher while studying under Gustav Droysen, August Böckh, Leopold von Ranke, and Friedrich von Raumer, among others. Nevertheless, the conflict thesis is part and parcel of a Protestant heritage. Despite their marked differences in temperament and background, Draper and White both promulgated a particularly English Protestant narrative of progress.

This narrative, of course, was not antireligious. Indeed, both Draper and White are clear on this. But by the nineteenth century, a new understanding of "religion" was emerging. Draper and White sought to salvage the cardinal values of their religious heritage by reconstituting them in a way that would make them intellectually acceptable, as well as emotionally pertinent. Perhaps one of the most neglected areas of research in current historiography is the failure to recognize the rapid and significant changes in religious thought that occurred during the nineteenth century. Throughout their writings, Draper and White relied heavily on the then-emerging comparative study of religion and the increasingly challenging historical-critical scholarship of the Bible. Numerous works appeared during the nineteenth century that ushered in dramatically new perspectives on religion in general and Christianity in particular. Much of this new scholarship depended on developments in the study of religion during the seventeenth and eighteenth centuries, which was largely produced by Protestant writers.[54] For Draper and White, almost every episode of conflict in their overarching narratives begins with some comment on comparative mythology, anthropology, or biblical criticism. Their dependency on these new scholarly traditions is clear throughout their work. By emphasizing these elements in the narratives of Draper and White, we shall see that the conflict thesis has always had unacknowledged antecedents in Protestant polemics. But by the mid-nineteenth century, religious challenges and reconceptualizations brought on by comparative religious studies transformed this essentially Protestant narrative into a more general antidogmatism. In short, the narratives constructed by Draper and White found a ready audience. By reexamining the origins, development, and dissemination of the nineteenth-century conflict thesis, I hope to demonstrate that Draper and White were not the "embattled founders" of this narrative, but rather inheritors and codifiers of an already existing narrative. The conflict thesis, in other words, was a received narrative.

In chapters 1 and 2, I offer an intellectual biography of Draper and White, one that pays particular attention to their religious background and development. I discuss the people who influenced them most—family, friends, and associates. With the exception of their biographers, few historians of science have carefully engaged with these sources of inspiration, and fewer still have shown how Draper and White constructed their narratives based on those sources.[55] In taking account of their intellectual and religious development, I also draw from new archival material that further redresses the gap in current scholarship.

After calling attention to the unique background, cultural heritage, and intellectual predecessors of Draper and White, in chapters 3 and 4 I place their thoughts in a wider historical context. These pages are devoted to accounting for the intellectual and religious changes in England and America, from the beginning of the Reformation to the end of the nineteenth century. The goal is not to trace the rhetorical genealogy of each particular episode of conflict recorded in their respective narratives. Rather, my aim is to explain how views like those found in Draper, White, and many others were even possible. Indeed, Draper and White should not be viewed as founders of the conflict thesis but rather as representatives of tensions within Protestantism. Whereas Draper's thoughts were rooted in the "rational" Christianity of seventeenth- and eighteenth-century English Protestant thinkers, White's ideas can be traced to more romantic and idealistic conceptions of religious progress found in the writings of early nineteenth-century German thinkers and in American transcendentalism. In short, certain changes or transformations in religious thought had occurred during the seventeenth and eighteenth centuries that enabled Draper, White, and others to construct such narratives, and their overarching arguments make sense only in light of these historical developments.

In chapter 5, I examine how Youmans disseminated and popularized the narratives of Draper and White. However, Youmans, who was also a friend and publisher of many of the scientific naturalists, including Huxley, Tyndall, and especially Spencer, ultimately appropriated the narratives of Draper and White for his own purposes. Thus an examination of their relationship with Youmans is crucial to understanding the emergence and popularization of the conflict thesis. Finally, in chapter 6, I examine the public reception of Draper and White. As we shall see, their redefinition of religion was deeply contested by readers. More importantly, it shall be evident that Draper and White lost control of their narratives to secularists, freethinkers, and atheists, who used their work to support their own efforts at secularizing society.

Draper and White are no doubt guilty of using ambiguous language that could easily be misconstrued and appropriated for purposes they never intended. This ambiguity has misled many readers, including modern historians of science, into believing that they, in fact, did posit a conflict between science and religion. But the ambiguity of their words was shared with many other liberal Protestants before and after them. By the nineteenth century, the divide between liberal and conservative Christians was becoming sharper, and the gulf between faith traditions would only widen in the early part of the twentieth century to become even more significant than other intra-Protestant denominational divisions in preceding generations. More progressive Protestants sought to formulate a version of Christianity adapted to the critical demands of the modern age—one that worked in consonance with, rather than in opposition to, Enlightenment philosophy, science, historical research, and culture as a whole. One crucial strategy liberal Protestants used was the history of science. They told narratives and popularized anecdotes of how traditional religious beliefs obstructed the progress of liberal ideas, whether in science or religion. As we shall see, this strategy ultimately backfired. In the late nineteenth century and early twentieth century, secularized versions of their narrative emerged, thus giving credence to the view that the conflict between science and religion is just a secularized polemics developed between Protestants and Catholics.

The more complicated reality is that the notion that "science" and "religion" are in a constant state of conflict or warfare is a more recent invention, found not in Draper and White but in their promulgators near the end of the nineteenth century and the beginning of the twentieth century, by those who appropriated liberal Protestant narratives in their campaign to discredit all religion and secularize society. A better understanding of the context of their work, including its subsequent reception, demonstrates that Draper and White envisioned a conflict not between science and religion, but one within the religious conscience, between what they praised as a progressive or diffusive Christianity against a more orthodox or traditional Christianity. The irony in all this is that the reconciliatory prescriptions offered by Draper, White, and many others to the tensions within the religious conscience at the end of the nineteenth century had the unintended consequence of undermining the very religion they maintained to preserve.

DRAPER AND THE NEW PROTESTANT HISTORIOGRAPHY

IN 1874, English-born American scientist, educator, and historian John William Draper (1811–1882) published his enormously popular book, *History of the Conflict between Religion and Science*. His title alone seems to suggest that religion and science are caught in an endemic battle. In his preface, Draper claimed that the history of science was "a narrative of the conflict of two contending powers, the expansive force of the human intellect on one side, and the compression arising from traditionary faith and human interests on the other." He went on to explain that "faith is in its nature unchangeable, stationary; Science is in its nature progressive; and eventually a divergence between them, impossible to conceal, must take place." Presumably, the "faith" Draper had in mind was primarily that of the "Roman Church," which he accused of subjecting men of science to social ruin, mental torment, physical torture, and even death. Indeed, he accused the Catholic Church of having hands "steeped in blood!" By contrast, he asserted that science "has never attempted to throw odium or inflict social ruin on any human being. She has never subjected any one to mental torment. She presents herself unstained by cruelties and crimes." With such aggressive rhetoric, it is unsurprising that most historians of science interpret Draper's book as a vehement dia-

Figure 1.1. John W. Draper, photograph by Edward Bierstadt (1821–1906). Courtesy of the Smithsonian Institution, National Museum of American History, Archives Center, Draper Family Collection.

tribe against the Roman Catholic Church. He explicitly stated that his exposition would have "little to say respecting the two great Christian confessions, the Protestant and Greek Churches."[1]

That Draper's book was an indictment against the Roman Catholic Church is undisputable. He asserted that "an impassable and hourly-widening gulf intervenes between Catholicism and the spirit of the age." Even a cursory reading of his book reveals a catalog of what would become the most prevalent myths about science and religion—that an angry mob of Christians murdered the brilliant fifth-century Alexandrian mathematician Hypatia; that St. Augustine did more than any other Church Father to bring science and religion into antagonism; that patristic theology had become a stumbling-block to the intellectual advancement of Europe, thus plunging mankind into mental darkness for a millennium; that medieval Christians believed in a flat earth; that a fearless Christopher Columbus proved it was a globe; that Copernicus's heliocentric claims demoted humanity by removing the earth from the center of the cosmos; that Galileo was tortured and imprisoned by the Inquisition; that Giordano Bruno was a scientific martyr; and that geology and evolution came into conflict with the biblical view of human origins.[2]

In his concluding remarks, Draper claimed that "it is not given to religions to endure for ever," and that it was "impossible that Religion and Science should accord in their representation of things." The final arbiter must be "reason," he declared, the "supreme and final judge." He predicted that a "time approaches when men must take their choice between quiescent, immobile faith and ever-advancing Science—faith, in its medieval consolations, Science, which is incessantly scattering its material blessings in the pathway of life, elevating the lot of man in this world, and unifying the human race." Ultimately the Church and science, Draper argued, are "absolutely incompatible; they cannot exist together; one must yield to the other; mankind must make its choice—it cannot have both."[3]

For historians of science, Draper's book still serves as the *locus classicus* of the conflict thesis. While I do not dispute the claim that his historical narrative led many to believe there was a conflict between religion and science, I challenge the imposed designation. Simply put, it is wrong. A closer reading demonstrates that Draper, the title of his book notwithstanding, did not in fact propose an inherent conflict between "religion and science." Draper's *History of the Conflict* was largely a compilation and condensation of other works he published earlier in his career. The *History of the Conflict* must be placed within the larger context of this oeuvre. Draper himself believed his work followed a coherent philosophical trajectory. Reflecting later in life, he claimed that each publication followed an important sequence, a legitimate offspring of the other, presenting an expansion of ideas that illustrated his belief in intellectual and social progress.[4] In other words, if we concentrate on the *History of the Conflict* alone, we miss important details of Draper's argument. We must also situate him within the larger context of contemporary thought. He produced a philosophy of history written from the perspective of a physical scientist who collated and used the available recorded "facts" of history to illustrate and explain his own theory of progress. Set within a wider context, we discover that his ideas reflected a number of disparate traditions, discussions, and themes prevalent among other authors—particularly among Protestant historians. Indeed, as we shall see, a closer inspection of his work reveals the continuing power and influence of his Protestant upbringing.

In this chapter, I take seriously the Protestant heritage in Draper's understanding of history. I argue that Draper looked back to the "rational religion" found among seventeenth- and eighteenth-century liberal Anglican intellectuals, who viewed new knowledge as evidence of the ineffable creative power of God. Rather than instigating a conflict, he believed his narrative would bring about a permanent solution to an already existing conflict between the religious consciousness and the spirit of the age. Like many before and after him, Draper redetermined the relationship between science and religion by redefining Christianity. He thus represented a synthesis of ideas, partly derived from a rich and complex Protestant legacy and partly from other emerging nineteenth-century conceptions of law, evolution, and progress. Draper, in short, was a somewhat belated heir of the "reasonable" Christianity of eighteenth-century "rational supernaturalism." Insufficient attention has been given to these Protestant elements of Draper's understanding of religion and science. This chapter, therefore, attempts to redress this problem by placing Draper firmly within the Protestant tradition.

A BIOGRAPHICAL SKETCH

To better understand how the son of an Englishman grew up to become an American citizen, scientist, educator, and then historian, it is important to know a few details about his life.[5] John William Draper was born in St. Helens, England, into a family of devoted Methodists. His father, John Christopher Draper (1770–1832), was a Roman Catholic who converted to Methodism during a revival meeting in England. After his conversion, John Christopher became a Wesleyan itinerant minister for the Sheerness circuit.[6] His mother, Sarah Ripley, was the daughter of the famous abolitionist and evangelist Dorothy Ripley (1767–1831). A number of Dorothy's ancestors had emigrated to America before the Revolutionary War and founded a Wesleyan colony in Christianville, Virginia. At the time of John William's birth, the elder Draper was a preacher at the Methodist Chapel in St. Helens. The young Draper was baptized by his father's friend, autocratic leader of the Wesleyan Methodists Jabez Bunting (1779–1858).[7]

At age eleven, Draper was sent to a Methodist boarding school at Woodhouse Grove in neighboring Yorkshire, presumably to follow in his father's footsteps and prepare him for the ministry.[8] John Christopher had a strong Nonconformist outlook and, like many other Dissenters in the late eighteenth and early nineteenth centuries, he combined his religious commitments with a penchant for scientific subjects, especially chemistry and astronomy.[9] According to an early biographer, for instance, John Christopher was "in the habit of amusing his lei-

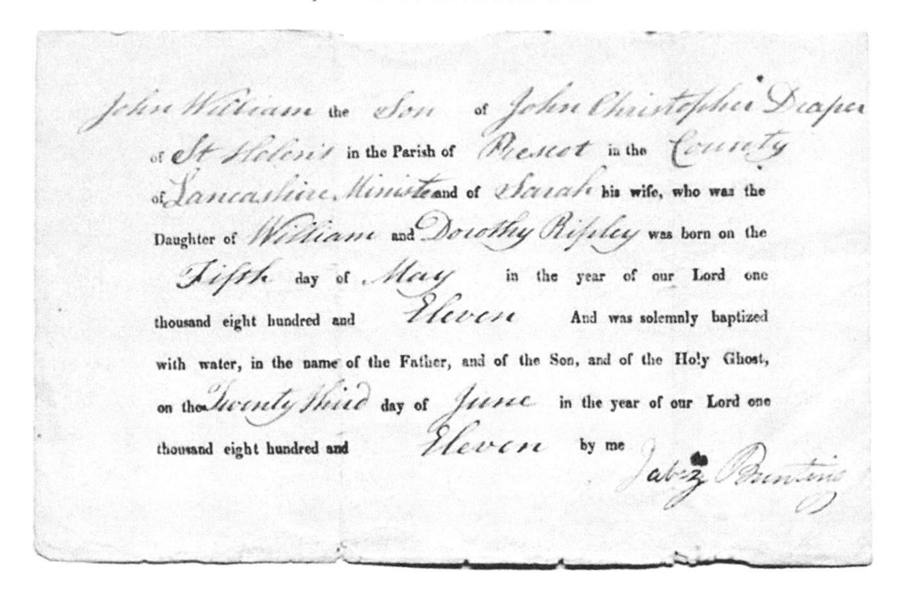
John William the Son of John Christopher Draper of St Helen's in the Parish of Prescot in the County of Lancashire, Minister and of Sarah his wife, who was the Daughter of William and Dorothy Ripley was born on the Fifth day of May in the year of our Lord one thousand eight hundred and Eleven And was solemnly baptized with water, in the name of the Father, and of the Son, and of the Holy Ghost, on the Twenty Third day of June in the year of our Lord one thousand eight hundred and Eleven by me Jabez Bunting

Figure 1.2. Birth certificate of John W. Draper, showing he was baptized by famous Methodist preacher Jabez Bunting (1779–1858). Courtesy of the Library of Congress, Manuscript Division, John William Draper Family Papers, Container 18.

sure by observing the heavens through a Gregorian telescope, and on one occasion his little son, then scarcely more than six years of age, was permitted to look through it at some of the heavenly bodies."[10]

Historians have pointed out that the religious revival at the turn of the century was a revolt against a cold, ineffective, and emasculated Church of England.[11] The serious decline in the public expression of Christian faith was matched by other depravities of English life. The English countryside, for example, ran rampant with old pagan beliefs, superstitions, and practices, and urban life had become increasingly violent, drunk, and debauched.[12] The Church of England, it seemed to many, had failed its spiritual mission. In his classic study, French historian Élie Halévy (1870–1937) summed up well the condition of established religion at the end of the century: "apathetic, sceptical, lifeless; sects weakened by rationalism, unorganized, their missionary spirit extinct. This was English Protestantism in the eighteenth century."[13]

Spiritual renewal began with the rise of the evangelical movement, and particularly with the ministry of John (1703–1791) and Charles Wesley (1707–1788) in the 1730s.[14] William Law's (1686–1761) *A Practical Treatise Upon Christian Perfection* (1726) and *Serious Call to a Devout and Holy Life* (1729) had inspired the Wesleys and many others to form the "Holy Club" while studying at Oxford University.[15] "These convinced me," John Wesley later wrote, "more than ever, of the absolute impossibility of being *half a Christian*."[16] As is well known, in 1738

Wesley felt his "heart strangely warmed" during a Moravian gathering on Aldersgate Street, and thereafter would inspire innumerable others (including Draper's father) to take up the call to minister the gospel.[17] The following year, Wesley published a tract that pithily summarized the "Character of a Methodist."[18] In it he emphasized the distinction between essential beliefs of the Christian faith and mere opinions. "The distinguishing marks of a Methodist," he wrote, "are not his opinions of any sort. His assenting to this or that scheme of religion, his embracing any particular set of notions, his espousing the judgment of one man or another . . ." Unless it directly strikes at "the root of Christianity," Wesley argued, "we think and let think." For Wesley, a Methodist was "one who has the love of God shed abroad in his heart by the Holy Ghost given unto him; one who loves the Lord his God with all his heart, with all his soul, and with all his mind, and with all his strength."[19] Intellectual historian James Turner has observed that theologians of the previous century had rationalized belief, exalting "reason at the expense of revelation"; but with the evangelical revivals in Britain and America, there emerged a "stress on religion as a matter of the affections"—a "religion of the heart."[20] By making religion a matter of the heart, Wesley believed that Christians could be united with a few essential doctrines and yet have differing opinions about a range of other, less important things. Draper's father John Christopher is said to have exemplified these beliefs.

Embracing a "religion of the heart" did not necessarily entail the rejection of learning, of course.[21] That the Wesleyan approach sought to bring the whole person—emotions, spirit, and mind—into the Christian life is undeniable. John Christopher's Gregorian telescope and his fascination with the natural world was a characteristic feature of many Dissenters and Nonconformists, including the evangelicals.[22] Draper's Methodist upbringing, then, had instilled in him a set of values that stretched back into English history and prepared him well for the next step in his academic career. In 1828, his father had purchased shares in support of establishing the new University of London, to which he sent Draper in 1829 to commence studies in chemistry and medicine.

Information on the intellectual influences upon Draper during his time in London is scanty. However, we do know that he fell under the influence of chemist Edward Turner (1796–1837). Indeed, of all his teachers, Turner seems to have struck Draper the most.[23] Turner wrote a very popular textbook of chemistry, *Elements of Chemistry* (1827), which went through eight editions during his lifetime. Educated in England, Scotland, and Germany, Turner was a fellow of both the Royal Society of London and of Edinburgh.[24] When the chair of chemistry was established at the University of London, Turner was selected after Michael Faraday (1791–1867) had turned down the appointment. According to an obituary printed in the *Gentlemen's Magazine*, "Dr. Turner was a member of the established Church of England, and a strict observer of its ordinances," but added that "his religion was perfectly free from bigotry or intolerance."[25]

In addition to working closely with Turner, Draper also took courses in legal philosophy from John Austin (1790–1859).[26] Raised in a middle-class Unitarian family, Austin became close friends with Jeremy Bentham (1748–1832) and his two disciples, James Mill (1773–1836) and John Stuart Mill (1806–1873).[27] In fact, the younger Mill and Draper likely sat in the same classroom during Austin's lectures on jurisprudence at the University of London.[28] Bentham was of course the intellectual leader of the utilitarians, and no one did more than James Mill to spread his philosophy. Austin, in turn, was greatly influenced by Bentham as well, acknowledging in his lectures that "I much revere [him], and whom I am prone to follow."[29] As will become apparent, Draper seamlessly combined aspects of utilitarianism and Methodism in his conception of scientific identity, natural law, and theory of progress.

While studying at the University of London, Draper boarded with a friend of his father, a Mrs. Mary Barker, aunt of Antonia Caetana de Paiva Pereira Gardner (1814–1870), Draper's future wife. Antonia was the daughter of the famous Daniel Gardner, the attending physician to King John VI of Brazil and Portugal (1767–1826) and Carlota Joaquina of Spain (1775–1830). Antonia's father had sent her to live with his sister in London, and it did not take long for Draper to propose to the young lady. Only a few months after Draper arrived, they were engaged.[30] Antonia's father, however, died in 1831. Tragedy piled upon tragedy when Draper's own father died the following year. Resolute in character, Draper nevertheless completed his certificate in chemistry, married Antonia, and then, urged by his relatives, emigrated to the United States, to the Wesleyan community in Virginia set up by his mother's ancestors.

As one of his biographers has observed, Draper "kept to the end of his life the sense of belonging to the European intellectual community, the sense, it would almost seem at times, of being in exile."[31] Shortly after arriving in America, Draper pursued a medical degree at the University of Pennsylvania in 1835, studying under chemist Robert Hare (1781–1858) and physician John Kearsley Mitchell (1798–1858). Hare, it should be noted, decried the sectarianism in his day, and once complained in a pamphlet that "orthodoxy, so-called, is productive of great injustice and oppression to many who, like myself, consider a book of no higher authority than the fallible men who wrote it."[32]

Draper took the doctorate in medicine in 1836 and was promptly appointed professor of Chemistry and Natural Philosophy at the Presbyterian affiliated Hampden-Sydney College, Virginia. A year later he took up an appointment as Chair of Chemistry at the University of New York, an institution that had been deliberately modeled after the University of London. He would spend the rest of his life in New York. By all accounts, he had a remarkably successful career. In 1841, he founded the School of Medicine at the University of New York and served as its president from 1850 to 1873. In 1869, he also became the founding president of the American Union Academy of Literature, Science, and Art. Draper was also elected the first president of the American Chemical Society in 1876.

FIXED AND IMMUTABLE LAW

Like many men of science in the nineteenth century, John William Draper imbued his scientific work with certain metaphysical speculations. One of the most prominent features of Draper's scientific writings was the very popular nineteenth-century belief in the absolute uniformity of nature. This principle stated that all natural phenomena are invariably governed by immutable laws. As one writer in the *Pall Mall Gazette* put it, the Victorian period was characterized by "an ever-increasing conviction of the uniformity of the operations of all physical law."[33] This conviction in the uniformity of nature was not, however, antireligious. Matthew Stanley has recently demonstrated that the principle of the uniformity of nature was shared by both scientific naturalists and theistic men of science.[34] From such men as Turner, Hare, Mitchell, and others, Draper imbibed a rationalistic approach toward religion. Historians of science have missed the fact that Draper's earliest scientific publications were coauthored with geologist William Mullinger Higgins (1809–1882).[35] Together they argued that "geology must be brought to the simple but noble condition of a practical science; and, like chemistry, watch over its accumulating facts, jealous of hypothesis." Geologists, they wrote, must be "cured of the mania which has seized them." This was an undercutting allusion to the so-called "scriptural geologists" of the time. They contended that geologists must "turn their powerful energies to the discovery of facts . . . for the geologist alone is capable of making those observations upon which rational opinions can be formed."[36] Higgins, however, was also the author of such popular science works as *Mosaical and Mineral Geologies* (1832), in which he attempted to reconcile Genesis with the principles of geology.[37] But in describing the geological process, Higgins was entirely naturalistic. He argued that the earth "was created at some period, and this we deduce from science, independent of revelation." According to Higgins, "the great object of the Divine Spirit under whose guidance he [Moses] wrote, was to detail the history of man, his character, condition, and prospects," the main concern being to demonstrate *why* God created and not *how*. He noted further, following a theory first articulated by Scottish divine Thomas Chalmers (1780–1847), that "this was done before the six days; how long, we are not informed, and are, consequently, at liberty to attempt to determine it by the assistance of science."[38]

Unsurprisingly, then, when Draper began lecturing at Hampden-Sydney College, he also displayed a number of strong affinities to a more rational theology. What is particularly noteworthy about these early lectures is that he was already making references to the progress of science. Thus, as early as the 1830s, Draper was instructing students that the Middle Ages was held captive by a "fanatical spirit," and that the "Mohammedans and Saracenic conquests" brought astronomy and algebra back to the West. At the close of the thirteenth century, he asserted, "the human intellect awoke from its sleep." Intellectual advance, however, was often opposed by "governments." The same government that "laid violent hands on Galileo," he averred, also burned books of astronomy and geography in England. In what would become a characteristic refrain in later writings, he declared that "the persecutions which were endured by philosophers from the malice of princes, could neither rein nor stop the progress of knowledge." Men of science had perceived "eternal decrees" in nature, the presence of law and order operating everywhere. All manner of existence is the "the result of simple and uniform laws." Significantly, Draper gave the study of these natural laws a religious character, musing that "there is something in the calm regularity of these laws, that persuades us to commit ourselves unreservedly to their operation."[39]

Upon taking up the Chair of Chemistry at the University of New York in 1837, Draper further elaborated on these themes. He repeated that "there is not a force in nature which does not affect us," and that all change is the result of "fixed and immutable laws." He extended these laws to individuals and society, arguing that as individuals pass away, so have "whole species, tribes, and genera." History attested, he claimed, to an organic process of life, death, and rebirth. Nearly two decades before Darwin published his *Origin of Species*, in these lectures Draper announced his support of developmentalism.[40]

But Draper also attempted to reassure students that the connection between the inorganic and the organic did not pose a problem for religious belief. Beyond the material world, he said, there exists "an intelligent principle with all its affections and feelings, and acquisitions and knowledge unaltered and untouched." Studying nature demonstrates, he claimed, the "peculiarity of the works of God."[41] Draper believed that the plan of nature reveals "not a God of expedients, but the master of endless resources." In a remarkable statement, he even claimed that to study nature was to elevate the "mind of a philosopher to a perception of the laws upon what it pleases God to govern the universe."[42]

While there are unmistakable and evocative hints of a natural theology in these early lectures, where the study of nature is likened to reading the mind of God, Draper had shunned Christian orthodoxy. For example, in a lecture on oxygen gas, he spoke very highly of English chemist, natural philosopher, and Unitarian minister Joseph Priestley (1733–1804), praising especially his religious opinions. "We must not impute it to mental weakness," he told students, "but rather to a pursuit of *truth,* that in succession [Priestley] passed through many phases of religious belief." In Priestley, Draper found a "deeply pervading and redeeming faith." He also explicitly rejected the "feeble contrivances" of incessant divine intervention in nature. A "reflecting mind," he argued, would not oppose the idea of a God acting "through ancient and self-imposed law." While it made no difference to him whether God was directly involved or had set in place certain laws, Draper maintained that experience demonstrates that nature and civilization operated by certain, fixed, immutable laws. He wrote that "a Pure Intelligence will rarely act by intervention, but always through law."[43] For Draper, then, God was the "great Agriculturist" who worked the farm of civilization by "rotations." Both seasons in nature and seasons in civilization "arise true to their times, as with the precision of clock-work." The same fundamental laws that regulate biology, he contended, operate in human social organization, in the life and death of men and empires. The lesson to be learned from these reflections, as Draper put it, is that "the physical laws of the world are universal and unvarying in their operation."[44]

While many of these themes reappear in Draper's popular textbooks,[45] they were expressed more fully in his *Human Physiology,* first published in 1856.[46] In this book Draper aspired to remove from physiology what he believed was an obstinate and pervasive "mysticism." He attempted to demonstrate that life was the result of chemical and physical processes rather than some otherworldly "vital force." At the same time, he also believed that the "right progress" of society depended on its "religious opinions." A "religion of superstition," he wrote, "is very liable to be connected with a life of evil works." Mankind needed a new guide to truth. That guide, Draper argued, was "Positive Science." The new positive science will inevitably reveal the truth about God, the human soul, immortality, the afterlife, and the nature of the cosmos. In short, this new guide, rather than renouncing religion, will ultimately reveal its most essential and reasonable tenets.[47]

It is important to point out that while most of *Human Physiology* dealt with chemistry, anatomy, and geography, Draper also dedicated several concluding chapters to examining the progress in history. He called his approach "comparative sociology," which he believed allowed him to see history as a science. The progress of mankind, he maintained, is like a "series of changes to those which have been traced in the psychical career of the individual, and this, whether we consider the progress in theology, policy, philosophy, or any other respect. It is a continued passage from the general to the special—from the homogeneous to the heterogeneous." Significantly, he went on to give a brief sketch of the history of religion, tracing its growth from fetishism to polytheism to a "purer" monotheism. The development of the religious consciousness, according to Draper,

DIFFERENCES IN MEN. 563

CHAPTER VII.

ON THE INFLUENCE OF PHYSICAL AGENTS ON THE ASPECT AND FORM OF MAN AND ON HIS INTELLECTUAL QUALITIES.

Differences in Form, Habits, and Color of Men.—Ideal Type of Man.—Its Ascent and Descent.—Causes of these Variations.

Doctrine of the Unity of the Human Race.—Doctrine of its Origin from many Centres.

Influence of Heat on Complexion.—Cause of Climate Variations.—Influence of Heat illustrated by the cases of the Indo-Europeans, the Mongols, the American Indians, and the Africans.—Distribution of Complexion in the Tropical Races.

Variations in the Skeleton.—Four Modes of examining the Skull.—Connection of the Shape of the Skull and Manner of Life.—Physical Causes of Variation of the Skull.

Influence of the Action of the Liver on Complexion.—Influence of the Action of the Liver on the Form of the Skull.—Base Form of Skull arising from Low as well as High Temperatures.—Disappearance of the Red-haired and Blue-eyed Men in Europe.

The Intellectual Qualities of Nations.—Synthetical Mind of the Asiatic.—Analytical Mind of the European.—Their respective Contributions to Human Civilization.—Spread of Mohammedanism in Africa.—Spread of Christianity in America.—Manner of the Progress of all Nations in Civilization.

THERE are great differences in the aspect of men.

The portrait of Newton is from the frontispiece of his immortal Prin-

ed the Marquis de l'Hôpital, himself a great contemporary French math-

Figure 1.3. Page from John W. Draper's *Human Physiology: Statical and Dynamical; or, The Conditions and Course of the Life of Man* (1856), where Draper argued, among other things, that environmental conditions helped explain the differences between men like Newton and the "Australian savage," who was lost, according to Draper, "in filth and vermin."

reveals that the human mind manifested "a premarked and predestined course in which it must go."[48]

In his *Human Physiology* Draper also offered an early critique of the "papal government," which he would repeat in later writings. The Church, he wrote, had become "the autocrat of Europe." But in her quest for power, he contended, the Church "forgot her duty." The central error of the "Italian government" was "the compression of human thought." Rome had misunderstood the qualities of the European mind, its progressive tendency and direction. "How different would it have been," he bemoaned, "if she had taken the lead, and directed the human mind in the channels through which it was destined to pass, instead of opposing herself as an obstacle! She might have guided, but she could not resist."[49]

EVOLUTIONISM

Another important feature of Draper's thought was what some scholars have called "evolutionary deism" or "providential deism," which was the popular belief that the developmental process was the unfolding of a divine plan instituted by God. This position first became popular during the late eighteenth century, among thinkers who hoped to reconcile belief in divine providence with the laws of nature. While God was separated from the cosmos, the natural order was nevertheless benevolently designed by him.[50] The origins of this view are typically traced to German transcendental biology, or *Naturphilosophie*, especially to the work of Johann Wolfgang von Goethe (1749–1832), Friedrich W. J. Schelling (1775–1854), Lorenz Oken (1779–1851), and Karl Ernst von Baer (1792–1876). As Nicholas Jardine has explained, such work attempted to comprehend nature in its totality, often using organic metaphors (growth, development, maturity, decay) to describe the activity of nature as a whole.[51] While this approach arose mainly in Germany, many ideas were shared with French natural historians. By the early nineteenth century, a great wave of transcendentalism had passed over biological thought, which centered on the idea that there existed a unique plan of structure between the development of the individual and the evolution of the species. Evolutionary deists thus often applied a wider teleological or progressive view towards society.

Draper's alma mater, the University of London, had from its inception deliberately emphasized science and philosophy from the Continent. Indeed, its curriculum had been influenced by German and French naturalists, especially Benoît de Maillet (1656–1783), George-Louis Leclerc, Comte de Buffon (1707–1788), Étienne Geoffroy Saint-Hilaire (1722–1844), Jean-Baptiste Lamarck (1744–1829), and Antoine Étienne Renaud Serres (1786–1868).[52] German and French evolutionary deists were thus highly influential on prominent British men of science. In his *Human Physiology*, for instance, Draper acknowledged his debt to such English writers as Humphry Davy (1778–1829), James Cowles Prichard (1786–1848), Robert Bentley Todd (1809–1860), William Benjamin Carpenter (1813–1885), James Paget (1814–1899), William Bowman (1816–1892), and William Senhouse Kirkes (1822–1864), men who were themselves greatly influenced by developmental theories from the Continent.

Draper himself also credited a number of Continental authors as well. He pointed his American readers to, for example, German explorer Alexander von Humboldt's (1769–1859) essays on the geographical distribution of plants. He also utilized French naturalist Louis-Jean-Marie Daubenton's (1716–1800) anatomical descriptions of quadrupeds; suggested readers follow the classificatory system of German anatomist Johann Friedrich Blumenbach (1752–1840) in examining skulls; and even found it useful to mention the work of German phrenologists Franz Joseph Gall (1758–1828) and Johann Gaspar Spurzheim (1776–1832). He also cited early embryologist Karl Ernst von Baer's "law of development" exactly, quoting him as stating that "the heterogeneous arise from the homogeneous by a gradual process of change."[53]

More telling still was Draper's reference to Parisian naturalist Comte de Buffon and his descriptions of the growth of man and the influence of climate. Buffon, a mentor of Lamarck, offered perhaps the most comprehensive and influential natural history of the eighteenth century. In his monumental *Histoire Naturelle, générale et particulière, avec la description du Cabinet du Roi* (1749–1804), which appeared in translated and abridged editions throughout the nineteenth century, Buffon envisaged the historical development of life as the result of uniform, natural processes, describing the whole of the natural world from its molten origins to the birth and organization of human civilization. While Buffon repudiated appeals to the "supernatural" in studying nature, he retained, as historian Peter Bowler observes, "a faith in the existence of an underlying pattern which constrains the activity of the material universe." That is, Buffon maintained the existence of an "internal mold" that directed development and acknowledged that the Creator had set things in motion.[54]

In later editions to his textbook on human physiology, Draper admitted that the original book "aimed at much more than was directly expressed upon its pages." He stated that "man must be studied not merely in the individual, but also in the race," and that there is an analogy between his advance from infancy to childhood, youth, manhood, and old age, and the progress of civilization. He announced that he had been working on a "companion" to *Human Physiology*, a book in which he would attempt to study the progress of civilization as a whole. That "companion" volume was his *A History of the Intellectual Development of Europe*, published in 1863.[55]

Before its publication, however, Draper shared portions of his thesis at two different venues. The first was an address at the opening ceremony of the Cooper Union for the Advancement of Science and Art in New York in 1859. In his speech Draper supported American industrialist, inventor, and Unitarian Peter Cooper's (1791–1883) educational reforms aimed at working-class men and women. Unsurprisingly, Draper believed education would improve and stabilize society in the rapidly changing environment of mid-nineteenth-century America. Draper declared that "the life of a nation is meant for intellectual development," and disclosed to his audience his desire "to see, feel, and understand and know that there are immutable laws designed in infinite wisdom, constantly operating for our good." According to Draper, God had placed the government of this world under the same laws as the development of man. History, then, demonstrated that "nations, like individuals, are born, run through an unavoidable career, and then die."[56]

Much of Draper's Cooper address reflected the work of educational reformer Henry Brougham (1779–1868), who was incidentally one of the leading founders of the University of London. To shed better light on the time and place of Draper's thoughts, it is important to briefly outline Brougham's own ideas. In 1802 Brougham and a group of other Whigs founded the *Edinburgh Review* as a platform for political and educational reform. Brougham believed strongly in the importance of scientific education for the lower classes. When he became Rector of the University of Glasgow in 1824, for instance, he delivered its inaugural address in which he offered guidelines for the "improvement of the mind." Brougham had basically separated secular and religious knowledge, declaring that:

> Real knowledge never promoted either turbulence or unbelief; but its progress is the forerunner of liberality and enlightened toleration. Whoso dreads these, let him tremble; for he may be well assured that their day is at length come and must put to sudden flight *the evil spirits of tyranny and persecution, which haunted the long night now down the sky*. As men will no longer suffer themselves to be led blindfold in ignorance, so will they no more yield to the vile principle of judging and treating their fellow creatures, not according to the intrinsic merit of their actions, but according to the accidental and involuntary coincidence of their opinions. The Great Truth has finally gone forth to all the ends of the earth, THAT MAN SHALL NO MORE RENDER ACCOUNT TO MAN FOR HIS BELIEF, OVER WHICH HE HAS HIMSELF NO CONTROL. Henceforward, nothing shall prevail upon us to praise or to blame anyone for that which he can no more change than he can the hue of his skin or the height of his stature.[57]

That same year, Brougham also published his *Practical Observations upon the Education of the People, addressed to the Working Classes and their Employers*. In this work he discusses the education of adults, particularly the working population of England. "People must be the source and the instruments of their own improvement," Brougham declared. He believed this was possible with the diffusion of cheap popular science books. To this end, Brougham founded in 1826 the London Society for the Diffusion of Useful Knowledge, perhaps the most prominent, influential, and controversial educational program in early nineteenth-century Britain, putting hundreds of books and periodicals into the hands of hundreds of thousands of readers of all classes. While his own efforts were initially met with resistance, Brougham proclaimed that there are now few "objections to the diffusion of science among the working classes." "Happily the time is past and gone," he says, "when bigots could persuade mankind that the lights of philosophy were to be extinguished as dangerous to religion; and when tyrants could proscribe the instructors of the people as enemies to their power. It is preposterous to imagine that the enlargement of our acquaintance with the laws which regulate the universe, can dispose to unbelief." Science, rather, is the cure for superstition and a remedy for intolerance: "a pure and true religion has nothing to fear from the greatest expansion which the understanding can receive by the study either of matter or of mind."[58]

As James Secord put it, Brougham's goal was "nothing less than the complete reformation of society through knowledge."[59] Then, in 1827, Brougham published his *Discourse on the Objects, Advantages, and Pleasures of Science* (1827), where he presented, as Draper himself would later present, a sweeping overview of all branches of knowledge.[60] Like Draper, Brougham's survey was less descriptive than prescriptive. His overview was intended to show how the study of science "elevates the faculties above low pursuits, purifies and refines the passions, and helps our reason to assuage their violence." Interestingly, the highest gratification, the highest pleasure, we may gain from contemplating the sciences, Brougham maintained, is that "we are raised by them to an understanding of the infinite wisdom and goodness which the Creator has displayed in his works." This indeed is the most consoling inference, to be able to follow "with our own eyes, the marvelous works of the Great Architect of Nature." As we shall see, Brougham was among many writers in the nineteenth century that claimed the progress of knowledge would not only have material benefits, but religious ones as well.

The second venue where Draper developed his own views was the thirtieth annual meeting of the British Association for the Advancement of Science (BAAS), held at Oxford in 1860. That meeting, of course, became legendary. According to the conventional view, Darwin's "bulldog," Thomas H. Huxley, and the Bishop of Oxford, Samuel Wilberforce, clashed over Darwin's theory of

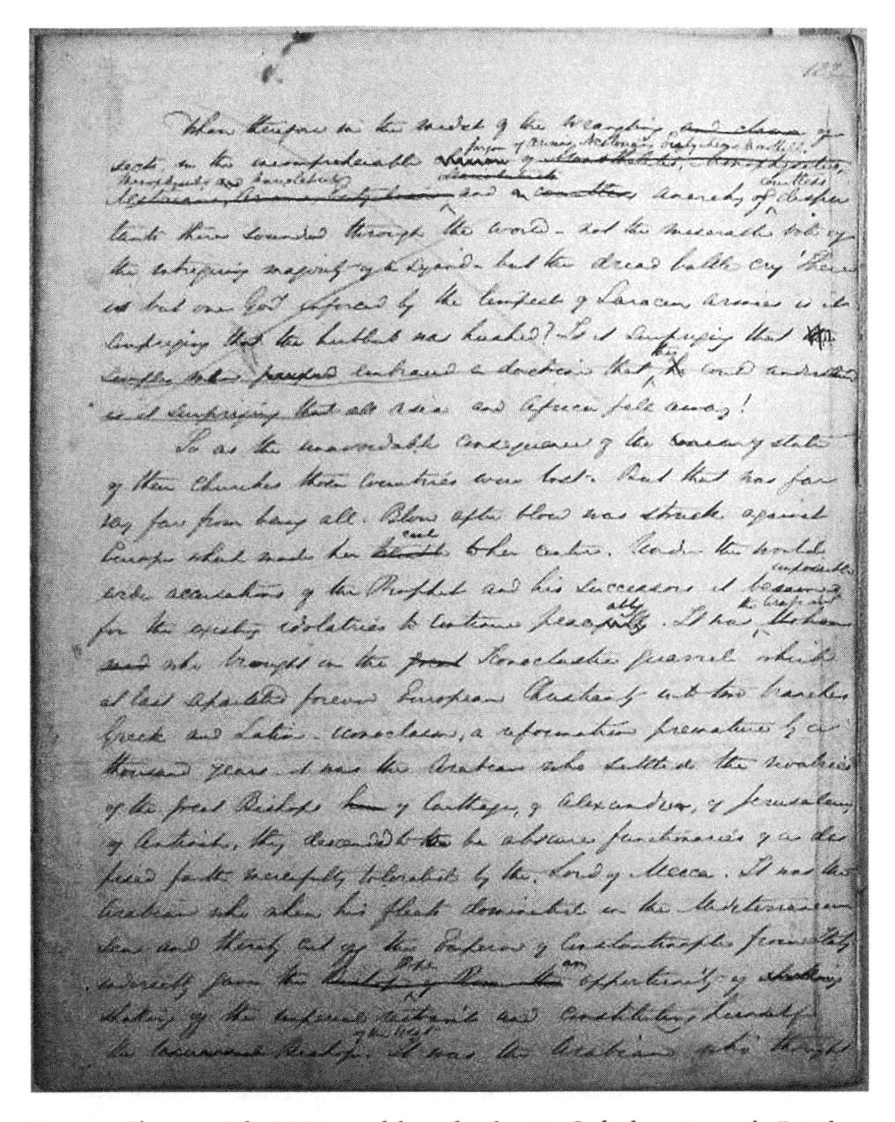

Figure 1.4. The paper John W. Draper delivered at the 1860 Oxford meeting at the British Association for the Advancement of Science. Courtesy of the Library of Congress, Manuscript Division, John William Draper Family Papers, Container 8.

evolution.[61] Since most of the primary and secondary literature has focused on this altercation, it is often forgotten that Draper had delivered one of the principal papers of that meeting. His paper was initially entitled "On the Possibility of Determining the Law of the Intellectual Development of Europe," but at some point Draper amended it to fit the occasion, retitling it "The Intellectual Development of Europe (considered with reference to the views of Mr. Darwin and others) that the Progression of Organisms is Determined by Law." Despite invoking Darwin's name, however, the paper was more a reflection of Draper's own conception of evolution. His central argument was the same one he had made since the beginning of his career—namely that "the progression of organisms is determined by law."[62] Draper was also more willing than Darwin was at the time to accept transmutation of species. For instance, he regarded humanity "as the first member of an infinite series of organisms all composed of the same elements[,] chemical and anatomical[,] subject to the same influences[,] governed by the same laws." From a microscopic speck to the ascension of "his humbler comrades" (i.e., apes) the study of humanity reveals a long career of submission to universal law.[63]

But rather than debating the details of evolutionary theory, Draper was more interested in arguing that "civilization does not occur accidentally or in a fortuitous manner, but is determined by immutable law." In other words, the same physical laws directing biological evolution also direct the progress of human society as a whole. "In this manner," he told his Oxford audience, "there emerges into prominence the noble conception that man is the archetype of Society, that individual development is the model of social progress." But like any other living organism, he argued, societies are "necessarily ephemeral" and "altogether transient." Death, in short, was a necessary condition of all life. History manifests this "perfectly definite course." Examining the history of civilization demonstrates that nations and empires run through the same cycles as human development. The rise, decline, and fall of a nation are determined by immutable laws.

Draper more explicitly outlined his views on evolution in 1877, when he was invited to give an address to a group of Unitarian ministers in Springfield, Massachusetts, on "Evolution: Its Origin, Progress, and Consequences."[64] Interestingly, he told the ministers that the idea of evolution had been discussed for centuries and that it was inextricably connected to questions of theology. In explaining the origin of life, for example, the issue had always been divided between the "creationists" and "evolutionists." On the one hand, the creationists asserted "that Almighty God called into sudden existence, according to his good pleasure, the different types of life that we see." On the other, the evolutionists claimed "that from one or a few original organisms all those that we see have been derived, by a process of evolving or development." The creationist sees each species as independent and divinely appointed by God. The evolutionist, by contrast, does not admit any divine intervention, sees each species as interrelated and constrained by the universal reign of law, and is ultimately silent about the origin of organisms.[65]

Ultimately, Draper seemed to promote a neo-Lamarckian version of evolution. He told the Unitarian ministers that nature presented itself in "geometrical forms," and that "every organism is the result of the development of a vesicle, under given conclusions, carried out in material execution." The "genealogy of any organism," he argued, manifests the "dominion of law." But while he maintained that the "capricious intrusion of a supernatural agency has never yet occurred," Draper believed that natural law nevertheless magnified "the unutterable glory of Almighty God." Indeed, he said the evolutionary hypothesis offered "nobler views of this grand universe of which we form a part, nobler views of the manner in which it has been developed in past times to its present state, nobler view of the laws by which it is now maintained, nobler expectations as to its future." But above all, evolution offered a more vivid conception of the "awful majesty of the Supreme Being." He reminded his audience that the "doctrine of evolution has for its foundation not the admission of incessant divine interventions, but a recognition of the original, the immutable fiat of God." History has demonstrated, he declared, that the "heaviest blow the Holy Scriptures have ever received was inflicted by no infidel, but by ecclesiastical authority." He thus emphatically urged members of the audience to find a means of reconciling evolution and Scripture.[66]

THE SPIRIT OF THE AGE

Draper's *Intellectual Development* was indeed an expanded version of his Oxford lecture. But before we examine that work in detail, we still need to further set its ideas in context. As already mentioned, Britain in the early decades of the nineteenth century had experienced a communications revolution. In 1831, Draper's University of London classmate philosopher John Stuart Mill wrote a series of articles in the weekly radical paper, *The Examiner*, on the "Spirit of the Age." Crucially, for Mill, the spirit of the age was an age of transition. The "march of the intellect," according to Mill, had pitted "men of the present age" against "men of the past." He pronounced that mankind had "outgrown old institutions and old doctrine," and that a "change has taken place in the human mind, a change which, being effected by insensible gradations, and without noise, had already proceeded far before it was generally perceived."[67]

The early decades of the nineteenth century had indeed witnessed the ever-expanding scope of popular education and the growing demand for the availability of scientific publication. Shortly after Brougham's campaign, for example, such publishers as Archibald Constable (1774–1827), John Murray (1808–1892), Thomas Norton Longman (1771–1842), and William (1800–1883) and Robert Chambers (1802–1871) began producing good, cheap literature to enlighten the middle and working classes.[68] Like Brougham, these publishers were particularly interested in spreading the latest scientific advances to a broad audience. As Bernard Lightman has shown, the emergence of the professional scientist created a golden age of popular science writers.[69] But with the organization of such scientific societies as the Royal Institution (1799), Geological Society (1807), Zoological Society (1826), and the BAAS (1831), among others, there was an increasing need to provide an overarching narrative of the growth, meaning, and place of science in modern life.

Near the end of his life, for example, the gifted electrochemical experimenter and arguably the most visible and influential practitioner of science in early nineteenth-century Britain, Humphry Davy—whom, recall, Draper had referenced in his *Human Physiology*—contemplated nature's harmony and its underlying unifying principles in his posthumously published *Consolations in Travel, Or the Last Days of a Philosopher* (1830). In this exceptionally popular book, Davy offers a history of the progress of science in order to understand the present, alluding to the "remarkable laws belonging to the history of society" from which we might be able to develop "higher and more exalted principles of being." This law of progress was providential, for it leads not only to material advance, but spiritual enlightenment. Christianity, Davy seemed to intimate, was one step in the progress of religion.[70] Thus, according to Davy, the progress of science and religion will be guided by a scientific clerisy who could take society in a new direction.[71]

Notions of progress were pervasive in the nineteenth century. Historians, theologians, and scientists maintained that history demonstrated inexorable laws of progress.[72] For instance, such notions were particularly representative among members of the "Cambridge Network," which included such English polymaths as Charles Babbage (1791–1871), John F. W. Herschel (1792–1871), William Whewell (1794–1866), and others. These "gentlemen of science" produced reflective scientific treatises that went beyond simply studying the natural world. They were participants of a much larger cultural project, which included promoting, supporting, delineating, interpreting, and popularizing science. Since many of them were clerics, they were also concerned with securing faith in God's dual revelations, the Book of Scripture and the Book of Nature. As such, they looked to the sciences as a tool of intellectual and religious reform. This often required accessing and evaluating its history. As Richard Yeo aptly summarizes, during the first half of the nineteenth century, "public discourse on science served at least two purposes: it not only conveyed scientific discoveries to the public, but also legitimated science as part of a cultural discourse." These "metascientists," as Yeo calls them, shaped the cultural meaning of science, and often published lengthy, progressive philosophies or histories of science. These histories were concerned with the nature of science broadly conceived: its history, ethos, metaphysical foundations, and—perhaps most importantly—relations to religion.[73]

It is important to note that this "old guard" of clerical naturalists, whose science was colored by a natural theology in which nature was indicative of divine design, were often theologically liberal. In his own scientific work, Draper made numerous references to the old guard of science, betraying a deep longing to be part of the scientific scene in England. In his *Scientific Memoirs* (1878), for instance, which were a collection of previously published articles on his experimental researches in radiation, Draper demonstrates a close and consistent reading of the BAAS reports.[74]

In his more popular work, Draper also recommended to his readers the work of Scottish minister and popularizer of science Thomas Dick (1774–1857), the author of *The Christian Philosopher, or, the Connection of Science and Philosophy with Religion* (1823).[75] Significantly, Dick accused Christians of overlooking the "visible manifestations of the attributes of the Deity." He even discussed the "bad effects of setting Religion in opposition to Science." The true "Christian philosopher," he insisted, seeks to "illustrate the harmony which subsists between the system of Nature and the system of Revelation." Dick warned his readers of the "pernicious effect on the minds of the mass of the Christian world, when preachers in their sermons endeavour to undervalue scientific knowledge, by attempting to contrast it with the doctrines of Revelation." The Christian philosopher who investigates the works of creation under the guidance of true science will expand his conceptions of the power, wisdom, benevolence and superintending providence of God. According to Dick, "it is now high time that a complete reconciliation were affected between these contending parties."[76]

As a Presbyterian minister, by "Christian" Dick meant, of course, Protestantism. He argued that natural philosophy enabled one to detect "pretended" miracles. Ignorance of the principles of natural science had allowed "pretenders to supernatural powers" to beguile humanity. These pretenders, according to Dick, were "Pagan and Popish Priests" who endeavored to "support the authority of their respective religious systems." But in the light of modern science, he said, "every species of degrading superstition, vanish into smoke." Dick also criticized Protestant Christians for taking a too literal interpretation of Scripture and feared the advances of science. In short, Dick wanted to reassure Protestants that science was safe. Despite what they might have read in tracts, or heard from the pulpit, Protestants ought to use science in support of "true religion," which was in essence "gratitude to the God of our life, and the Author of our salvation."[77]

Perhaps the most striking parallel between Dick and Draper was his *On the Improvement of Society by the Diffusion of Knowledge* (1833). Here Dick argued that the historical record revealed a "melancholy scene of intellectual darkness." There were a few "rays of intellectual light" in ancient civilization, but these were "doomed to be speedily extinguished." While the hordes of barbarians almost "annihilated every monument of science and art which then existed," the "debasing superstitions of the Romish church, the hoarding of relics, the erection of monasteries and nunneries, the pilgrimages to the tombs of martyrs and other holy places, the mummeries which were introduced into the services of religion, the wild and romantic expeditions of crusaders, the tyranny and ambition of popes and princes and the wars and insurrections to which they gave rise, usurped the place of every rational pursuit, and completely enslaved the minds of men." A "night of ignorance" thus ensued, in which both civil and ecclesiastical governments worked together to retard "the progress of the human mind." Like Draper, Dick spoke of the "laws uniformly operating" in nature as evidence of the Power, Intelligence, Wisdom, and Benevolence of the Divine Being.[78]

Remarkably similar perspectives also existed between Draper and a number of "gentlemen of science," including Babbage, Herschel, and Whewell. Known for his magnificent calculating engine, Babbage, for example, had criticized the Established Church for constantly interfering with the progress of science. He called on Protestants to consider the merits of a more "Natural Religion," and thus avoid imitating "the bigoted Romanists who imprisoned Galileo." He bitterly remarked that while the "best informed and most enlightened" clergymen see no danger in the advance of science, a significant portion "protests against anything which can advance the honour and the interests of science." According to Babbage, "all established religions are, and must be in practice, political engines—they have a strong tendency to self-aggrandisement. Our own is by no means exempt from this very natural infirmity."[79]

Herschel, whose reputation in England is difficult to overstate, had in fact corresponded with Draper in 1840 about his daguerreotypes, in which he improved on Louis Daguerre's (1787–1851) process for photography.[80] During the BAAS meeting at Oxford in 1860, Draper had hoped to visit him at his residence in Slough.[81] In his enormously influential *A Preliminary Discourse on the Study of Natural Philosophy* (1830), Herschel maintained that the study of nature leads man "to the contemplation of a Power and an Intelligence superior to his own," thus rendering "doubt impossible." In later editions, he put it even more strongly, insisting that science "places the existence and principal attributes of a Deity on such grounds as to render doubt absurd and atheism ridiculous."[82]

But while natural philosophy affords religion testimony, Herschel argued, it should never be bent or twisted to fit "narrow interpretations of obscure and difficult passages in the sacred writings." These were the "bigots" and "dreamers" who persecuted Galileo and who continue to obstruct the progress of science in the present day. The study of nature should foster humility and unfetter the mind from "prejudices of every kind." In an oft-quoted passage, one that Draper paraphrased in his early lectures at Hampden-Sydney College, Herschel declared that "There is something in the contemplation of general laws which powerfully induces and persuades us to merge individual feeling, and to commit ourselves unreservedly to their disposal; while the observation of the calm, energetic regularity of nature, the immense scale of her operations, and the certainty with which her ends are attained, tends, irresistibly, to tranquillize and re-assure the mind, and render it less accessible to repining, selfish, and turbulent emotions."[83]

Though Herschel saw an alliance between natural philosophy and belief in the "Divine Author," he was indeed among the more theologically moderate of the "old guard." For instance, when chemist Capel H. Berger (1839–1868) petitioned him to sign the "declaration of students of the natural and physical sciences," which called on scientists to publicly declare "that it is impossible for the Word of God, as written in the book of nature, and God's Word written in Holy Scripture, to contradict one another, however much they may appear to differ," Herschel responded in an open letter in the literary magazine *Athenaeum* that to sign such a document would be "an infringement of that social forbearance which guards the freedom of religious opinion in this country with especial sanctity." He added that he considered "this movement simply mischievous, having a direct tendency (by putting forward a new Shibboleth, a new verbal test of religious partisanship) to add a fresh element of discord to the already too discordant relations of the Christian world."[84] Early in life, moreover, Herschel had made a distinction between a "religion established by law" and one "established by nature," and even expressed hostility toward orthodox Christianity to his Cambridge friend John William Whittaker (1790–1854).[85] This distinction allowed Herschel to bewail the "bigots" and "dreamers" who persecuted Galileo, and claim that there was a "great eclipse of science" that lasted for nearly eighteen centuries, "till Galileo in Italy, and Bacon in England, at once dispelled the darkness."[86]

Herschel shared the same intellectual context in which both Whewell and geologist Charles Lyell (1797–1875) worked. Unsurprisingly, then, both Lyell and Whewell constructed similar histories of science. Lyell, an ardent liberal Protestant who attended the services of Unitarian minister James Martineau (1805–1900), rejected all attempts at harmonizing Genesis and geology. In his *Principles of Geology* (1830–33), for instance, he wrote that "the progress of geology is the history of a constant and violent struggle between new opinions and ancient doctrines, sanctioned by the implicit faith of many generations, and supposed to rest on scriptural authority."[87] In his *History of the Inductive Sciences* (1837), Whewell also demanded strict separation of science from theology, and similarly spoke of the history of science as a drama of "conflict," "war," "contest," and "crisis" between the new knowledge and theological dogmatism. Like Herschel and Lyell, Whewell warned against the error of seeking "a geological narrative in theological records." Relying on "a few phrases of the writings of Moses," he advised, will only lead to "arbitrary and fantastical inventions." This hostile attitude toward natural philosophy, according to Whewell, was present throughout the history of Christianity. "During a considerable period of the history of the Christian church," he wrote, "the study of natural philosophy was not only disregarded but discommended."[88]

While Draper was undoubtedly concerned with the cultural meaning of science, he should not be mistaken as one of the "gentlemen of science," however much he desired to be part of the British scene. Rather, Draper was likely more deeply influenced by a number of science popularizers of his own generation. Perhaps most obvious, a great affinity existed between him and Scottish publisher Robert Chambers. Chambers first met Draper at the 1860 BAAS meeting in Oxford. He was so impressed with Draper's speech that he invited him to his home in Edinburgh. Later, Draper had sent Chambers a copy of his *Intellectual Development*. In a remarkable 1864 letter, Chambers thanked Draper for the volume, declaring it a brilliant work. He was particularly impressed with Draper's "view of human progress and human destiny." Chambers especially admired Draper's point that theology has always obstructed real knowledge and that the "pretension to infallibility, whether in a pope or a book," is the "master evil of the world." Biblical criticism and science, he asserted, are "working a vast change" in the history of humanity. However, Chambers denied that this "vast change" was in any way inimical to "religion." "After all that we have learned of the natural system of the world," he wrote, "there appears to me ample scope and range for the religious feelings in regard to the divine author and ruler, and in the concerns of the immaterial principle which works within us and is, as we hope, destined to survive the frail body."[89]

Chambers was of course known posthumously as the author of the sensational *Vestiges of the Natural History of Creation*, first published in 1844.[90] In his book, Chambers advanced a theory of progressive development based on uniform natural laws. Like Draper, he suggested that the stages of progress in nature were "as if, seeing a child, a boy, a youth, a middle-aged, and an old man together."[91] Chambers also believed his *Vestiges* presented a case for a new natural theology, and made constant reference to "God," "Divine Author," "Great Ruler of Nature," "Author of Nature," and "Creator," to mention but a few epithets.[92] When numerous naturalists criticized *Vestiges*, Chambers reminded readers in a "sequel" published in 1846 that he never intended to write a scientific work, and that the *Vestiges* sought to articulate a new worldview rather than a scientific theory.[93]

Another important Scot was phrenologist George Combe (1788–1858), whom, incidentally, Chambers greatly admired.[94] His famous *Constitution of Man considered in Relation to External Objects* (1828) promoted the idea that the laws of nature are universal, unbending, and progressive. Significantly, while he rejected

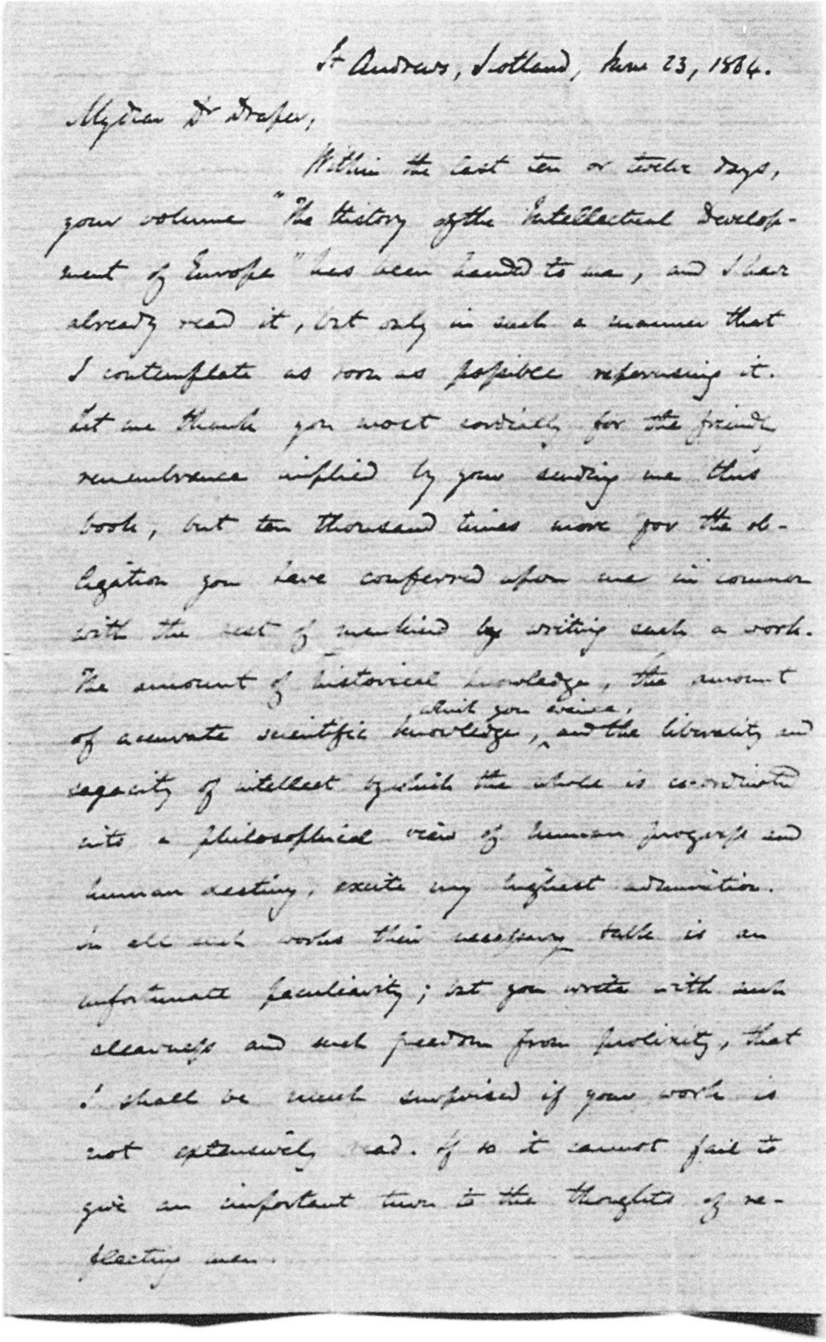

St Andrews, Scotland, June 23, 1864.

My dear Dr Draper,

Within the last ten or twelve days, your volume "The History of the Intellectual Development of Europe" has been handed to me, and I have already read it, but only in such a manner that I contemplate as soon as possible reperusing it. Let me thank you most cordially for the friendly remembrance implied by your sending me this book, but ten thousand times more for the obligation you have conferred upon me in common with the rest of mankind by writing such a work. The amount of historical knowledge, the amount of accurate scientific knowledge which you evince, and the liberality and sagacity of intellect by which the whole is co-ordinated into a philosophical view of human progress and human destiny, excite my highest admiration. In all such works their necessary bulk is an unfortunate peculiarity; but you write with such clearness and such freedom from prolixity, that I shall be much surprised if your work is not extensively read. If so it cannot fail to give an important turn to the thoughts of reflecting men.

Figure 1.5. Scottish publisher and anonymous author of the sensational *Vestiges of the Natural History of Creation* (1844), Robert Chambers (1802–1871) thanking John W. Draper in 1864 for his copy of *A History of the Intellectual Development of Europe*. Courtesy of the Library of Congress, Manuscript Division, John William Draper Family Papers, Container 2.

the doctrines of orthodox Christianity, Combe was no atheist.[95] He argued that the laws of nature attested to the wisdom and benevolence of its Creator, and even regarded his *Constitution* as a work of "natural theology." In discussing the "relations between science and scripture," he wrote that in all ages "new doctrines have been charged with impiety." But he contended that opposition between science and revelation was "impossible." If the "facts in nature are *correctly* observed, and divine truth is *correctly* interpreted," there should be no conflict between religion and science. According to Combe, revelation was never "intended to supersede the necessity of all other knowledge."[96]

Combe also published a remarkable treatise entitled *On the Relation between Religion and Science* (1847), a document almost entirely ignored by historians of science. He explained that the work of natural theologians convinced him that God reigns through fixed, immutable natural laws. Interestingly, he also argued that the Reformation remains to be completed, equated progress in religion with progress in knowledge, and even accused "religious professors" of atheism when they denied the laws of nature. What needs to occur, according to Combe, is a second or "new Reformation."[97]

While men like Combe and Chambers rejected orthodox Christianity, they nevertheless drew from a nineteenth-century natural theological tradition that claimed moral and spiritual value for the study of the laws of nature. This belief, as we shall see later, had a long precedent among English Protestant natural and experimental philosophers. But natural theology always had protean qualities.

THE TYRANNY OF THEOLOGY OVER THOUGHT

Whether or not Draper was directly influenced by these men, the similarities between them are indicative of a much wider currency of thought. As we have seen, one of the defining characteristics of both gentlemen of science and science popularizers was the use of history to support a new, emerging scientific worldview. Draper no less took to surveying the historical record in support of his conceptions of scientific and religious progress in history. If we turn to his *Intellectual Development*, for example, we notice immediately that Draper, like other early nineteenth-century British natural philosophers, proposed an "intellectual history" of Europe that explicated the changes in its philosophy, science, literature, religion, and government. Change, according to Draper, occurred not by "incessant divine interpositions" or "by miracles and prodigies"—rather, all affairs followed each other in the relation of cause and effect. Such a conception of history revealed its "majestic grandeur," and ultimately disclosed the very mind of God. These sentiments should be unsurprising. Nineteenth-century intellectual history was not isolated from moral discourse. As we have seen and shall see again later, many other contemporary historians offered similar holistic visions.[98]

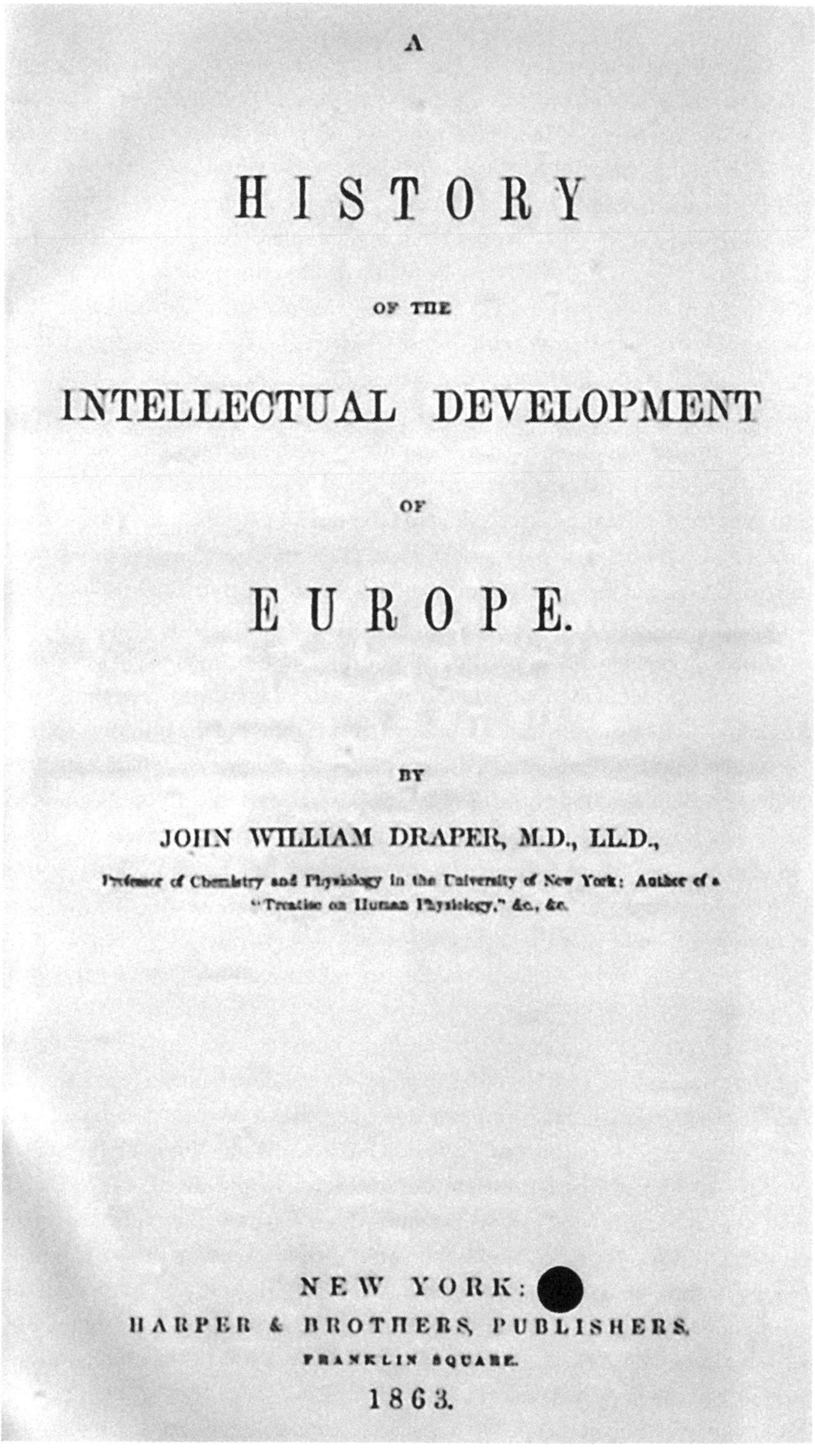

A

HISTORY

OF THE

INTELLECTUAL DEVELOPMENT

OF

EUROPE.

BY

JOHN WILLIAM DRAPER, M.D., LL.D.,

Professor of Chemistry and Physiology in the University of New York: Author of a "Treatise on Human Physiology," &c., &c.

NEW YORK:
HARPER & BROTHERS, PUBLISHERS,
FRANKLIN SQUARE.
1863.

Figure 1.6. Title page for John W. Draper's *A History of the Intellectual Development of Europe,* which he completed in 1861 but due to the American Civil War could not publish until 1863.

Surveying multiple countries, centuries, governments, wars, philosophies, religions, and scientific theories, Draper's *Intellectual Development* offers a grand overarching narrative of inevitable progress—a progress he considered to be driven by natural law. But what is also evident upon a closer inspection of Draper's *Intellectual Development* is its theological character. He begins, for example, by discussing the origins, development, and extinction of Hindu, Egyptian, Greek, and Roman religions. When he finally reaches the topic of the intellectual development of Europe, he offers nothing less than a history of Christianity. In fact, the bulk of his narrative was concerned with Christian history. At the outset of this history, he made an important distinction that many historians of science have ignored. He argued that the "pure doctrines" of the primitive church were debased by the "commingling" of "ceremonies of the departing creed." That is, the dying creeds of the Greco-Roman world had been absorbed by early Christianity and thus "paganized." According to Draper, this paganization corrupted the original, pure, and simple message of Christ. He wrote: "that I may not be misunderstood, I here, at the outset, emphatically distinguish between Christianity and ecclesiastical organizations. The former is a gift of God; the latter are the product of human exigencies and human invention, and therefore open to criticism, or, if need be, to condemnation." In other words, the theological system of the Catholic Church was categorically different from the teachings of the primitive church. He argued that the surrounding circumstances forced primitive Christianity to undergo a dramatic transformation. It was "contaminated." "In the beginning," he declared, "the creed and the rites were simple; it was only necessary to profess belief in the Lord Jesus Christ, and baptism marked the admission of the convert into the community of the faithful." As his narrative progresses, it becomes more and more apparent that the distinction was, more precisely, between what he called "true religion" and all later theological interpretations, Catholic and Protestant alike.[99]

More than any other figure, Draper blamed the Roman Emperor Constantine for contaminating or "materializing" primitive Christianity. He argued that the "fatal gift of a Christian emperor had been the doom of true religion." He also faulted the Alexandrian or Platonic Christians for adulterating the simple message of Christ. The Trinitarian controversies of the fourth century concluded with the orthodox party victorious, which was politically enforced by the emperor. But the "horrible bloodshed and murders attending these quarrels in the great cities," he asserted, "clearly showed that Christianity, through its union with politics, had fallen into such a state that it could no longer control the passions of men." With the succession of Constantine's sons, "religion had disappeared, and theology had come in its stead."[100]

Draper proclaimed that those "who had known what religion was in the apostolic days, might look with boundless surprise on what was now ingrafted upon it, and was passing under its name." As it gained greater power and authority, the Church became increasingly more corrupt. It was, as he put it, the "tyranny of Theology over Thought." But soon a collected revolt emerged against the "Italian system," beginning first with the Germans, and then spreading throughout Europe. According to Draper, Latin Christianity had to be destroyed and built up anew. He pointed to thirteenth-century Calabrian abbot Joachim of Fiore (ca. 1135–1202), for example, and his doctrine of the "everlasting gospel" as displaying an "enlarged and masterly conception of the historical progress of humanity." He cited his triadic division of history, tacitly proclaiming that Joachim was correct in predicting that the final stage will usher in a "new time" of "wisdom and reason" to replace ecclesiastical institutions and clergy.[101]

According to Draper, a growing coalition of intellectuals began asserting reason and the right of private judgment in the individual, a condition, he insisted, "utterly inconsistent with the dominating influence of authority." This principle, he claimed, was at the heart of the Reformation. The monk Martin Luther (1483–1546) had brought out the "grand idea which had hitherto silently lain at the bottom of the whole movement"—namely, "the right of individual judgment." But by the time we reach Genevan reformer John Calvin (1509–1564), however, the movement was already coming to an end. Alliances were divided and contending Protestant groups began attacking each other. Nevertheless, the Reformation "introduced a better rule of life, and made a great advance toward intellectual liberty." But while the Reformation ceased to exist as a movement in Europe, he argued, it found its fulfillment in America.[102]

It is only near the end of the volume that Draper finally begins outlining what later historians of science have called episodes of "conflict" between "religion and science." But these were not Draper's terms. It must be emphasized that he continued to make a distinction between "ecclesiastical power and authority" and "true religion." When the geocentric and heliocentric theories in astronomy were fiercely debated, for example, Draper argued that the Church was constrained to uphold the former since it was the long-held "ecclesiastical view." He also pointed out that adherents of heliocentrism were attempting to elucidate the works of God, and were therefore not irreligious. The mechanical philosophy, the discoveries of resistless laws and the uniformity of nature, universal gravitation, and the nebular hypothesis that appeared during this period were likewise not necessarily irreligious. But one thing was perfectly clear: any anthropocentric view of our existence, of God, or of nature is no longer tenable. While scientific advance was not in itself "inconsistent with the admission of a Providential guidance of the world," he warned that man "is not always a reliable interpreter of the ways of God." He thus believed the chief characteristic of an individual's maturity was "intellectual advancement and improvement" in science *and* religion.[103]

THE CIVIL WAR AND THE FUTURE POLICY IN AMERICA

Long before he published his *History of Conflict*, then, Draper had already turned his thoughts to history, politics, sociology, philosophy, and religion. He continued to address these issues in other writings that historians of science ignore or count as irrelevant.[104] In the aftermath of the American Civil War, for instance, Draper turned his thoughts away from Europe and to the *Future Civil Policy in America*, which was originally a course of lectures he delivered before the New York Historical Society in 1864. In this work, he argued that the progress made in America was not an isolated event but one event in a "chain of Empires." In describing this "chain of development," he reasserted many of his earlier positions, especially his belief that the "paganization" of Christianity had led to conflicts with the new learning. He confidently declared that history demonstrated the consequences of adhering too tightly to the "dead weight" of an ossified theology. In those countries where the "ecclesiastical system" still dominated, he asserted, "every onward step that science makes implies a conflict." In America, however, there was no such "dead weight"—if only the American clergy could avoid the same great and mortal mistake of the Catholic Church. Citing the classic two-books metaphor, he called on his American audience to "remember that the Reformation remains only half completed, until to the free reading of the *Book of God* there is added the free reading of the *Book of Nature*."[105]

Over the course of the next decade, Draper would further develop this theme of "conflict." In his three-volume *History of the American Civil War* (1867–70), he posited that the course of the war "had been influenced by uncontrollable causes," and that it "exemplified the great truth that societies advance in a preordained and inevitable course." The war was thus an inescapable effect of physiological factors resulting from the operation of universal laws. The differing climates of North and South, he argued, instigated sectional disputes that eventually led to conflict. Whereas the climate of the North produced freedom, the climate of the South encouraged slavery. The resolution of this conflict was "knowledge," according to Draper, and particularly knowledge of natural law. He refused to accept the assumption that the conflict was incorrigible, though, and believed that a better understanding of the laws of nature would eliminate the antagonism on both sides.[106]

As his narrative of the Civil War unfolds, it becomes clear that when Draper spoke of the conflict between the North and South, he meant more than just geographical regions. The difference in philosophical and even theological outlooks also led to the growing divide. In other words, this was a conflict between opposing worldviews. In strikingly familiar language, he argued that the "South mistook the spirit of the times. She did not recognize that modern civilization is adverse to her institution." In short, Draper interpreted the conflict between the North and South as the same conflict between ecclesiasticism and the modern spirit of the age. One side inspired the "right of private judgment," the other the blind adherence to "traditionary authority." Draper believed that the victory of the North was largely the result of their knowledge, education, and, ultimately, science. Posterity will look back on the war and not see it simply as a victory of the North, he said, but "as the fiat of God." This recognition, he believed, would ultimately lead to reconciliation.[107]

A HISTORY OF CONFLICT

What most historians of science have not realized is that this same principle of reconciliation guides Draper's last major work, his *History of the Conflict*. Indeed, the book was his attempt at condensing and popularizing ideas he had elaborated at greater length in earlier publications. It is important to note that Draper prefaced the volume with a warning: "whoever has had an opportunity of becoming acquainted with the mental condition of the intelligent classes, must have perceived that there is a great and rapidly-increasing departure from the public religious faith." As someone who had indeed become "acquainted" with the intellectual development of civilization, he insisted that "science" was not the cause of this recession. Rather, it is "the continuation of a struggle that commenced when Christianity began to attain political power." In other words, the decline of religious faith is a direct consequence of a politicized or "materialized" Christianity, not science. His chief indictment was thus against ecclesiasticism—or more precisely, orthodoxy, and its claim of divine authority. Such authority had exercised, he argued, a restraining and retarding force upon progress. Science had been and must continue to be the chief source of progress. Nothing else has so effectively broken down the strictures and bonds of human advancement. For Draper, the history of human progress was both the history of scientific advance and of obtaining truer conceptions of God. Thus in this work he more explicitly focused on the conflict between the "expansive force of the intellect" and the "compression arising from traditionary faith and human interests."[108]

While Draper believed that "faith" was in its nature "unchangeable," he also implicitly called on enlightened religious thinkers to save religion from ignorant men: "when the old mythological religion of Europe broke down under the weight of its own inconsistencies," he wrote, "neither the Roman emperors nor the philosophers of those times did any thing adequate for the guidance of public opinion. They left religious affairs to take their chance, and accordingly those affairs fell into the hands of ignorant and infuriated ecclesiastics, parasites, eunuchs, and slaves." Thus, seen in the context of his entire corpus, what Draper understood by "religion" in the title of his book was the theological dogmatism of a corrupt, defunct, and dying creed. This distinction allowed him to express concern over the capitulation of "religious faith" among the "intelligent classes," which he interpreted as an impending religious crisis in history. This distinction also better explains why Draper began his *History of the Conflict* with an account of the religious condition of the Greeks, four centuries before the rise of Christianity. Conflict existed, in short, since the dawn of human civilization, between those who embraced new knowledge and those who held tightly to old traditions.

As he had done in earlier work, Draper then offered a condensed account of the origins and spread of Christianity. He recapitulated his idealized image of primitive Christianity, where the demand of faith was simply the veneration of God, purity of life, and social benevolence. But according to Draper, this simple, pristine faith did not last. Once it became popular, it was adopted out of expediency. "Crowds of worldly persons," he decried, "who cared nothing about its religious ideas, became its warmest supporters." It thus relapsed into various forms of paganism, which was then subsequently incorporated as dogma. By the time of Constantine, Draper argued, Christianity had become thoroughly "paganized." It was this "transmuted" and "debased" religion, he explained, that eventually came into conflict with science.[109]

Once again Draper argued that "Mohammedanism" rekindled the expiring embers of knowledge and religious purity. This "Southern Reformation," as he called it, was an attempt to purify and reform the abuses and wickedness of the Catholic Church. Although not uncritical, Draper nevertheless believed Islam produced "more correct" views, particularly in rejecting the notion of incessant divine intervention. In the same way they "vindicated the majesty of God," he averred, the followers of Mohammad embraced learning and the "scientific truth."[110] Again, one should not confuse Draper's remarks on Islam as endorsement; he was simply following a commonplace historiography that emphasized Arab nations as collectors and codifiers of as well as commentators upon ancient Greek philosophy. The polymath Whewell, for instance, argued that the "Arabs bring to the cultivation of the science of the Greek their own oriental habit of submission, their oriental love of wonder; and thus, while they swell the herd of commentators and mystics, they produce no philosophy. Yet the Arabs discharged an important function in the history of human knowledge, by preserving and transmitting to more enlightened times, the intellectual treasures of antiquity."[111]

More importantly, many eighteenth- and nineteenth-century scholars who commented on Islam did so mainly as a foil to attack Christianity—or at least traditional, institutional Christianity, particularly the Roman Catholic Church. Crucially, this was especially true among Protestant historians. While Protestants did not consider themselves friends of Islam, they saw elements in that religion that they could use in their conflicts with Catholics. In such historical surveys, Islam is broadly represented as a pure Semitic monotheistic religion that had resisted external influences in stark comparison to the darkness that lulled contemporary Europe, for which Catholicism was responsible.

In his discussion on Islam, for example, Draper mentions two writers of special interest. First, he offers several lengthy excerpts from French scholar Ernest Renan's (1823–1892) "Historical Essay on Averroism," which was likely his *Averroès et L'Averroïsme* (1852).[112] Renan is famous, above all, as the author of the wildly popular *Vie de Jésus*, first published in 1863. In this book, Renan combined critical scholarship with novelistic aesthetic appeal to write an historical account of the *human* Jesus. For Renan, Jesus was not the Son of God but a humble visionary, a "noble initiator" who believed he could hear the word of God within him. Importantly, Renan argued that Jesus's religion was devoid of priests and external observances.[113] In his earlier study on the Muslim philosopher Ibn Rushd (1126–1198), or Averroes, Renan similarly depicted his subject as a champion of a pure monotheistic philosophy that contradicted religious orthodoxy and was persecuted for doing so. For Renan, this was the victory of the "religious party" over the "philosophical party."[114] In short, for Renan both Averroes and Jesus were pure monotheists who faced opposition from the reigning religious orthodox authorities.

One can clearly see why Renan would have appealed to Draper. But in his discussion on the "Southern Reformation," Draper also depended on the work of English historian Edward Gibbon (1737–1794).[115] In his classic *History of the Decline and Fall of the Roman Empire* (1776–89), the amateur *literati* Gibbon noted the contributions of Muslims to astronomy, chemistry, and medicine.[116] That Draper read Gibbon is revealing indeed. Gibbon, of course, described his narrative as the "triumph of barbarism and religion." According to Gibbon, the historical development of human societies since the second century after Christ was a retrogression for which Christianity was mainly to blame. More precisely, and of central importance for our purposes, Gibbon argued that it was "the union and discipline of the Christian republic," with its Church offices, ecclesiastical councils, and collection of revenue, that eventually damaged the welfare of the Empire.[117]

While he vehemently rejected revealed religion, it is important to note that Gibbon's narrative on the origin, development, and corruption of Christianity derives largely from the work of eighteenth-century, antidogmatic Protestant historians, particularly the German Lutheran church historian Johann Lorenz von Mosheim (1693–1755), English clergyman Conyers Middleton (1683–1750), Bishop of Gloucester William Warburton (1698–1779), and English church historian John Jortin (1698–1770).[118]

We will return to some of these authors below, for Draper himself credits them for his understanding of the history of Christianity. It is enough to say here that for Draper, the first major "conflict" was not between "religion and science," but between the theological doctrines of Catholicism and Islam, between the Trinitarian dogmatism of the ecclesiastical system and the unity of God promulgated by Mohammed and his followers. Draper went on to narrate a series of "conflicts" that historians of science now more commonly refer to as "myths about science and religion." But Draper also addressed a series of "controversies"—namely, the so-called geological controversies regarding the age of the world and the antiquity of man. The substitution of "conflict" with the term "controversy" was deliberate, for Draper argued that "the Church did not exhibit the active resistance she had displayed on the former occasion." That is, the belief that the earth was created in six days, and that the human race is but six thousand years old, was relatively unproblematic, for by the nineteenth century there was a growing conviction that Scripture, rightly apprehended, did not teach these doctrines.[119]

Significantly, according to Draper, what made the religious mind more open to new areas of scientific investigation was the Protestant Reformation, or what he called the "Northern Reformation." Draper claimed that the nature of the Reformation was the contention over the "criterion of truth." When Christianity attained imperial power, "truth" was enforced by civil authority. Despite such coercive efforts, a "dismal disbelief stealthily pervaded all Europe—a denial of Providence, of the immortality of the soul, of human free-will, and that man can possibly resist the absolute necessity, the destiny which envelops him." But then the reformers called into question the authority of the Church and replaced it with the Bible as the sufficient guide for every Christian.

Draper believed the Reformation was progress. But in denouncing the papacy, he argued, Protestants had imputed to Scripture the sum and substance of all knowledge, including the interpretation of nature. "The Reformers," he wrote, "would tolerate no science that was not in accordance with Genesis." But new developments in history and philosophy brought the Pentateuch under scrutiny, exposing its "inconsistencies, incongruities, contradictions, and impossibilities." Moreover, comparative religious studies revealed that most of the Old Testament narrative derived from Mesopotamian and Assyrian sources. While science had established a new criterion of truth, it did not mean, Draper argued, that the Bible should be discarded. Indeed, the comparative history of creation legends was first exposed by "churchmen." With Scripture rightly interpreted, and the truths of science well ascertained, there could be no antagonism.

Draper then turns to the "controversy over the government of the universe." Again, the choice of terms is important, for Draper believed the issue of whether the universe was governed by "incessant divine intervention, or by the operation of unvarying law" had already been settled by more advanced theologians. Historically, he argued, priests repressed the notion of unvarying law, for it "seemed to depreciate their dignity" and "lessen their importance." Nevertheless, with the progress of the sciences there emerged a growing conviction of the causal and natural efficacy of secondary causes. Indeed, from Brahe to Kepler to Newton and others, the investigation of secondary causes, as established by God, came to be seen as a legitimate means of studying natural phenomena.

But while the exclusion of God from the natural order of the universe does not itself entail impiety, the principle of government by law was in absolute contradiction to "Latin Christianity." "The history of this branch of the Christian Church is almost a diary of miracles and supernatural interventions," Draper contended. But Protestants had destroyed ecclesiastical miracles at the onset of the Reformation. Remarkably, Draper sided with Calvin's "stoical austerity," particularly his doctrine of predestination, for he believed that "God has from all eternity decreed whatever comes to pass." But since Protestants in his own day continued to struggle with the nature of divine agency, Draper reminded them that the whole point of the Reformation was a "protest against the Catholic doctrine of incessant divine intervention in human affairs." The spirit of scientific investigation, in short, derived from the Protestant spirit. The progress of science was the byproduct of this religious revolution.[120]

In his conclusion Draper returned to his earlier remarks in the preface, stating that "No one who is acquainted with the present tone of thought in Christendom can hide from himself the facts that an intellectual, a religious crisis is impending." He maintained that "it is not given to religions to endure forever." Indeed, all religions must "undergo transformation with the intellectual development of man." Science, he proposed, could and should transform religion because it offered nobler and truer conceptions of God. Disbelief, however, was not the result of science—rather, it was the result of the unreasonableness of what religious men and women are required to believe. Draper argued that the Catholic Church, rather than addressing the real concern, responded in typical fashion when it promulgated the Syllabus of Errors (1864) and the declaration of papal infallibility at the First Vatican Council (1869–70). But in denouncing "liberty of conscience," he contended, the Church failed to appreciate the demands of the spirit of the age. Draper did not limit the coming crisis to Catholics alone. The tendency is also found among Protestants, he argued, claiming that the recent Evangelical Alliance meeting in New York also failed to appreciate that modern science "is the legitimate sister—indeed, it is the twin-sister—of the Reformation." If churches, both Catholic and Protestant, wished to remain relevant, if they wished to continue to exist, they need to be open to the discoveries of science. Draper believed that there were "formidable, perhaps insuperable obstacles in the way" for Catholicism and science to be reconciled. Such was not the case with Protestantism, for it "is not only possible, but would easily take place, if the Protestant Churches would only live up to the maxim taught by Luther." In this way, the "friendship" between science and religion could be restored, which "misunderstandings have alienated."[121]

If this sketch of Draper's argument seems remarkable, we may turn to one last publication that further confirms this analysis. In an article published in the *Princeton Review* in 1879, one of the last published pieces of his career, Draper addressed the "Political Effect of the Decline of Faith in Continental Europe." The decline of faith, according to Draper, has given rise to such troubling movements as nihilism, socialism, and communism. The rise of such movements, he argued, puts society itself in great peril. While he still spoke bitterly of the medieval Church, he also recognized the benefits the Church conferred on society, instilling an important sense of justice and hope that held in check the passions of men. The Church prevented the development of that spirit of unrest among the lower classes that in his time had taken shape in these new troubling movements.

But in characteristic fashion, Draper claimed that "the plain and simple demands of primitive Christianity had been burdened with many pagan fictions, or with legends that outraged common-sense." This "fraudulent" religion was attacked by that "great political event, the Reformation." With the reformers, progress was made. Thousands of "vulgar impostures" had disappeared. By the nineteenth century, however, many men and women had taken an extreme position, relating all religion as fiction. He writes, "we have come to the conclusion that the whole, from the beginning to the end, was a deception." The result, he said, is the "wide-spread religious unbelief of so many thousands of men."

Thus, according to Draper, the birth of nihilism, socialism, and communism came with the extinction of religious belief. "With no spiritual prop to support them, no expectation of an hereafter in which the inequalities of this life may be adjusted, angry at the cunningly-devised net from which they have escaped, they have abandoned all hope of spiritual intervention in their behalf, and have undertaken to right their wrongs themselves." But while the ecclesiastic blamed the rise of nihilism, socialism, and communism on the Reformation, Draper contended that the blame should fall squarely on "those who invented these [theological] delusions, persuaded humanity to accept them, and reaped vast benefits from them." Draper, in short, argued that the lack of faith in his time was entirely due to "ecclesiastical impostures," those who had mixed Christianity with paganism.

Draper therefore imputes the origins of these troubling movements to the repressive policies of the Church, and cited the striking fact that they do not seem to flourish in Protestant nations—"Hence it may be said that the existence of these dreaded societies is a consequence of the failure of the Reformation to establish itself in the countries in which they are found."[122]

THE PROGRESS OF HISTORY

It is crucially important to see Draper, like many of his contemporaries, as a man of science who spoke not only on science but also on its history and its relation to religion. Scholars familiar with nineteenth-century literature and science will no doubt recognize many of the arguments found in Draper's writings. Tracking down his sources, however, is a process fraught with difficulty, for there is not a single footnote in any of his publications. Many historians of science who have found it difficult to locate the provenance of his ideas usually cite French positivist philosopher Auguste Comte (1798–1857) as a source of inspiration.[123] But upon closer inspection, as we have already seen, Draper does refer to a variety of authors, and sometimes even quotes directly from their work. What is clear from our reading is that Draper never once quotes, mentions, or even alludes to Comte.

Admittedly, Draper did use the phrase "positive science" in his *Human Physiology*. At the same time, he never again used the phrase in subsequent work. Moreover, the phrase is frequently found in medical treatises in the first half of the nineteenth century that opposed "vitalism," the belief that living organisms are actuated by some immaterial "vital principle" or force, or some such higher power other than mere physical processes.[124] At the 1834 meeting of the BAAS held in Edinburgh, for instance, anatomist William Clark (1788–1869) reported on the latest advances in animal physiology, in which he declared that "physiology, as a positive science, can only be founded in observation and experiment."[125] Similarly, Thomas Hun (1808–1896), Dean of Albany Medical College, argued before the medical college that "systems and vague speculation have now passed away from medicine, and have been succeeded by positive science, founded on observation and experiment."[126] Incidentally, Hun, like Draper, had been a medical student at the University of Pennsylvania during the 1830s. Perhaps most curiously, Scottish country gentleman Patrick Edward Dove (1815–1873) declared in his *Theory of Human Progression* (1851) that the progress of humanity coincided with the advances of "positive science." More remarkable still, Dove later argued in his *Romanism, Rationalism, and Protestantism, Viewed Historically* (1855) that the history of modern Europe has been a "warfare" between "the Church of Rome and her antagonists." "Romanism," Dove contended, has always been a "despotic power, whose whole energies are bent on the abolition of liberty."[127]

Even if we would admit the influence of Comtean positivism on Draper, the most we can say is that it was indirect, as what he had likely encountered at the University of London or as he continued to monitor its development across the Atlantic. But as a number of scholars have observed, while Comte's writings attracted much attention on both sides of the Atlantic, most nineteenth-century English writers "fused" Comte's thought with parts of other systems.[128] That is, the popularity of Comte's philosophy among British and American readers was largely the result of it acquiring broader meaning, particularly in the case of a number of important Protestant theologians in the century.[129]

Perhaps the most instructive example is Draper's old University of London classmate, John Stuart Mill. By the 1830s, Mill had grown dissatisfied with the radical Benthamism of his upbringing. He began reading the writings of the French philosopher Claude Henri de Rouvroy comte de Saint-Simon (1760–1825) and his followers. He was particularly struck, Mill later wrote, "with the connected view which they for the first time presented to me, of the natural order of human progress; and especially with their division of all history into organic periods and critical periods."[130] Like many other French intellectuals after the anarchy of the French Revolution, Saint-Simon and his followers turned to historical theorizing to find meaning and hope for the future. They taught that history was not static but dynamic, and that institutions appearing to be corrupt were not originally so and had once served a progressive social purpose. Interestingly, they believed that civilization moved in cycles of organic and critical epochs. They also believed men of science could provide a new unified worldview, grounded in physical science, and extended to the social world. In his *Nouveau Christianisme* (1825), Saint-Simon had even proclaimed the establishment of "new Christianity," where scientists would function as priests rather than clergymen. [131]

The most ambitious Saint-Simonian was, of course, Saint-Simon's *protégé*, Auguste Comte. But where Saint-Simon and some of his followers still saw value in religion, including aspects of orthodox Christianity, Comte saw religion as a primitive stage through which humanity must pass to achieve maturity. He aspired to replace traditional religion with a new religion of science, progress, and humanity. In his *Cours de philosophie positive* (1830–42), which first appeared in a condensed English translation in 1853, Comte infamously argued that each branch of human knowledge "passes successively through three different theoretical conditions: the theological, or fictitious; the metaphysical, or abstract; and the scientific, or positive." Individual development parallels social development, and thus the progress of society may be allegorically referred to as passing through childhood (theology), youth (metaphysics), and manhood (natural philosophy). Moreover, in tracing the progress of social systems, Comte contended that all such phenomena are subject to invariable natural laws. The present age was in fact transitioning to the higher positive stage of scientific development.[132]

Mill was deeply impressed by Comte's stadial scheme, and claimed that he "had contributed more than any one else to make his speculations known in England."[133] While Mill had indeed published in 1865 an entire treatise on *Auguste Comte and Positivism*, the reality is that he, too, sought to reconcile Comte's philosophy with English thought, particularly English Protestantism.[134] As Timothy Larsen has convincingly demonstrated, the "sea of faith was full and all around" Mill throughout his life, evident in his family, friends, and even published works.[135] Indeed, religion is surprisingly pervasive in all his major publications, including *A System of Logic* (1843), *On Liberty* (1859), *Utilitarianism* (1863), and other writings.[136] Above all, and most important of our purposes, in his *Three Essays on Religion*, which was published posthumously in 1874, Mill presents his own version of Comte's "Religion of Humanity" as a viable replacement for orthodox Christianity.[137] He follows Comte closely in describing the path by which men progressed from primitive polytheism to the more advanced conceptions of monotheism. But here the agreement between Comte and Mill ends, for while he believed that the progress of the physical and natural sciences had made orthodox Christianity untenable, Mill nevertheless granted that there is nothing in scientific experience inconsistent with the belief that the laws of nature are "themselves due to a divine will." According to Mill, a God who governs through universal law, through a "general Providence," is far more coherent than a God who "acts of variable will." "Science contains nothing repugnant to the supposition that every event which takes place results from a specific volition of the presiding Power, provided that this Power adheres in its particular volitions to the general laws laid down by itself," Mill wrote. Rather than eschewing religion before the progress of positive science, Mill believed a more rational faith could continue to thrive in the future. Ironically, and perhaps fittingly, Mill's final words on religion turn to Christ rather than God. Our need for a concrete model of moral perfection is found in the teaching and life of the "Prophet of Nazareth." While historical Christianity has warped and distorted morality, Mill argued, Christ, "probably the greatest moral reformer" who "ever existed upon earth," is the "ideal representative and guide of humanity." Mill's "Religion of Humanity," in short, is an ethical Protestantism without doctrinal Christianity.[138]

Mill's theological views had much in common with other Protestant intellectuals, particularly liberal Anglican Broad Churchmen.[139] Similarly, Draper's own view of history had a long and complex religious pedigree. For example, Draper referred to a number of Catholic writers for his view on historical progress, suggesting that his alleged anti-Catholicism was not as total as historians of science believe it to be. As we have already noted, he mentioned the twelfth-century Italian mystic and theologian Joachim of Fiore and his "Everlasting Gospel," which, according to Draper, displayed a "masterly conception of the historical progress of humanity." Joachim and his followers equated the papacy with the Antichrist, as did the Fraticelli or Spiritual Franciscans, which subsequently influenced Protestant reformers. Joachim also envisioned human history as ascending through a triadic division of Father, Son, and Spirit, where the final spiritual stage would be a time when ecclesiastical authority would be no longer needed, when mankind reached a great and pure "contemplative" stage. As a number of scholars have suggested, Joachim's vision of historical progress not only influenced Renaissance thinkers and Protestant reformers but also radical secularists.[140] Draper also applauded sixteenth-century French political philosopher Jean Bodin (1530–1596) for his emphasis on climate and topography as factors in the advance or decline of empires and peoples. While Draper only referred to his *Les Six de la République* (1576), Bodin's discussion of the origin and development of human society in his *Methodus ad facilem historiarum cognitionem* (1566), where he posited a cyclical view of history, with its constant rise, fall, and rebirth, with each cycle representing a new, higher level than its predecessor, is strikingly similar to his own view.[141]

Draper's own philosophy of history thus seemed to have been a combination of a linear or teleological view of history with a cyclical view, which emphasized waves or pulses of change that had cumulative effects, where certain repetition occurs but each new cycle constantly ascends to higher levels. This view of history, according to Draper, was exemplified in the rise and fall of nations and civilizations. Just as every human life exhibits development from infancy to old age, so every civilization manifests a similar pattern of development. At the point where a stage of development reaches maturity, the process of decay and decline begins. This course of development is then repeated in other periods—not in the same events, but in the general pattern. But there can be no progress in history unless history is moving toward some goal. For Draper, that was the "liberty of conscience" and "freedom of thought."

Draper's idea of progress thus follows closely what M. H. Abrams calls the "romantic spiral," a historiographical interpretation that fuses narratives of historical decline and progress to suggest that the goal of history is a return of the past in a higher synthesis.[142] It is, perhaps, no coincidence that German philosopher Georg W. F. Hegel (1770–1831) and Draper both referred to Protestantism as the "Northern Reformation." In his vast historical and philosophical synthesis, Hegel argued that history reveals the progressive manifestation of *Geist* (spirit) in the world. History, according to Hegel, propels itself forward through a dynamic process, where each successive age "resolves and synthesizes" conflicts of earlier eras. In his phenomenology of the human spirit, Hegel envisioned the human race passing through successive stages of religious life, each one reflecting a necessary moment in the development of spiritual consciousness and adding to a deeper, fuller comprehension of truth. Indeed, Hegel's "trinitarian dialectic" of world history, which covered many of the same countries, centuries, governments, wars, philosophies, religions, and scientific theories as Draper did in his work, advanced conflict as positively necessary for improvement or change, where alienation or opposition produces a higher synthesis or new unity. This dialectical progression, from abstraction, negation, and synthesis is not the vanquishing of a side by the other in a progressive movement away from religion, but rather a process in which both elements of the antithesis are preserved in a new, higher form.[143]

Draper's views had been influenced by German thought in other ways. In a remarkable exchange in 1875, for example, he disclosed to a reader that in early life he had sincerely believed the "miraculous acts of Jesus Christ, those of his disciples and followers, as well as those of the prophets and patriarchs," but when he became "familiar with what the great German writers had done," he could no longer trust the biblical miracles.[144] While Draper does not name which "great German writers" he had read, even a cursory reading of his oeuvre reveals deep debts to nineteenth-century historical-critical scholarship. In his *Intellectual Development*, for instance, he claimed that the Age of Reason had been preceded by the rise of "European criticism," and that the "rise of many-tongued European literature" corresponded to the decline of "papal Christianity." The ecclesiastic, he asserted, "soon learned to detect a heretic from his knowledge of Greek and Hebrew, just as is done in our day from a knowledge of physical science." According to Draper, with the rise of criticism a new appreciation for the nature of evidence emerged. "We see it in such fact," he wrote, "as the denial that a miracle can be taken as the proof of any thing else than the special circumstances with which it is connected." The invention of printing in the fifteenth century, moreover, "gave an instant, a formidable rival to the pulpit." It was, after all, Luther's critical study of the Bible that led him to the "grand idea" of "the right of individual judgment." The Reformation, according to Draper, "had been, to no small extent, due to the rise of criticism, which still continued its development, and was still fruitful of results." In the aftermath, classical authors were compared against each other, contradictions were pointed out, errors exposed, weaknesses detected, and new views offered.[145]

Moreover, in his *History of the Conflict*, in a section questioning the authorship of the Pentateuch, Draper referred to the critical work of Dean of Norwich Humphrey Prideaux (1648–1724), particularly his *The Old and New Testament Connected in the History of the Jews and Neighbouring Nations from the Declension of the Kingdoms of Israel and Judah to the Time of Christ* (1716–18). More contemporaneously, Draper suggested readers turn to the Anglican Bishop of Natal John William Colenso's (1814–1883) controversial *The Pentateuch and Book of Joshua Critically Examined* (1862); German Lutheran churchman Ernst Wilhelm Hengstenberg's (1802–1869) *Dissertations on the Genuineness of the Pentateuch* (1847); the Duke of Argyll, George Douglas Campell's (1823–1900) *Primeval Man* (1869); and Protestant German orientalist Hermann Hupfeld's (1796–1866) *Über Begriff und Methode der sogenannten biblischen Einleitung* (1844).[146] In acknowledging the diverse authorship of the Pentateuch, and how most of its narrative derived from other Ancient Near Eastern legends, Draper maintained that many Christian doctrines, such as the Fall and Atonement, could no longer be supported. But this did not trouble Draper, for he believed that Christianity, in its earliest days, when it was converting and conquering the world, knew little or nothing about those doctrines. Indeed, he contended that such doctrines had "originated among the Gnostic heretics."[147]

Furthermore, Draper and Broad Church liberal Anglicanism seem to have been influenced by a number of the same group of writers. For his general understanding of church history, for example, he turned to eighteenth-century Göttingen church historian Johann Lorenz von Mosheim, who, as we mentioned earlier, had also deeply influenced Gibbon. First published in 1726, Mosheim's *Institutes of Ecclesiastical History* was a representative example of "Protestant Enlightenment" thought, replacing theology with "the history of theology."[148] Incidentally, John Wesley, who was obviously central to Draper's own Methodist upbringing, had published an abridged version as *A Concise Ecclesiastical History* in 1781. Interestingly, here and in many of his sermons Wesley had constantly lamented the "Constantinian fall of the church."[149] J. G. A. Pocock observes that Mosheim believed that "the immoderate love of philosophy did violence to the holy scriptures," and "inflamed by philosophic zeal, men laboured with extraordinary subtlety to extract what dogmas they thought reasonable from the writings of inspired authors; and if any revelation in the sacred text seemed contrary to philosophical doctrine, they twisted and misinterpreted them desperately."[150] Draper had referred to Mosheim to support his own claim that the ecumenical councils had corrupted the original teachings of Christianity.[151] He also found inspiration from the Latitudinarian ecclesiastical historian John Jortin, another important source for Gibbon. Jortin, whom Draper referred to as "a very astute ecclesiastical historian," loathed theological systems. He had argued in his *Remarks on ecclesiastical history* (1751–73) that true Christianity, plain and simple in its original principles, had been paganized by the early councils—that "a clear and unpolluted fountain, fed by secret channels with the dew of Heaven, when it grows a large river, and takes a long and winding course, receives a tincture from the various soils through which it passes."[152] This was proof, according to Draper, that primitive Christianity had been corrupted.[153] Draper also referred to English universalist Bishop of Bristol Thomas Newton's (1704–1782) *Dissertation on the prophecies, which have been remarkably fulfilled, and are at this time fulfilling in the world* (1754), from which he quoted a long passage that further buttressed his argument that early Christianity had quickly become paganized with the rise of ecclesiastical power.[154]

While Draper also mentioned Whigs Henry Hallam (1777–1859)[155] and Thomas Babington Macaulay (1800–1859)[156] in several places, he seems to have been more deeply impressed by the Anglophile French scholar François Guizot (1787–1874). Indeed, he was naturally drawn to Guizot's *General History of Civilization in Europe* (1828), which argued that diversity of thought had kept European civilization steadily progressive throughout the centuries. Raised under an austere Protestantism, Guizot retained an unmistakable belief in Providence. At the same time, he proposed that it is humanity who works out God's plan for civilization. Draper affirmed Guizot's "profoundly religious and philosophical" perspective and quoted at length a passage where Guizot spoke of God's providential plan and humanity's responsibility in fulfilling it.[157]

In discussing the Reformation in England, one other important historian Draper mentioned was James Anthony Froude (1818–1894).[158] Froude was a central figure in Victorian thought, and Draper was very likely influenced by many of his writings. He memorably expressed his religious doubts in his semi-autobiographical novels *The Shadows of the Clouds* (1847) and *Nemesis of Faith* (1849). These were more than thinly disguised autobiographies, however—they were also Froude's thoughts on intellectual history, ranging from German historical-critical scholarship, theology, religious history, to the contemporary state of Anglicanism. He scandalously condemned the English clergy for strangling religion with dogma and expressed disgust and horror at the ethical implications of orthodox beliefs. Catholicism was anathema to Froude as much as Protestantism.[159]

But Froude was neither an atheist nor an agnostic. However unorthodox his views, Froude always claimed to be Protestant. Indeed, many accused his massive twelve-volume *History of England* as a Protestant polemic. But it needs to be emphasized that for Froude the Protestantism of his day was not Protestantism at all.[160] As Michael Madden observes, Froude thought nineteenth-century Protestantism was a "mere shadow of the movement begun at the Reformation."[161] Protestantism, according to Froude, placed its faith in reason and free inquiry, not authority. He thus criticized nineteenth-century Protestants for failing to live up to the original principles of the Reformation. Indeed, he said that the Reformation was "the cause of humanity against theology, the cause of God against the devil." He believed that the Reformation was a conflict between the laity and ecclesiastical authority, between the individual inquiring mind and the establishment. This was the same conflict he himself had experienced, and one he believed other contemporaries were experiencing. The Reformers and all true Protestants, then, were martyrs of reason and free inquiry.[162]

Perhaps most important, Froude also viewed Protestantism as a revivalist and corrective force in world history. Protestantism was not just a sixteenth-century movement—it was part of a general law, a regenerative principle throughout the course of history. In this sense, doctrinal formulations paralyzed religious progress. Religion, according to Froude, was "not a series of propositions or a set of outward observances of which the truth or fitness may be properly argued; it grows with the life of a race or nation; it takes shape as a living germ develops into an organic body."[163] It should be no surprise, then, that Draper also spoke of multiple "reformations," believing strongly that if the principles of the Reformation could be revived in his own day, if Protestants could return and adhere to the right of private judgment, liberty of conscience, and free interpretation of both Scripture and nature, then there would be no conflict between religion and science.

CONCLUSION

Like many nineteenth-century Protestants, Draper expressed the necessity of returning to the founding principles of the Reformation, which he believed could effect a reconciliation between religion and modern thought. But for Draper, the Reformation was not just a single event in history. There were numerous attempts at "reformation" throughout the record of European civilization.[164] The sixteenth-century Protestant Reformation was simply another occasion when "pious men" recognized that "Religion was not accountable for the false position in which she was found, but that the misfortune was directly traceable to the alliance she had of old contracted with Roman paganism."[165] The reformers had thus called for a "return to primitive purity." But rather than returning to the simple message of Christ, Protestants replaced the authoritarianism of the Roman Church with a hermeneutical authoritarianism. This eventually led, according to Draper, to the conflict over the "criterion of truth." This conflict, in other words, was a conflict within the religious conscience. Nevertheless, Draper retained the belief that "a reconciliation of the Reformation with Science is not only possible, but would easily take place, if the Protestant Churches would only live up to the maxim taught by Luther."[166] As we have seen, this maxim, as Draper defined it, consisted of three interrelated parts: the right of private judgment, liberty of conscience, and free interpretation of both Scripture and nature.

Draper appears to have neither struggled with belief nor ever proposed an alternative to Christianity. As we shall see more clearly in another chapter, Draper's religious views looked back to an older tradition of "rational religion" found among seventeenth- and eighteenth-century liberal Protestant thinkers who had subordinated revelation to human reason and science. Philosopher Charles Taylor has called these intellectuals "providential deists." These Protestants believed that an impersonal Creator had established a benevolent order that humans could not only grasp but obey. The regularities of nature and the laws of motion were designed for reason to see, and "by reason and discipline, humans could rise to the challenge and realize it." But according to Taylor, this shift in religiosity towards providential deism entailed the demise in the perceived need of God's intrusion. "God is relying on our reason to grasp the laws of his universe, and hence carry out his plan." Thus there is no need for work on God's part to carry out his purpose for human life. Indeed, such an intrusion was seen as circumventing the order of the plan.[167] Many of these thinkers believed the Reformation remained unfinished, that Luther and other reformers had only begun the task of emancipating reason and enlightening humanity. It was this more diffuse and rational Christianity, Draper believed, that would ultimately bring peace between these two contending powers and, more importantly, resolve the coming political and religious crisis. Draper, of course, was not alone in making these claims. As we have seen, countless other nineteenth-century scientists, historians, philosophers, and even theologians, of various shades of orthodoxy, made similar statements.

While most historians of science have characterized Draper's narrative as an endless series of conflicts between religion and science, a more accurate description would be to call it a "new Protestant historiography," a recasting of history that began with the Protestant Reformation itself. In his various writings, the antagonism arose not between religion and science but between divergent worldviews, between certain epistemological assumptions implicit in a progressive liberal Christianity and a more traditional conservative Christianity. This was, in short, a Protestant narrative about contending Christian traditions.

Despite these nuances, Draper's combative tone, selective handling of evidence, tendentious judgments, and disregard for the variety of opinions led many to interpret his writings as instigating conflict rather than bringing about a reconciliation. But if we take seriously the Protestant pedigree of Draper's ideas, it becomes undeniable that the Protestant conscience was a decisive solvent of Christian orthodoxy. The Protestant reformers attempted to replace the traditional teaching of the Church with their own faith. They based their doctrines on their interpretations of Scripture and the supposed beliefs of the early Christian communities. As we shall see, in addition to Scripture, their strategy was to appeal to history and rational argument. During the late seventeenth century and early eighteenth century, Protestants began subjecting all religion to intense scrutiny.

Thus, from the moment Luther began his protest, Protestants have constantly challenged what had hitherto been accepted as settled questions. The diversity of opinions they provoked, as we shall see later, led freethinkers and skeptics to conclude that many of them must be mistaken.

WHITE AND THE SEARCH FOR A "RELIGION PURE AND UNDEFILED"

JOHN William Draper believed that his work was only a forerunner "of a body of literature, which the events and wants of our times will call forth."[1] Indeed, in 1869, about a decade after Draper lectured on educational reform at the Cooper Union, Samuel D. Tillman, secretary of the American Institute of New York for the Encouragement of Science and Invention, invited American professor of history and first president of Cornell University Andrew Dickson White (1832–1919) to stand behind the same lectern and deliver an address on the "defects in the old system of education."[2] White, however, took it as an opportunity to reply to religious critics of his new university, which refused to apply religious exams to staff, faculty, and students alike. He entitled his lecture "The Battle-Fields of Science," which was printed in full the following day in the *New York Daily Tribune*, a newspaper edited by Horace Greeley, who also happened to be a Cornell trustee. In an oft-cited passage, White boldly proclaimed that "In all modern history, interference with Science in the supposed interest of religion—no matter how conscientious such interference may have been—has resulted in the direst evils both to Religion and Science, and *invariably*. And on the other hand all untrammeled scientific investigation, no matter how dangerous to religion some of its stages

Figure 2.1. Undated photograph of Andrew D. White, published in *Cornell Alumni News*, vol. 21, no. 7 (1918).

may have seemed, temporarily, to be, has invariably resulted in the highest good of Religion and Science." He then went on to survey the "great sacred struggle for the liberty of science," reviewing one by one the battles fought in cosmography, astronomy, chemistry, anatomy, and geology. White cited, for example, the numerous "mistakes of the church"—its battle over the rotundity of the earth and the debate over the existence of the antipodes; the position of the earth among the heavenly bodies and the persecution of scholars dedicated to the cause of free inquiry, such as Copernicus, Bruno, Galileo, and Kepler; the objection to and rejection of chemistry, physics, anatomy, medicine, and technology; and the humiliating trials faced by geologists earlier in the century. White concluded his onslaught by alluding to the criticism from sectarian clergy over the establishment of Cornell University.[3]

For the next thirty years, White doggedly pursued his thesis. He repeated his lecture at multiple venues in Boston, New Haven, Ann Arbor, Rhode Island, and elsewhere, before publishing an expanded version in two articles for the *Popular Science Monthly* in 1876.[4] These articles were revised and published later that same year under the title *The Warfare of Science*. Significantly, the edition published in England carried a preface by Irish physicist and scientific naturalist John Tyndall. Despite its expansion, White's general thesis remained unchanged. White continued developing his original Cooper Union lecture into a series of articles entitled "New Chapters in the Warfare of Science," which were again published in the *Popular Science Monthly* between 1885 and 1895.[5] These articles were collected, reorganized, and amplified into a massive two-volume *History of the Warfare of Science with Theology in Christendom*, published in 1896.[6] In these ponderous volumes, White described the disastrous role theologians played in the progress of science. Much like his original lecture, its twenty chapters teem with examples and illustrations of the warfare between science and theology, including battles over cosmology, biology, geography, astronomy, geology, chronology, anthropology, meteorology, chemistry, physics, medicine, philology, mythology, political economy, and even biblical criticism.

As we have already noted, many historians of science recognize that White's target was not "religion" per se, but "ecclesiasticism" or "dogmatic theology," and that he was, in general, a great deal more conciliatory in his approach than Draper. This was indeed a distinction White himself made. At the same time, it should be noted that in his earlier *Warfare of Science*, White frequently cited Draper's scientific and historical work as an important source.[7] Only later, in his *History of the Warfare*, does he distinguish himself from Draper. He wrote, for instance, that while Draper "regarded the struggle as one between Science and Religion[,] I believed then, and am convinced now, that it was a struggle between Science and Dogmatic Theology."[8] The staying power of this rhetorical strategy has been extraordinary, as most historians of science have repeated this distinction ever since. The distinction, however, is not entirely accurate.

It was not until 1896 that White clearly and unambiguously differentiated between "dogmatic theology" and "religion." This rhetorical strategy enabled him to assert that "Science, though it has evidently conquered Dogmatic Theology based on biblical texts and ancient modes of thought, will go hand in hand with Religion." He contended that "theological views of science" have "without exception . . . forced mankind away from the truth, and have caused Christendom

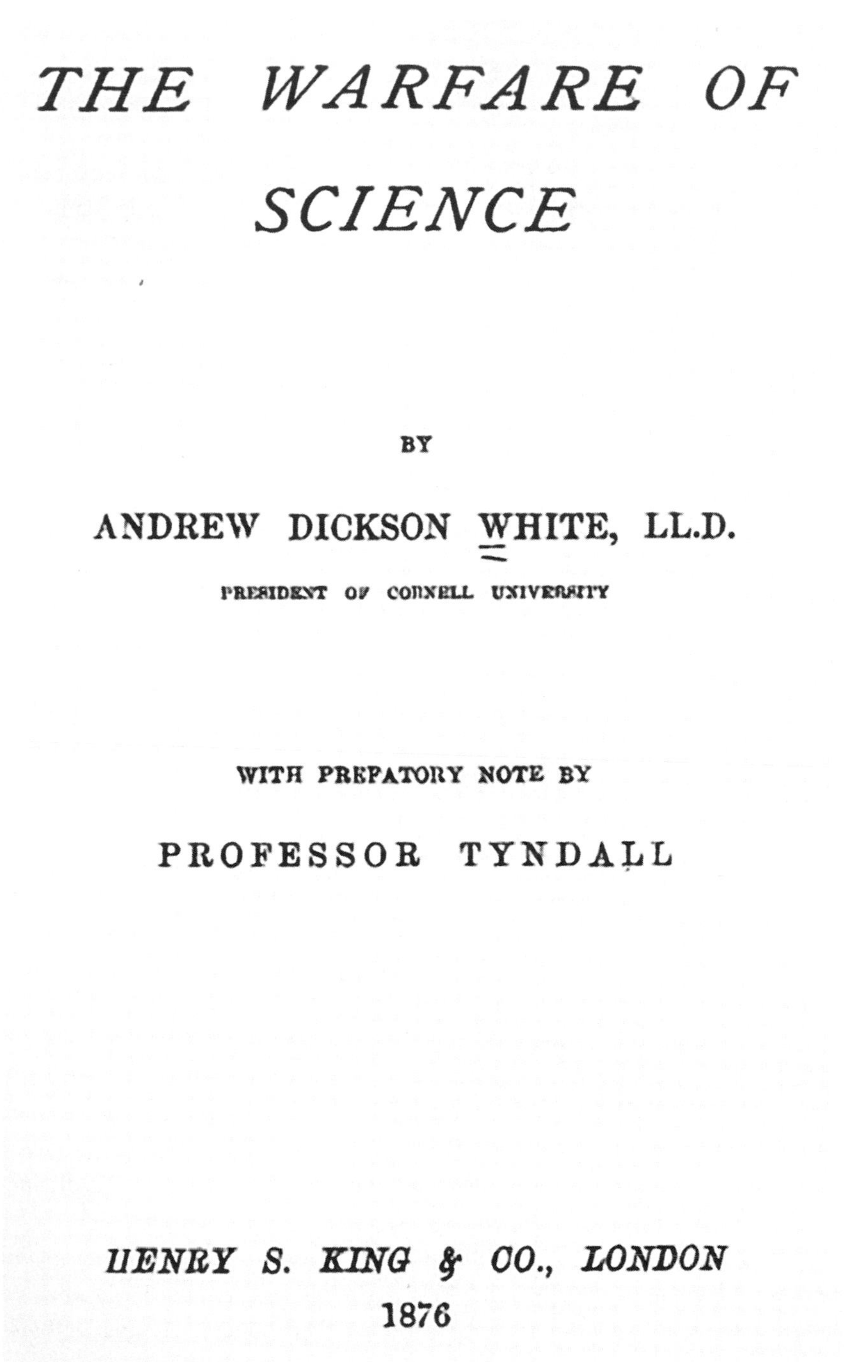

THE WARFARE OF SCIENCE

BY

ANDREW DICKSON WHITE, LL.D.

PRESIDENT OF CORNELL UNIVERSITY

WITH PREPATORY NOTE BY

PROFESSOR TYNDALL

HENRY S. KING & CO., LONDON

1876

Figure 2.2. The London edition of Andrew D. White's *The Warfare of Science*, published with a preface by Irish physicist John Tyndall.

to stumble for centuries into abysses of error and sorrow." From this persp he wrote that "the atmosphere of thought engendered by the developm all sciences during the last three centuries" had successfully dissolved the masses of myth, legend, marvel, and dogmatic assertion." Throughout his *History of the Warfare,* he referred constantly to "the theological view" and "the scientific view." Separating "theology" from "religion," then, allowed White to confidently conclude that "accounts formerly supported to be special revelations to Jews and Christians are but repetitions of widespread legends dating from far earlier civilization." For White, science was an aid to religion, encouraging its "steady evolution" into more purified forms. He thus denounced the "most mistaken of all mistaken ideas" as the "conviction that religion and science are enemies." "True religion," he argued, was consistent with science, and it was only the false teaching of theologians that came into conflict with its discoveries.[9]

Despite his stated religious purpose, White's general narrative was little different from Draper's. In fact, at times White was even more critical, severely undercutting Catholics and Protestants alike. Indeed, he frequently pointed out that Protestantism was just as destructively opposed to modern science as the Catholic Church. Most of his chapters took the form of narrating how some dogma has been fought over savagely, how the theologians feebly attempted some compromise, and how ultimately various scientific disciplines and other schools of modern thought were liberated when an intellectually decrepit theology finally gave way to the unstoppable advances of science. His assessment of historical Christianity is, in short, simply macerating. As biographer Glenn Altschuler aptly put it, despite White's claims to the contrary, his *History of the Warfare* seemed to "devour Christianity, core and all."[10]

In the preceding chapter, we saw how Draper appropriated an already existing historiography in his attempt to reconcile religious belief and modern thought. Influenced by a number of traditions in English and German liberal Protestantism, he rationalized Christian faith and reduced it to what he regarded as its most foundational, rationally justifiable elements. White took a similar approach. He claimed to be a theist and denied that he ever intended to attack religion as such. While perhaps no one did more to instill in the public mind a sense of warfare between science and religion, this was precisely the opposite of what he intended. Indeed, White consistently stressed that his work would ultimately lead to a reconciliation between science and religion.

But unlike Draper, White did not look to the Protestant Reformation as a guide. Rather, he followed other contemporaries in conceptualizing (or reconceptualizing) religion itself. Religion is found, White believed, in moral conscience, intuition, and sentiment. This definition of religion was, of course, not new. Indeed, it exemplified essential elements of the Romantic movement, which had become by the late nineteenth century a central component of American liberal Protestant thought. A number of ideological currents had thus converged to make White's position possible. At the turn of the century, a new generation of religious thinkers roundly condemned the rational religion of their predecessors as too arid, sterile, and devoid of feeling and personal experience. The rationalists of the previous generation alienated those who believed that morality and emotion were the very heart of religion. The romantics thus focused on themes of divine incarnation over atonement, immanence over transcendence, morality over doctrine, personal experience over authoritarian claims, and human goodness over human depravity. There was thus an important shift from the objective study of God as manifest in nature to an emphasis on subjective religious experience.

By the late nineteenth century, many liberal Protestants believed a new, purer Christianity was emerging. These theological currents precipitated not only a quest for new religious experience but a drive to reinvent religion. The basis of this new religion was not found in doctrinal rules or even the Bible but in an instinctive quality in humanity. The overwhelming purpose was not to destroy religion but to save it and make it relevant to a new setting. As Protestants, these religious thinkers protested what they believed to be a corrupt and spurious Christianity—which was any form of institutionalized Christianity. In Victorian Britain, perhaps the most outspoken figure of nineteenth-century liberal Protestantism was poet and literary and cultural critic Matthew Arnold. As we shall see in more detail below, Arnold's conception of God as the "Eternal power, not ourselves, that makes for righteousness" became White's own definition of divinity, and thus played a central role in his understanding of the relationship between science and religion.

This chapter argues that White must be located within this wider liberal Protestant tradition. While he embraced the ideas of the rationalists, he followed the liberal Protestant plan of strengthening Christianity by giving it a new basis in the moral conscience. Thus, unlike Draper, who attempted to restore what he understood to be the original and founding principles of Protestantism, White sought to establish Christianity on a new basis—in the recognition of "a Power in the universe, not ourselves, which makes for righteousness," and, more broadly still, as "the love of God and of our neighbor." By locating religion in feelings and religious experience, White was confident that it would never come into conflict with science. As we shall see, White's redefinition of religion and Christianity exemplified a strategy that became increasingly common among liberal Protestants, including many scientists, in America and Britain at the end of the century.

THE MAKING OF A PRESIDENT

White was born in Homer, New York, in 1832, the same year Draper and his young family were making their voyage to America from England. His father, Horace White, was one of the most successful bankers of Syracuse. His mother, Clara Dickson White, was the daughter of the cofounder of Cortland Academy, a prominent preparatory school in Homer. White later credited his maternal grandfather as the impetus behind his own future career.[11]

Both parents believed White was destined for the pulpit. His father thus sent him to the small Episcopalian-controlled Geneva College, later renamed Hobart College. It was at Hobart where White first began dreaming of establishing a university of his own.[12] But White found the curriculum at Hobart irrelevant and his classmates too rowdy and undisciplined. More importantly, his unpleasant experiences there encouraged his growing belief that a university "should be under control of no single religious organization; it should be forever free from all sectarian and party trammels; in electing its trustees and professors no question should be asked as to their belief or their attachment to this or that sect or party."[13] He ran away and enrolled at a Moravian academy until his father agreed to send him to Yale College. His father relented. After completing his studies, White and his Yale classmate, future first president of Johns Hopkins University Daniel Coit Gilman (1831–1908), embarked on a three-year European grand tour. White visited Oxford and Cambridge, attended lectures at the *Sorbonne* and *Collège de France*, and continued his studies at the University of Berlin. He spent several more months exploring Switzerland, Austria, Italy, and southern France before returning to America in 1856 to complete postgraduate study at Yale.

The following year, at the age of twenty-five, he was appointed to a professorship in History and English Literature at the University of Michigan. During the next six years in Ann Arbor, White lectured widely, covering an incredible array of subjects, including historiography, the fall of the Roman Empire, the rise of cities, the Crusades, monasticism, the growth of papal power, medieval Christianity, Islam, parliamentary power in England and France, the revival of learning and art during the Renaissance, the Reformation, the Jesuits, the Thirty Years' War, Louis XIV, XV, and XVI, the French Revolution and the *philosophes*.[14]

At the outbreak of the Civil War, White was forced to resign his appointment at Michigan after being unexpectedly nominated and elected for New York State Senate. In 1864 he became chairman of the Education Committee, where he first became acquainted with Ezra Cornell (1807–1874). Cornell had made a fortune in the telegraph business, but as a devoted Quaker he cared very little for wealth. His success had made him a renowned philanthropist, and in 1863 he was elected

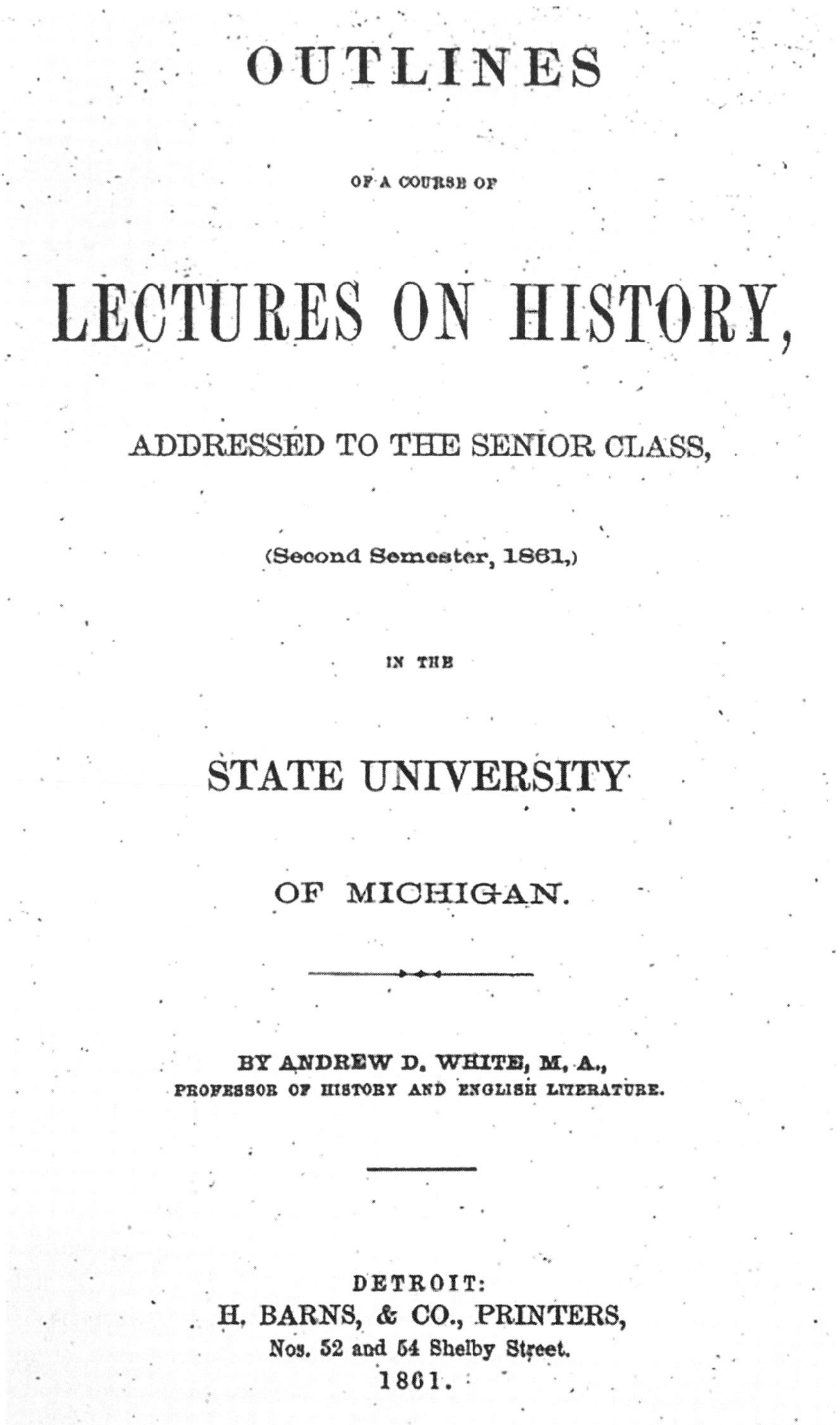

OUTLINES

OF A COURSE OF

LECTURES ON HISTORY,

ADDRESSED TO THE SENIOR CLASS,

(Second Semester, 1861,)

IN THE

STATE UNIVERSITY

OF MICHIGAN.

BY ANDREW D. WHITE, M. A.,
PROFESSOR OF HISTORY AND ENGLISH LITERATURE.

DETROIT:
H. BARNS, & CO., PRINTERS,
Nos. 52 and 54 Shelby Street.
1861.

Figure 2.3. Title page of Andrew D. White's *Outlines of a Course of Lectures on History, Addressed to the Senior Class.* Courtesy of the Division of Rare and Manuscript Collections at Cornell University Library, Andrew Dickson White Papers, Reel 139.

to the Senate.[15] In 1864 Cornell proposed a bill to endow a public library in Ithaca, an act that greatly impressed White. The two became friends and began planning together a foundation for a new university. The Morrill Land Grant Act of 1862 gave White and Cornell an opportunity to allocate their funds for the purpose of building an innovative, modern, and nonsectarian university in Ithaca. After much political maneuvering, the White-Cornell bill was passed in 1865. The following year White was unanimously elected president of the new university, and in 1868 it opened its doors to the public. It must be pointed out, however, that White was largely an absentee president. His post was interrupted by a Presidential commission to Santo Domingo in 1871, another European tour from 1876 to 1878, and by a diplomatic post as United States Minister to Berlin from 1879 to 1881. White eventually resigned the presidency in 1885, when he decided to focus exclusively on his research and writing.

THE IDEA OF A (NONSECTARIAN) UNIVERSITY

While the idea of a "more stately, more scholarly, more free" university first came to White as an undergraduate, it was not until his time at the University of Berlin that he found an ideal model to follow. He later wrote, "There I saw my ideal of a university not only realized, but extended and glorified."[16] A German education had become quite fashionable for many American students in the middle of the nineteenth century. While the Victorians, as we shall see in more detail later, drew much from German thought, greater still was the influence of German ideas on American intellectuals and educational reformers. Many studies confirm that throughout most of the nineteenth century, young American males who desired and could afford higher education sailed abroad to study at one of the German universities.[17]

The "free" University of Berlin, which was established under the leadership of Prussian philosopher Wilhelm von Humboldt and theologian Friedrich Schleiermacher in 1810, became paradigmatic. As Thomas A. Howard observes, Berlin's guiding principles were "in many respects antithetical to the lingering confessionalism that had justified theology's institutional centrality in the post-Reformation period." Berlin infused the historical sciences with the air of historicism or "developmentalism."[18] This view of history, and especially its nonsectarian approach, greatly appealed to White. In discussing the guiding principles of Cornell in the *Report of the Committee on Organization* (1866), for example, White declared that "we have under our charter no right to favor any sect or promote any creed. No one can be accepted or rejected as a trustee, professor or student, because of any opinions or theories which he may or may not hold." Its essential principles were founded on "freedom" of education and religion, and White resolved that Cornell should be above the restricting influences of parties, creeds, and dogma. By instilling such principles in its charter, Cornell would ensure that "whatever their individual theories on this or that dogma," students will work for "the glory of God and the elevation of man."[19]

But for White, nonsectarianism neither implied irreligion nor secularism. He amplified this point in its *First General Announcement* (1868), where he claimed that while Cornell University was nonsectarian, its highest aim "seeks to promote Christian civilization."[20] This idea was further developed during its inauguration celebrations, where Ezra Cornell himself spoke of making "true Christian men, without dwarfing or paring them down to fit the narrow gauge of any sect." In his inaugural address, White boasted that he had assembled the best "Christian Faculty" available, arguing that "perhaps no one thing has done more to dwarf the system of higher education than the sectarian principle." Sectarianism, according to White, was an "evil growth from an evil germ." Indeed, while strictly nonsectarian, the newly established university "shall not discard the idea of worship. This has never been dreamed of in our plans. The first plan of buildings and the last embraces the university chapel . . . From yonder chapel shall daily ascend praise and prayer. Day after day it shall recognize in man not only mental and moral but religious want. We labor to make this a Christian institution—a sectarian institution may it never be."[21] White's severe indictment against sectarian power in higher education demonstrates his determination to make Cornell free from the influence of any one particular religious creed.

White was therefore deeply disturbed when his university encountered strident religious opposition. In 1869, he confided to Ezra Cornell that "the papers, addresses, and sermons on our unchristian character are venomous."[22] In the wake of such criticism, he continued to take every opportunity to stress the "Christian" character of his university. In his annual report to the board of trustees, White stressed that Cornell "is now, and always has been, an institution managed by Christian trustees, under a Christian faculty, laboring for the advancement of Christian civilization."[23] In other words, he did not believe that his nonsectarian university subverted Christianity. He reminded the public in speeches and interviews that the university began each working day with prayer. He hoped that Cornell, as a "Christian institution," would equip young men to "go forth from these hills to do battle for truth and right in the Christian ministry as in other professions."[24]

These religious principles were also highlighted in White's annual speeches to students, where he often admonished them to develop themselves physically, mentally, morally, and religiously. For instance, on one occasion he urged students that "to be a complete man or woman the religious side must also be developed." Indeed, a "man is wretchedly off who does not feel that there is a great power in this Universe which shapes for good."[25] These exhortations were given in Sage Chapel at Cornell, a "non-sectarian pulpit" for the university that White planned to establish from its inception. He desired that students begin each day with a simple religious service, though participation was not mandatory.[26] While these were voluntary services, he exhorted students to "take advantage of these sermons. One single hour—the time of a single lecture—is all that is needed. What is gained is something in the long run which is of great importance to every one. Even from the intellectual side it is well worthy of your attention."[27] In the formation of character, White thus recognized the importance of religious development in university education. However, consistent with his idea that Cornell should remain nonsectarian, he made no attempt to promote any particular religious doctrine or denomination.

THE GREAT "SACRED STRUGGLE" OF HISTORY

The point that needs emphasizing here is that while historians of science argue that White came to see a "warfare" between science and theology after he faced religious criticism of his university, they ignore the fact that White's antipathy toward religious sectarianism, theology, and dogma, and, more importantly, his own conception of historical progress, predated the founding of his beloved university. While the university no doubt faced suspicion and scorn from many quarters, White consoled himself that "slander against the university for irreligion was confined almost entirely to very narrow circles, of waning influence."[28] In other words, the opposition he received merely confirmed what he already believed. By looking more closely at his work, including his less-known writings, we shall gain a better understanding of White's interpretation of history and its intimate connection to his views on science and religion.

It should be noted at the outset that in his early education White encountered many works that developed both his religious and historical perspective. At the Moravian academy, for instance, he was encouraged to read the Genevan historian Jean-Henri Merle d'Aubigné's (1794–1872) *History of the Reformation of the Sixteenth Century*, which was first published in Paris beginning in 1835. "No reading ever," White later reflected, "did a man more good."[29] Merle d'Aubigné, a pastor and professor of historical theology at Ecole de théologie de Genève who was influenced by the German historical tradition and popular romantic historiography, treated the Reformation as the key event in the rise of liberty and conscience. He strongly believed that history was guided by the hand of divine providence, stressing—as White himself stressed—the struggle between good and evil, in which great individuals, in spite of fear and persecution, made lasting contributions to human progress. The purpose of history, according to Merle d'Aubigné, is to reveal the universal, eternal principles that guide its development.[30]

Like many educated American Protestants in the nineteenth century, White developed an obsession with the Reformation. This obsession led to his amassing an extensive collection of books on the Reformation, which, by the 1880s, had accumulated to about thirty thousand volumes, including pamphlets and manuscripts. He eventually donated this entire collection to the Cornell Library.[31] During his first European tour, White also visited Wartburg castle where Frederick the Elector of Saxony had sheltered Martin Luther from persecution. He noted in his diary that he "lingered for a long time at [Luther's] window, looking at the wild scenery at which *he* once looked. I can see that a great change has come over me in the things that I love to see and to linger over. At my first sight, seeing it was all castles and abbeys, regardless of their tenants in great measure. Now I ask more—ask for places where something has been done for the race, for men as well as monks, and of these man-monks Luther is chief."[32] While studying at Berlin, moreover, White attended lectures by pioneering German scholars Carl Ritter (1779–1859), Friedrich Ludwig Georg von Raumer (1781–1873), August Böckh (1785–1867), and Leopold von Ranke (1795–1886). He read the works by Gotthold Ephraim Lessing (1729–1781), Johann Wolfgang von Goethe (1749–1832), Johann Christoph Friedrich Schiller (1759–1805), Friedrich D. E. Schleiermacher (1768–1834), Friedrich Christoph Schlosser (1776–1861), Christian Charles Josias von Bunsen (1791–1860), Karl Gutzkow (1811–1878), and other eminent German intellectuals. In this heady environment, White began formulating a conception of history that was both providential and progressive.[33]

It is important that we pause for a moment to reflect on a few of the figures mentioned in White's reading list and consider how important they were to his intellectual and religious development. German literary critic Gotthold Ephraim Lessing's influence, for example, was extraordinary, profoundly affecting White's religious outlook and, ultimately, much of modern Protestant thought.[34] As is well known, Lessing published Hermann Samuel Reimarus's (1694–1768) *Fragments of an Unnamed Author* between 1774 and 1778. This document, which were portions of Reimarus's enormous 4,000-page *Apology for or Defense of the Rational Worshipper of God,* cast doubt on historical revelation and argued that the Bible was a confused collection of ambiguous and contradictory documents. During the controversy that ensued after the publication of the *Fragments,* Lessing responded that his primary objective was to uphold the right of free discussion even of the most sacred subjects. He contended that while the *Fragments* might embarrass the theologian, the "true Christian" is safe. The teachings of Christ, according to Lessing, teach an "inward truth" that existed long before the Bible, and this inward truth can be known only by "spirit and power" or personal experience. Accordingly, while Christianity may have failed, the "religion of Christ" remained. Thus, objections to the literary details of the Bible were not objections to the spirit of "true religion." Christianity, Lessing stressed, should not be equated with the Bible. Lessing even appealed to Luther, "the Great Misunderstood," claiming to be following the reformer's spirit if not his writings. Significantly, Lessing believed that the leading characteristic of Luther's theology was his proclamation of the individual's total freedom in matters of religious opinion, unfettered by ecclesiastical interference, and contrasted this with the theological straitjacket imposed by not only Roman Catholic writers but the "bibliolatry" of orthodox Protestants.[35]

Lessing's theological position was further developed in the theater, in his play *Nathan the Wise* (1779). Set in the time of the Crusades, the play was a dramatic plea for liberty of opinion in all matters of theology. Nathan, the hero of the play, is a projection of Lessing's own religious convictions—a pure, tolerant religiosity independent of historical revelation. Lessing pleaded for tolerance of all religions, provided they show love to all. He presented an argument that many, including White, would later embrace, that "true religion" consists of love of God and man, and that the true seeker can find God equally through any of the great faiths.[36]

Perhaps even more influential on White was Lessing's *The Education of the Human Race* (1780). Lessing argued that no dogmatic creed is final, for religion is continually evolving. Those who do not allow for development cannot truly appreciate the growth of religion. According to Lessing, every historical faith has played its part in developing the spiritual life of mankind. The Old Testament is seen as an "elementary primer" in the education process. But just as children outgrow their first toys and books, after a time a maturing humankind sensed the inadequacy of the Old Testament. With the New Testament, humankind begins moving toward "the time of a new eternal Gospel," the last age of man, the "time of perfection." For Lessing, Jesus was "the first reliable, practical teacher" who preached "an inward purity of heart." History thus reveals a definite law of religious progress. According to Lessing, the age of reason and the self-realization of humanity were the fulfillment, not the contradiction, of Christian revelation.[37]

From Lessing one may naturally pass to Friedrich Schleiermacher, another important author White read while studying in Berlin.[38] Schleiermacher's vision was the confluence of a number of conflicting thoughts—evangelical piety, rationalism, idealism, romanticism, and patriotism all left their mark on his new theology. His theology was, as Karl Barth put it, a "theology of feeling" or "awareness."[39] While he grew up in a Reformed Calvinist home, Schleiermacher was educated among the Moravian Brethren, and thus came under the influence of a German pietism that called for an intimate relationship with Christ and emphasized one's personal experience with God. Indeed, the movement's founder, Count Nicholas Ludwig von Zinzendorf (1700–1760), was convinced that the essence of religion was "something very different than holding an opinion," and that religion was to be "grasped by sensation alone, without any concepts." Anticipating Schleiermacher's own ideas, Zinzendorf argued that humans had a *sensus numinis*, a feeling of absolute "dependence on something superior."[40] But after completing his education at the University of Halle, Schleiermacher seems to have been influenced by both rationalistic and romantic writers. In 1776 he was called to serve as the chaplain at a hospital in Berlin, where he began associating with the city's literary and intellectual circles. He soon helped found a university and became head of its theology faculty. During these years in Berlin, Schleiermacher lectured and wrote on an astounding array of topics, including historical, philosophical, and practical theology, the New Testament, hermeneutics, and psychology.

At the end of the eighteenth century, many Berlin intellectuals had rejected the miracles of Christ's incarnation and resurrection as crass superstition. In this cultural milieu, Schleiermacher was a puzzle. Here he was, a Reformed pastor eagerly associating with believers and skeptics who had rejected organized religion as irrelevant and constraining. While he shared many of their values, he still proclaimed that he was committed to his Christian religion. In turn, Schleiermacher attempted to reconcile what he thought the best of Christianity with modern skeptical thought. He constructed a theological position intended to disarm critics by placing greater weight on the religious *experience* of the believer than on the cognitive, presuppositional claims of theologians. In his *On Religion: Speeches to its Cultured Despisers* (1799), Schleiermacher called on readers to "become conscious of the call of your innermost nature and follow it." For Schleiermacher, the essence of religion was not dogma but intuition and feeling. He repudiated the idea that religion is a compendium of fixed doctrines that must be accepted on faith.

Most of the "cultured despisers" of Christianity equated religion with the institutional church, its hierarchy, and dogma. But according to Schleiermacher, our most profound encounters with God have little to do with any of these things. Our awareness of God is immediate, a "feeling of absolute dependence." This is an awareness of our finitude before the Infinite and Eternal, the universal and ineffable mystery that surrounds us. At the root of Schleiermacher's theological vision, then, was the redefinition of religion, neither as morality nor as belief or knowledge, but as an immediate self-consciousness or feeling of absolute dependence on God. This premoral and precognitive "religious consciousness" is common to all people, though recognized and expressed in a great variety of forms. From this religious consciousness of the individual stemmed religious communities, bodies of people who recognized in each other similarities in religious emotion, awareness, or intuition, and in doing so began a communion and a church.

Perhaps more important is Schleiermacher's contention that "ecclesiastical society" had become an obstacle to perceiving the "religious consciousness." Despite its origins in this feeling of total dependence, religion was never "found among human beings . . . undisguised," has never appeared in the "pure state." A mere glance at ecclesiastical history, he argued, is "the best proof that it [the church] is not strictly a society of religious men." In other words, the "religious consciousness" almost immediately became shrouded by human contrivances. He wrote, "You are right to despise the paltry imitators who derive their religion wholly from someone else, or cling to a dead document by which they swear and from which they draw proof. Every holy writing is merely a mausoleum of religion, a monument that a great spirit was there that no longer exists . . . it is not the person who believes in a holy writing who has religion, but only the one who needs none and probably could make one for himself."[41] Consequently, what the cultured despisers despise, according to Schleiermacher, is not religion but theological dogma. But this is only the husks and not the kernel of religion, Schleiermacher contended. In this dramatic separation of religion from theology was contained the ecumenical promise that appealed to many in his generation and the next. Christianity was being prepared for a new basis, a Christianity free from theology and free from the Bible. In 1829, Schleiermacher declared to a friend that his goal was to "create an eternal covenant between the living Christian faith and an independent and freely working science, a covenant by the terms of which science is not hindered and faith not excluded."[42] Science and religion have an "eternal covenant," according to Schleiermacher, because each uses its own methods to undertake different tasks that cannot overlap. The rise of science and the rise of the historical consciousness, therefore, does not threaten Christian faith, for both are distinct spheres and thus independent of each other.

Another significant source of inspiration for White was German historian Leopold von Ranke. In his *Autobiography*, White told an amusing story of studying under Ranke. He admitted that he had a difficult time understanding the German historian. In his lectures, he reported, Ranke "had a habit of becoming so absorbed in his subject, as to slide down his chair, hold his finger up toward the ceiling, and then, with his eye fastened on the tip of it, to go mumbling through a kind of rhapsody, which most of my German fellow-students confessed they could not understand."[43] But while White was confused by Ranke the man, he embraced the Rankean ideal of *wie es eigentlich gewesen*. Rankean "scientific history" was a romantic revolt against the rationalism of an earlier generation. Indeed, while Ranke accepted a division between analytical and empirical knowledge, he cautioned against regarding history as simply a collection of historical facts. "I believe," he wrote, "that the discipline of history—at its highest—is itself called upon, and is able, to lift itself in its own fashion from the investigation and observation of particulars to a universal view of events, to a knowledge of the objectively existing relatedness."[44]

History is, in Ranke's view, a guiding theme, a dramatic plot. The political predicament of Germany in the first half of the nineteenth century undoubtedly influenced Ranke to characterize history as the struggle for identity and power between nation-states. But this position is easily generalized. Through this struggle, human life discovers its potential and forms its very self. According to Ranke, this struggle bears the imprint of the hand of God. Critically, Ranke's conception of God, like many of the other German Protestant theologians and historians at the end of the century, was emancipated from "certain narrow theological notions."[45]

Ranke, of course, was not alone in conceiving history as struggle between competing people, groups, or ideas. We shall discuss in more detail later the extraordinary changes in religious thought in nineteenth-century Germany. For the moment it is enough to point out that the dramatic interplay of ideas and events between the "competing truths" of Catholic Austria and Protestant Prussia captured the imagination of the *Bildungsbürgertum* or educated middle class in the latter half of the nineteenth century. In the work of German liberal Protestant theologian Albrecht Ritschl (1822–1889), for instance, the church was conceived analogously to a struggling nation-state that creates itself in the crucible of contending historical forces. History is the story of striving, willing humanity. Jesus was the founder of a community that is the vehicle of freedom and purpose for those who belong to it. While White never cited or referred to Ritschl in his published work, he no doubt found common cause with Ritschl's emphasis that God is known by practical engagement in human affairs. For Ritschl, knowledge of the divine is moral knowledge. "Religion springs up as faith in the superhuman spiritual powers," he proclaimed, "whose help the power of which man possesses of himself is in some way supplemented, and elevated into a unity of its own kind which is a match for the pressure of the natural world." Ritschl conceived Christianity both as a relative phenomenon of history, ever changing the particular forms of its faith, and as a transcendent reality that perseveres in its identity through history in the battle against contrary forces. Importantly, Ritschl also argued that conflicts between science and religion arise only when people fail to properly distinguish between theoretical knowledge and religious knowledge. Scientific knowledge, he asserted, strives for pure theoretical, disinterested, objective cognition of things in themselves. Religious knowledge, by contrast, consists of value judgments about reality.[46]

German historians, philosophers, and theologians thus effected a separation between reason and religion, opening a space in which religion could be articulated on a new basis—one of immanence, feeling, and personal experience. This new basis inevitably challenged religious orthodoxy by placing the authority of religion in the experience of the individual. German thinkers also aspired to replace theological dogmatism with a more organic, evolutionary view of religion. Religion changed over time, and the history of religious evolution demonstrated humanity's progress towards greater maturity and self-knowledge. Change in religion was thus seen as the unfolding of a providential plan. The human race passed through successive stages of religious life, each one reflecting a necessary moment in the development of spiritual consciousness.

It should be no surprise, then, that White's first serious publication as a historian was "Glimpses of Universal History," a review article on German historical scholarship. In this article White bemoaned how "sham history" had become so popular in his day. "Sham" historians, he explained, see history as a "mere game at cross-purposes, a careless whirl; and in sequences of national virtue and vice, barbarism and civilization, death of great states, and birth of great principles,—sometimes fate, sometimes caprice." In other words, such history was mere chronology and desultory, without any kind of guiding principle, meaning, or purpose. Good "universal history," on the other hand, was guided by a philosophy, and served "as a foundation for all special study of history." Universal history, or what the Germans called *Universalgeschichte,* was necessary to "lift the new race of young men above the plane of our old demagogues, so that whether with pen or voice, they may attack them from above and at advantage." According to White, this type of history is exemplified in Friedrich Schlosser's work, particularly his *Weltgeschichte für das Deutsche Volk, unter Mitwirkung des Verfassers bearbeitet von Dr. G. L. Kriegk* (1844–57). Interestingly, Schlosser viewed history as a kind of theodicy to humanity, in which man steadily evolved in a teleological fashion toward a more rational worldview. White greatly admired this conception of history, and believed that such a philosophy could provide students with an overarching narrative, giving them the ability to "grasp at the meanings of great events, sequences and changes—seeking to fit their fragments of historic knowledge, and to bind them together,—longing to know something of the way the world has gone and goes." Such moralistic history, he declared, bears witness to a God "ever giving growth through all new light from new history." White believed that history was "providential," and that the "pursuit of historical truth revealed something about God's plan" in history.[47]

Prior to his presidency at Cornell, White spoke often of great obstacles to this moralistic vision of human progress. He referred to it as a "sacred struggle and battle" in history, a war against an "aristocracy based upon habits or traditions of oppression."[48] His early lectures on history at Cornell also attest to the fact that he had already designed the basic structure of his later *magnum opus.*[49] In an essay lamenting the rapid industrialization of American society, White warned that the "mercantile spirit" was undermining those essential elements of human civilization—society, education, science, literature, art, and especially religion. He thus called on the next generation to place these elements on "higher" ground. The next generation, he declared, must "form an idea of religion higher than a life devoted to grasping and grinding and griping with a whine for mercy at the end of it. They must form an ideal of science higher than increasing the production of iron or cotton goods. They must form an ideal of art higher than pandering to the prejudice or whimsy of to-day, and of literature better than anything known to the cynical prigs. And they must form an ideal of man himself worthy of that century into which are to be poured the accumulations of this."[50]

Unsurprisingly, he encouraged readers to look to Germany for a "better theory of life."[51] From the Germans, White learned that the "highest effort and the noblest result" of historical study was a "philosophical synthesis" that leads to a "large, truth-loving, justice-loving spirit." Historical study, he insisted, must seek to perfect "man as man" and "man as a member of society." The historian, in other words, had a decisive role in the development and progress of civilization. As a universal panacea, the work of the historian

> show[s] us through what cycles of birth, growth, and decay various nations have passed; what laws of development may be fairly considered as ascertained, and under these what laws of religious, moral, intellectual, social, and political health or disease; what developments have been good, aiding in the evolution of that which is best in man and in society; what developments have been evil, tending to the retrogression of man and society; how various nations have stumbled and fallen into fearful errors, and by what processes they have been brought out of those errors; how much the mass of men as a whole, acting upon each other in accordance with the general laws of development in animate nature, have tended to perfect man and society; and how much certain individual minds, which have risen either as the result of thought in their time, or in spite of it—in defiance of any law which we can formulate—have contributed toward this evolution.[52]

Once again, Germany provided White with an exemplary model. "Germany," he wrote, "has surpassed other modern nations not only in special researches, but in general historical investigations."[53]

These ideas culminated in his late essay on "Evolution and Revolution," published in 1890, six years before his final version of the *History of the Warfare* was published. White argued that the sciences have demonstrated the "upward evolution" of humanity. By "evolution," White meant the "regular, natural processes," a "growth in the main, quiet, steady, and peaceful." Humanity could avoid repeating the "catastrophic" and violent revolutions of the past—the Fall of Rome, the

Crusades, the Thirty Years' War, the American and French Revolutions, the Civil War—by striving for a "steady, healthful evolution." But like Merle d'Aubigné, White believed that progress or evolution was not some haphazard process, but ultimately the "unfolding" of a providential plan. This was only possible, however, if mankind participated in God's plan for history. Historians have demonstrated, White believed, that religion had evolved from "fetichism [*sic*], shamanism, idolatry, monotheism, through an allegiance to tribal gods, to monotheism under one God who is over all." Religion must thus continue to advance, not by revolution, as it was done in the past, but by a "steady, healthful evolution." The advance in modern history by a more steady evolution, he asserted, is seen in the efforts of "Melanchthon, Contarini, and Cranmer, of the Wesleys, Edwards, Bishop Butler, and Channing, of Emerson, Theodore Parker and Newman, of Arnold, Maurice, and Robertson [Smith]." These men, according to White, taught the "idea that dogmas and metaphysics are but the mere husks and rinds enclosing the precious kernel of truth." He explained that mankind must avoid the approaches of the religious revolutionaries, the "ultraists," both conservative and radical alike. On the one hand, the conservatives believed that "all truth is at last reached" and thus "all progress must now stop." The radicals, on the other, were "wild schemers," "dreamers," and "scoffers," denouncing and insisting on the abolition of all that was held sacred. White pointed to a *via media*, to "those who are laboring for a more quiet, beautiful, and effective evolution of religious thought and effort." Mankind must follow, he proclaimed, the "blessed example of the Master," that is, Jesus Christ, who cared little for dogma and doctrine and promoted instead "devotion to truth as truth, casting aside more and more the mere husks and rinds of religious truth, developing more and more the ethical contents of those forms of religion in which they find themselves, caring little for theories as to the origin of evil, devoting themselves to the evolution of good." He called on his audience to devote themselves to these "great vitalizing truths," which were, according to White, "the essentials of religion."[54]

TOWARDS A NEW RELIGIOUS EPOCH

Having more clearly defined his conception of history, we are now in a better position to evaluate White's views on science and religion. While his reconstruction of the past is questionable, it is more important to understand the function this narrative performed, and how, exactly, it fit into his understanding of religion and its place in history. As early as his Cooper address, for instance, White declared that "God's truth must agree, whether discovered by looking within upon the soul, or without upon the world." This is an interesting new formulation of the "two books" metaphor. Rather than pointing to the book of Scripture, White argued that the truth written upon the "human heart" cannot be at any variance with the truth written upon the "fossil." While the speech made the case for a university free of sectarian control, White defined "true religion" as something far nobler than narrow theological dogmatism. Like Draper, then, White wanted to demonstrate how organized or institutionalized religion was the enemy of both science *and* religion. Throughout the speech, he maintained that science offered a "far more ennobling conception of the world, and a far truer conception, and more devout reliance upon Him who made and sustained it." "Which is more consistent with a great, true religion," he added, the cosmography of the Church Fathers or Newton? Indeed, he bemoaned that epithets of "godless," "infidel," "irreligious," "unreligious," and "atheist" were hurled as weapons at men of science who were also "pious" Christians. The result of such antagonism "made large numbers of the best men in Europe hate" Christianity. The intelligent classes were turning away from Christianity because it was made "identical with the most horrible oppression of the mind." This led, according to White, to the "most unfortunate of all ideas—the idea that there is a necessary antagonism between science and religion."

Nevertheless, White explained how this great struggle had also led to "cheering omens." The struggle, in fact, encouraged men of learning in theology and natural science to find new ways of uniting all truth. "The greatest and best men in the churches," he observed, "are insisting with power, more and more, that religion shall no longer be tied to so injurious a policy—that searches for truth, whether in Theology or Natural Science, shall work on as friends, sure that, no matter how much at variance they may at times seem to be, the truths they reach shall finally be fused into each other. No one need fear the result." New knowledge of the natural world provides new ideas about the Creator and the "relations between his creatures." This was, according to White, a providential result—"the very finger of the Almighty has written on history that Science must be studied by means proper to itself, and in no other way." Although in the past warfare "has brought so many sufferings," it now "shall bring to the earth God's richest blessings."[55]

When White expanded this address for his *Warfare of Science* in 1876, he emphasized that the work of Christianity has been "mighty indeed." In particular, Christianity had aided the advance of science by imparting a "feeling of self-sacrifice for human good," which was an essential feature of the scientific endeavor. But like Draper, whom he approvingly cited throughout, he blamed "historical" Christianity and its theologians for obstructing "independent scientific investigation." One of the most notable features of White's *Warfare of Science* was his constant refrain that "religion" had always benefited from the triumphs of science. It was the theologians who drove "from religion hosts of the best men in all those centuries," instilling in the minds of so many generations that "Science and Religion are enemies." Men of science, on the other hand, have always offered a "more ennobling conception of the world, and a far truer conception of Him who made and who sustains it." He envisaged a day when "Religion and Science shall stand together as allies" against the enemies of truth and justice. He encouraged his readers to seek the "living kernel of religion rather than the dead and dried husks of sect and dogma," and declared that "the great powers, whose warfare has brought so many sufferings, shall at last join in ministering through earth God's richest blessings."[56]

As already mentioned, over the course of the next three decades White continued to expand his original Cooper Union lecture. In its early stages of development, he told his old college friend Gilman that he was reading with interest American geologist Alexander Winchell (1824–1891), writing that he lamented the controversy he faced over his views of evolution at Vanderbilt University. "What was tragical in Galileo's case is farcical in this," he opined. He also reported to Gilman that he found Winchell's *Reconciliation of Science and Religion*, which was published in 1877, particularly "valuable."[57]

In 1879, White wrote to Charles Kendall Adams (1835–1902), his successor at Cornell, that he was writing a book on "Ecclesiasticism and Science." Its intended purpose was to demonstrate that "Religion has never had any trouble or cause for trouble with Science, but that ecclesiasticism, which up to a certain point, has been necessary thus far to give protection and form to Religion in its worldly relations, had constantly developed in a way contrary to the best science and philanthropy, and of course the object is to show that ecclesiasticism must be steadily held in its proper place, and not allowed to encroach in fields which should be regulated by scientific thought."[58] Moreover, even before his "New Chapters" appeared in the *Popular Science Monthly*, White sat down with the New York *Evening Mail* for an interview in which he announced that "science is not antagonistic to religion, no matter what some person may say." On the other hand, religion was retarded by the constant bickering and quarrels between the "dogmatic assertions of the ministers of religion." White proclaimed confidently that "science will never destroy religion."[59] Much later, in his *Autobiography*, he stated that his aim, from the beginning, was to strengthen science *and* religion— "My aim in writing was not only to aid in freeing science from trammels which for centuries had been vexatious and cruel, but also to strengthen religious teachers by enabling them to see some of the evils in the past which, for the sake of religion itself, they ought to guard against in the future."[60]

Despite White's rhetorical distinctions, his 1896 *History of the Warfare* seemed to "devour Christianity, core and all," as Altschuler put it. But this was exactly what White intended to do. He believed that historical Christianity needed to be replaced. The conflict was between "two epochs in the evolution of human thought—the theological and the scientific." White took it upon himself to shine the light of truth "into that decaying mass of outgrown thought which attached the modern world to medieval conceptions of Christianity." The defunct theological world, in short, had been defeated by the modern scientific world.[61]

But in defeating theology, White believed science had only strengthened true religion. He defined such a religion as "the recognition of 'a Power in the universe, not ourselves, which makes for righteousness,' and in the love of God and of our neighbor." He hoped that this religion, "pure . . . and undefiled," would grow "not only in the American institutions of learning but in the world at large." Thus, most of his *History of the Warfare* consists of chapters showing how theological dogmatism suppressed progress in science *and* religion. As he put it, theological dogmatism had been the "deadly foe not only of scientific inquiry but of the higher religious spirit itself." It was the theologians who "wrought into the hearts of great numbers of thinking men the idea that there is a necessary antagonism between science and religion," who thrust "into the minds of thousands of men that most mistaken of all mistaken ideas: the conviction that religion and science are enemies." Seen in the context of his general philosophy of history, this great "sacred struggle" between an obstinate and villainous theology against an open and heroic science illuminated the deeper truths of religion as much as it revealed the laws governing the universe.[62]

Perhaps unsurprisingly, then, White, as a professor of history and literature, offered the fullest statement of his views in chapters dealing more specifically with the biblical text. In his chapter on "From Babel to Comparative Philology," for instance, he argued that "among the sciences which have served as entering wedges into the heavy mass of ecclesiastical orthodoxy—to cleave it, disintegrate it, and let the light of Christianity into it—none perhaps has done more striking work than Comparative Philology."[63] Depending mostly on the work of German philologist Friedrich Max Müller and English cleric Frederic Farrer, who incidentally were both inspired by romantic and idealistic movements of the century, White contended that the comparative study of languages "helped" Christianity rather than hurt it. Philology demonstrated that language was the result of an evolutionary process, and though it has undermined orthodoxy, the "essentials of Christianity, as taught by its blessed Founder, have simply been freed." White wrote that

> Darwin changed the whole aspect of our Creation myths; that Lyell and his compeers placed the Hebrew story of Creation and of the Deluge of Noah among legends; that Copernicus put an end to the standing still of the sun for Joshua; that Halley, in promulgating his law of comets, put an end to the doctrine of "signs and wonders" . . . Our great body of literature is thereby only

> made more and more valuable to us: more and more we see how long and patiently the forces in the universe which make for righteousness have been acting in and upon mankind through the only agencies fitted for such work in the earliest ages of the world—through myth, legend, parable, and poem.[64]

White then turned to "the Dead Sea Legends to Comparative Mythology," where he offered a diffuse account of how, "in the natural course of intellectual growth, thinking men began to doubt the historical accuracy of these myths and legends." The most notable feature of comparative mythology, according to White, is that "it shows ever more and more how our religion and morality have been gradually evolved, and gives a firm basis to a faith that higher planes may yet be reached." The growing skepticism over such myths and legends during the eighteenth century and especially the nineteenth century, White explained, "rendered a far greater service to real Christianity than any other theologian had ever done." According to White, since "all truth is one," comparative studies in philology and mythology could work "together in the steady evolution of religion and morality."[65]

In his last and perhaps most revealing chapter, White attempted to trace the transition "From Divine Oracles to Higher Criticism." He argued that "no matter how unhistorical" the great sacred books of the world were, they are still "profoundly true." In words strikingly similar to Draper, he believed that they embodied "the deepest searchings into the most vital problems of humanity in all its stages: the naïve guesses of the world's childhood, the opening conceptions of its youth, the more fully rounded beliefs of its maturity."[66] He derived this evolutionary perspective of religious belief from German Protestant historians, including those Anglo-American writers who were also influenced by their work.[67] He praised these historians and theologians for demonstrating that the "Bible is not a *book,* but a *literature*; that the style is not supernatural and unique, but simply the Oriental style of the lands and times in which its various parts were written." White claimed that these authors "rehabilitated" the Bible for him, "showing it to be a collection of literature and moral truths unspeakably previous to all Christian nations and to every Christian man."[68] Turning to the controversial liberal Anglican manifesto *Essays and Reviews,* which was first published in 1860, White praised the authors as a "new race of Christian scholars" who were bringing about an evolution in Christian thought.[69] But perhaps the most effective "Christian scholar," according to White, was Matthew Arnold. "By poetic insight, broad scholarship, pungent statement, pithy argument, and an exquisitely lucid style," White opined, the writings of Arnold "aided effectually during the latter half of the nineteenth century in bringing the work of specialists to bear upon the development of a broader and deeper view."[70]

Indeed, during the height of the religious crisis in England, Matthew Arnold (1822–1888) offered a solution. His most concentrated work in this area was *St. Paul and Protestantism* (1870), *Literature and Dogma* (1873), and *God and the Bible* (1875).[71] In these volumes, Arnold proposed to reestablish Christianity on a new basis. That is, he did not wish to extinguish Christianity, but rather to mediate between radical, freethinking liberals and what he believed was an ossified orthodoxy. As a mediator, Arnold sought a reconciliation between science and religion. This reconciliation, however, depended entirely on theological accommodations. Wanting to save religion from the dogmatists, Arnold warned believers that it was no longer possible to resist the demands of science, and that miracles and metaphysics and the popular fairy tales of a "materialized" theology must be abandoned. These volumes were thus designed to salvage religious belief from what Arnold believed was a narrow biblical literalism. The problem, then, was to devise a new interpretation of the Bible that all believers might embrace.

Beginning with *St. Paul and Protestantism,* Arnold argued that Protestants had misunderstood and even perverted the theology of the apostle Paul, and thus needed to be rescued. Ignoring Paul's "oriental" utterances, Protestants had wrongly interpreted his writings as "formal scientific propositions." This twisted "the master-impulse of Hebraism," which was, according to Arnold, "*the desire for righteousness.*" He urged readers to shake off these Protestant misinterpretations of Paul and embrace the universal order of righteousness, which will ultimately lead humanity towards its perfection. "In this conformity to the *will of God,* as we religiously name the moral order, is our peace and happiness." For Arnold, religion is that "which binds and holds us to the practice of righteousness," faith the "holding fast to an unseen power of goodness." He subsequently provided a historical narrative, not unlike those we have already encountered with Draper and White, looking back on the history of Christianity and arguing that "it was inevitable that the speculative metaphysics should come, but they were not the foundation."[72]

Moreover, in *Literature and Dogma* and *God and the Bible,* Arnold broadened his scope to include the reinterpretation of the Bible as a whole. Arnold argued for a return to the "essential" teachings of the Bible, which he believed was "righteous conduct." Significantly, he blamed Protestant sectarianism as the main obstacle to religious progress, arguing that religion must be understood as "morality touched with emotion." The books of the Bible, he maintained, were written by men, and thus reflected external circumstances. He argued that the "received theology of the churches and sects" hindered a proper understanding of the Bible. He postulated periodic invasions of what he called *Aberglaube,* or those "extra-beliefs" beyond what was certain and verifiable, such as superstitions and miracles. The Jews had lost their original, pure, and intuitive perception of God by replacing him with a "mere magnified and non-natural man." In time they also constructed the *Aberglaube* of a Messiah who would come to "judge the world, punish the wicked, and restore the kingdom to Israel." According to Arnold, Christ called for a return to the original experiential religion of the Hebrews. But in time this spiritual truth was once again weighed down with *Aberglaube*, as his own disciples had misunderstood him. Arnold thus spoke disparagingly of the Nicene Creed as "learned science," and characterized the Athanasian Creed as "learned science with a strong dash of violent and vindictive temper."[73] He made the arresting observation that "at the present moment two things about the Christian religion must surely be clear to anybody with eyes in his head. One is, that men cannot do without it; the other, that they cannot do with it as it is." He contended that "popular Christianity at present is so wide of the truth, is such a disfigurement of the truth, that it fairly deserves, if it presumes to charge others with atheism, to have that charge retorted upon itself." The God of popular religion was thus nothing more than a "fairy tale" in metaphysical dress. According to Arnold, if religion is to endure, it must be "transformed." Arnold called on the intelligent classes to "put some other construction on the Bible than this theology puts . . . find some other basis for the Bible than this theology finds . . . if we would have the Bible reach the people." His aim was therefore to supply a new and indestructible basis for Christianity, so that when the inevitable religious crisis emerges, it should not lead humanity to reject Christianity as a whole but only its *Aberglaube*. For Arnold, Christianity was simply the "eternal power, not ourselves, which makes for righteousness."[74]

Many liberal Protestants agreed with Arnold that the essence of Christianity is experience, not doctrine. Right religion consists of righteous living, or "goodness," after the example of Jesus of Nazareth rather than the adhering to a certain set of abstract, narrowly defined propositions about the deity and the condition of mankind. The Bible, then, must be read as literature, not as science. The language of the Bible, in short, is poetic and figurative. Arnold ridiculed the "pseudo-science of dogmatic theology" and sought to separate Christianity from the metaphysical assumptions enshrined in its historic creeds. Arnold, therefore, attempted to rescue Christianity by stripping away the *Aberglaube* and establishing its theology on a new basis.

In a strikingly similar way, White's rhetorical strategy of separating "theology" from "religion" enabled him to announce a new and peaceful religious epoch. In his estimation, "the atmosphere of thought engendered by the development of all sciences during the last three centuries" successfully dissolved what he called "vast masses of myth, legend, marvel, and dogmatic assertion."[75] But while the sciences had dissolved dogma, "it has also been active in a reconstruction and recrystallization of truth." All these sciences had given "a new solution to those problems which dogmatic theology has so long laboured in vain to solve." Thus, White urged his readers to go beyond the "reconcilers." The advance of the sciences confirmed for White the "evolution of morals and religion in the history of our race."[76] This evolution in religious thought was only possible "under that divine light which the various orbs of science have done so much to bring into the mind and heart and soul of man—a revelation, not of the Fall of Man, but of the Ascent of Man—an exposition, not of temporary dogmas and observances, but of the Eternal Law of Righteousness—the one upward path for individuals and for nations."[77] In short, White held that once science had completely dethroned dogmatic theology, a new religious epoch, "pure and undefiled," would come to prevail in America and all over the world.

THE NEW RELIGION OF THE "NEW THEOLOGY"

White's personal religious commitments played a central role in his historical work. Indeed, these commitments controlled his entire narrative. He was convinced that God's providential plan in history was revealed in the evolution of religious ideas. Science played an instrumental role in that plan, for it gradually dissolved dogmatic theology and allowed the "stream of 'religion pure and undefiled'" to "flow on broad and clear, a blessing to humanity." The warfare of science, therefore, was integral to the "great sacred struggle" of history.

We are fortunate to have White's own account of those religious ideas and figures who inspired him to take such views. From his *Autobiography*, we learn that White was troubled by orthodox Christianity at a young age. He was "disgusted" by sectarian preachers, who reminded him of rival sarsaparilla salesmen, "each declaring his own concoction genuine and all others spurious, each glorifying himself as possessing the original recipe and denouncing his rivals as pretenders." At some point during his early education, he broke with orthodoxy completely. While the incompatibility of the genealogies of Jesus given in the Gospels of Matthew and Luke disturbed him greatly, he decried the harsh Calvinism of his youth and cited it as eventually leading him to abandon such doctrines of hell and eternal punishment. He read apologetical tracts on "Christian evidences," but this seemed only to strengthen his doubts. His wide reading of other religious traditions, with all their own miracle stories, convinced him that all miracles were equally unbelievable. He turned to Bacon's pithy statement to explain his wavering religious beliefs during this period—that "it were better to have no opinion of God at all, than such opinion as is unworthy of Him."[78]

Despite his doubts regarding orthodoxy, White never ceased to call himself a Christian. He wrote, "I have never had any tendency to scoffing, nor have I liked scoffers."[79] While he refused to recite the creed, he continued to attend Episcopalian services throughout his life. He recognized that the "simple religion

Figure 2.4. Portraits of William Ellery Channing (1780–1842), Theodore Parker (1810–60), Horace Bushnell (1802–76), and Henry Ward Beecher (1813–87), religious leaders who greatly influenced Andrew D. White's religious outlook.

of the Blessed Founder of Christianity has gone on through the ages producing the noblest growths of faith, hope, and charity," but he was just as equally convinced that "many of the beliefs insisted upon within the church as necessary to salvation were survivals of primeval superstition, or evolved in obedience to pagan environment or Jewish habits of thought or Greek metaphysics or medieval interpolations."[80] Thus while White rejected most traditional conceptions of Christianity, he remained a firm believer in a benevolent creator God who acted and continues to act in history for the good of humanity. As he told Cornell librarian Willard Fiske (1831–1904), "whatever my heterodoxies I do believe in a higher power which controls human affairs for good."[81]

White was very familiar and even personally acquainted with many of the most prominent American liberal Protestant ministers of the nineteenth century, including William Ellery Channing (1780–1842), Horace Bushnell (1802–1876), Theodore Parker (1810–1860), Henry Ward Beecher (1813–1887), Thomas W. Higginson (1823–1911), Theodore T. Munger (1830–1910), Lyman Abbott (1835–1922), Minot J. Savage (1841–1918), Newman Smyth (1840–1925), and others. Indeed, his own writings reflected the ideas of this group who became known as proponents of the "New Theology." Differences certainly existed between them, but what impressed White was their shared ideological commitments, particularly their view that religion evolves.[82] Even before he had published his *Warfare of Science* in 1876, White had been popular among liberal Protestants. American Unitarian minister Thomas W. Higginson, for example, invited him to the Free Religious Association in Boston, telling him that "we are not all disreputable, you know, as R. W. Emerson, George W. Curtis, Prof. [Edward L.] Youmans etc., are our Vice Presidents."[83] White declined because of other obligations, but he did reassure Higginson that "if I could be with you in Boston at the time named in your letter, and take part in your Convention, I would."[84] White later sent a copy of his *Warfare of Science* to Higginson, who wrote in response: "Thank you much for your little book, the 'Warfare of Science,' which cover the points as able as Draper, I should say, and is much more just and judicial."[85] In White's later years, he was flooded with requests from liberal Protestant ministers and church members to speak on "the coming unification of science and religion, [and] what kind of religion will receive support and inspiration of science."[86]

While American liberal Protestants proclaimed Christianity, they often ignored or actively criticized many of its historical doctrines. As a consequence, they also sought to separate "religion" from "theology," assailing the latter as the true source of rising unbelief—essentially the same strategy White took in his own historical writings. Liberal Protestantism emphasized divine immanence over transcendence, morality over doctrine, and personal experience over authoritarian claims. Because they all shared a persistent optimism that human society was advancing toward the realization of the kingdom of God, they sought to accommodate or adjust Christianity to modern thought, to integrate Christianity within the broader confines of modern culture. They thus encouraged the free use of reason, insisting that God intended faith and knowledge to be complementary.[87]

The belief in divine immanence was also crucial to how liberal Protestants understood the course of history. Almost uniformly, liberals embraced a teleological view of history, a shared conviction that the historical process disclosed the gradual unfolding of a divine plan for the redemption of humanity. Thus, when conflicts or warfare emerged in history, it was interpreted as a consequence of humanity's inability to carry out the divine plan. This is why education and the sciences became so central to liberal Protestant theology. As we shall see later, religious liberalism infiltrated many prominent educational institutions in the second half of the nineteenth century. Schools such as Harvard, Yale, Princeton, Columbia, Johns Hopkins, Chicago, Stanford, Michigan, and others became influential centers of liberal theological ideas.

The influence of German mediating theology, or *Vermittlungstheologie*, on American Protestant theology is particularly evident in the work of learned Swiss-German émigré theologian and church historian Philip Schaff (1819–1893).[88] Arriving in America in 1844, Schaff noted America's flair for religious innovation and for altering old creeds and practices. He wrote that "America seems destined to the Phenix [sic] grave . . . of all European churches and sects." From the conflict of the American religious groups "something wholly new will gradually arise," Schaff predicted, and thus there was something positive in the sectarian conflicts often lamented by European observers.[89] This Hegelian dialectic model of history is pronounced throughout his multivolume *History of the Christian Church* (1858–1892), where he developed a philosophy of church history as an organic, developmental process toward consummation in Christ according to God's providential activity. In 1894, Schaff further observed that the influence of German theological thought in America "is both negative and positive; it tends to undermine old foundations, and aids in building up new constructions."[90]

As founder of the American Society of Church History, Schaff was a central figure in the transmission of German philosophical and theological ideas to the United States.[91] Schaff, as an advocate of a modern, theological science free from prejudice and sectarianism, was, perhaps surprisingly, one of White's great heroes. Indeed, White gave Schaff the honor of uniting the "scientific with the religious spirit," for "killing" old theories of biblical myths, and thus for rendering a greater service to religion than to science by "doing away with the enforced belief in myths as history which has become a most serious danger to Christianity."[92]

It is no coincidence that the golden age of liberal Protestant theology in America corresponded with the publication of many of White's views on the relationship between religion and science. A more intimate influence upon White was Unitarian minister William Ellery Channing. White was introduced to the work of Channing by one of his instructors at Syracuse Academy.[93] While Unitarianism was rooted in anti-Trinitarian European Protestantism, American Unitarianism emerged in the late eighteenth and early nineteenth centuries as a protest against New England Calvinism. Nascent American Unitarianism pivoted on Channing. In his famous Baltimore sermon of 1819, for example, Channing defined and defended Unitarianism against Calvinist critics. The Unitarian, he argued, believed that "God never contradicts in one part of Scripture what He teaches in another; and never contradicts in revelation what He teaches in his works and providence." The Unitarian is thus permitted to reject any interpretation of the Bible that violated reason and the laws of nature. Channing, believing that Unitarians were on the side of enlightenment and progress, called on his audience to resist the "gloomy, forbidding, and servile religion" of Calvinism. For Channing Scripture was authoritative only in the sense that it could be corroborated through reason, conscience, and experience. In closing, Channing instructed Unitarians to remember the "darkness which hung over the Gospel for ages," and that Christianity is still "dishonored by gross and cherished corruptions." There is much stubble "yet to be burned; much rubbish to be removed." He called for a new and "glorious reformation" of the church.[94]

Channing's assertion that liberal Protestantism was leading to a new reformation undoubtedly resonated with White. Any other belief that privileged creed over direct, personal revelation was, for Channing, akin to the "old theology" of Calvinism or the even older theology of Catholicism. Spreading "religious freedom," he insisted, would finally "redeem the Christian world from the usurpations of Catholic and Protestant infallibility."[95] In works like "Unitarian Christianity," he allied the old Reformation with a new American "glorious reformation" as he asserted the need for rooting out the "Papal dominion [that] is perpetuated in the Protestant churches."[96] Thus while obviously anti-Catholic in sentiment, liberal Protestants like Channing also aimed to critique the Protestant tradition itself. Calling for that "glorious reformation," and criticizing both Catholicism and Protestantism, Channing assumed the reformational mantle and considered the reforms advocated by liberal Unitarians to be contemporary manifestations of the Protestant reformers.

While studying at Yale, White also encountered many clergymen "outside the orthodox pale."[97] Perhaps the most important for White at this time was transcendentalist Theodore Parker. White credited Parker for checking his "inclination to cynicism and scoffing," and for strengthening his "theistic ideas," indeed for stopping "any tendency to atheism." There are many striking similarities between White's religious vision and Parker's. Parker, an avid reader of Schleiermacher, repudiated the idea of an external authority, and felt himself guided by a profound conviction that "true religious faith" was lodged in the intuition of the human spirit. In one of his most controversial sermons, for example, Parker examined the *Transient and Permanent in Christianity* (1841). "It must be confessed," he wrote, "that transient things form a great part of what is commonly taught as Religion." There was, according to Parker, too much doctrine and not enough of the "absolute, pure Morality; absolute, Pure Religion." Rigid adherence to a particular set of written tenets shackles the minds of men and restrains them from further religious progress. Parker was even willing to consider that if "Jesus of Nazareth had never lived, still Christianity would stand firm, and fear no evil." The only creed necessary "is the great truth which springs up spontaneous in the holy heart—there is a God."[98] According to Parker, "primitive Christianity was a very simple thing," with only two essential doctrines: to love man and to love God.[99] In his most substantial work, *A Discourse of Matters Pertaining to Religion* (1843), Parker asserted that Christian "theology is full of confusion," and that what passed for Christianity "in our times is not reasonable." He argued that historical Christianity had corrupted the simple message of Christ and that Unitarianism was carrying out the principles of the Protestant Reformation to completion.[100] White no doubt would have been greatly impressed with Parker's proclamation that "men of science, as a class, do not war on the truths, the goodness, and the piety that are taught as religion, only on the errors, the evil, the impiety, which bear its name. Science is a natural ally of Religion."[101]

In this New England atmosphere, White also became familiar with the work of liberal Congregational minister Horace Bushnell. Although there were important differences between Bushnell and New England Unitarians, he also stressed the importance of the experiential over the doctrinal and shared with them an optimism about human progress. Influenced by New England Puritan theology but also German romanticism, especially Schleiermacher, Bushnell criticized the prevailing orthodox doctrine of sin and hell, and demanded parents discontinue teaching their children Scripture history in his *Discourses of Christian Nurture* (1847). He explained that individuals gradually developed their Christian character, and therefore parents must stop enforcing "religious" or "theological" instruction onto their children.[102] In his *God in Christ* (1849), Bushnell also demanded that the Bible must be read "not as a magazine of proposition and mere dialectic entities, but as inspirations and poetic forms of life; requiring also, divine inbreathings and exaltation to us, that we many ascend into their meaning." Indeed, the very nature of language, its proximate and relative character, mitigates all dogmatic assertions, according to Bushnell.[103]

Perhaps his most important treatise, however, was *Nature and the Supernatural* (1858), which was an attempt to reconcile Christianity and science. According to Bushnell, there is a continuity between the supernatural and the natural, for both "have their common root and harmony in God." Science and religion are complementary if divine agency is reinterpreted as an intuitive sense of God's immediacy. "God's system is nature," Bushnell wrote, "and it is incredible that the laws of nature should be interrupted." Reconciliation would only occur, in other words, when religion adapted to the "sturdy facts of science." Science must correct revelation.

Significantly, Bushnell also wrote about the "dark ages," when the church was "unable to stay content with the humble guise and the simple doctrine of the cross." The Roman Emperor Constantine, he asserted, transformed Christianity "into a rank of political ascendency; which is the same as to say he dooms it, for ages to come, to be the mother of all unholy arts and oppressions, and the source of unspeakable public miseries." The Protestant Reformation, while helping to put Christianity back on track, was nevertheless "a thing too incomplete and partial to be allowed a sweep of universal triumph."[104]

But the greatest influence on White was no doubt Congregational minister Henry Ward Beecher. Indeed, White claimed that "nothing could exceed his bold brilliancy," and that he was "a man of genius." Beecher had even occasionally visited White at Cornell.[105] As a liberal Protestant, Beecher believed "true religion" was an "inward state," and that doctrine was secondary. He appealed to a growing number of religious believers who embraced progress in religion.[106] As historian Gary Dorrien observes, by midcentury liberal Protestantism had "seeped into American culture," and thus by the time Beecher took the prestigious pulpit in Brooklyn's Plymouth Church in 1847, he found a ready middle-class audience who soaked up his ideas. Dorrien writes, "Beecher made a singular contribution to the development and legitimization of American liberal Protestantism as a whole. In his journalism, lecturing, and preaching, mid-Victorian middle-class Americans took heart that religion, science, and American progress all worked together."[107]

In his lectures and preaching Beecher supported many of the emerging physical and historical sciences, particularly evolution. He was, indeed, one of the chief American disciples of the British scientific naturalists, such as Thomas H. Huxley, John Tyndall, and especially Herbert Spencer. It is important to recognize, however, that the kind of evolution Beecher popularized in late nineteenth-century America was infused with teleological speculations. In 1883, for example, Beecher delivered a lecture at the Cooper Union on "Evolution and Revolution." He opened his address with the assertion that "a greater change has taken place within the last thirty years, probably, than ever took place in any former period of five hundred consecutive years." This revolution had shifted the "old notion of creation" from the "instantaneous obedience of matter to the divine command" to the belief that the divine method of creation was "gradual, and as the result of steadily acting natural laws through long periods of time." In short, Beecher believed evolution was "God's way of doing things." While he was not prepared to accept the transmutation of man from lower animals, even if that were the case, it did not make him doubt any less the "agency of a divine, omnipotent, omnipresent God." He maintained that "a man may be an evolutionist and believe in God with all his heart and strength and soul." But more than this, he believed that evolutionary theory was a new revelation that revealed "true religion." "I thank God," he concluded, "for the growing life and power of the great doctrine of Christian Evolution."[108]

Two years later, Beecher published a series of sermons on *Evolution and Religion* (1885).[109] These sermons purported to discuss the bearings and applications of evolutionary philosophy to the Christian faith. As in his Cooper Union lecture, Beecher welcomed the "universal physical fact of evolution" as the "Divine method of creation." "For myself," he declared in the preface, "while finding no need of changing my idea of the Divine personality because of new light upon His mode of working, I have hailed the Evolutionary philosophy with joy."[110] In these sermons, Beecher confidently asserted that "Evolution is accepted as the method of creation by the whole scientific world and that the period of controversy is passed and closed." He assured audiences however that "Evolution is substantially held by men of profound Christian faith," and that while "Evolution is certain to oblige theology to reconstruct its system, it will take nothing away from the grounds of true religion." Like White, Beecher argued that "men are continually confounding the two terms, religion and theology. They are not alike." Indeed, Beecher believed religion needed to be emancipated from the "unbearable systems of theology." He poured contempt on conservative reservations about human evolution: "As it is now, vaguely bigoted theologists, ignorant pietists, jealous churchmen, unintelligent men, whose very existence seems like a sarcasm upon creative wisdom, with leaden wit and stinging irony swarm about the adventurous surveyors who are searching God's handiwork and who have added to the realm of the knowledge of God the grandest treasures. Men pretending to be ministers of God, with all manner of grimace and shallow ridicule and witless criticism and unproductive wisdom, enact the very feats of the monkey in the attempt to prove that the monkey was not their ancestor." To resist this progressive evolutionary theology was, according to Beecher, tantamount to opposing the very hand of God. This evolutionary theology "will obliterate the distinction between natural and revealed religion," and remove the "sands which have drifted in from the arid deserts of scholastic and medieval theologies."[111]

Beecher insisted that Christianity must be recast into the new revelation of evolution. This new revelation teaches that "God created through the mediation of natural laws; that creation, in whole or detail, was a process of slow growth, and not an instantaneous process." For Beecher, evolution was a grand metaphysical vision of progress. Creation, he wrote, was "moving onward and upward in determinate lines and directions so that the whole physical creation is organizing itself for a sublime march toward perfectness."[112] In his grand vision of recasting

Christianity in terms of a progressive evolutionary philosophy, Beecher envisaged the coming of a new and better Christianity:

> I believe that there is rising upon the world, to shine out in wonderful effulgence, a view of God as revealed in the history of the unfolding creation that men will not willingly let die—partly through a better understanding of the nature of God in Christ Jesus, and partly also through a growing knowledge of the universal God, the all-present God, the spiritual God, pervading time and space and eternity. The frowning God, the partial God, the Fate-God, men would fain let die; but the Father-God, watching, caring, bearing burdens, whose very life it is to take care of life and bring it up from stage to stage—that thought of God will quench utterly the lurid light of Atheism. We are coming to a time when we shall be so assailed by Atheistic philosophy, that men will be forced back upon this nobler view of God; and so, indirectly, God will restrain the wrath of men, and cause the remainder thereof to praise him.[113]

White shared with Beecher and other liberal and more radical Protestants an evolutionary perspective on religion. It became more and more clear to him, he later recalled, "that ecclesiastical dogmas are but steps in the evolution of various religions, and that, in view of the fact that the main underlying ideas are common to all, a beneficent evolution is to continue." The sermons of the new theologians revealed to him "a beauty in Christianity before unknown to me," and strengthened his own "liberal tendencies." All of this, he claimed, was an "aid in a healthful evolution of my religious ideas." He came to the conclusion that the world needed more "religion," not less. He wrote, "Whenever I spoke of religion, it was not to say a word against any existing form; but I especially referred, as my ideals of religious conduct, to the declaration of Micah, beginning with the words, 'What doth the Lord require of thee?'; to the Sermon of the Mount; to the definition of 'pure religion and undefiled' given by St James; and to some of the wonderful utterances of St Paul."[114] These authors and others had increased White's "distrust for the prevailing orthodoxy." Thus, like many of his contemporaries, White concluded that while the simple "religion of the Blessed Founder of Christianity has gone on through the ages producing the noblest growths of faith, hope, and charity," there also emerged a counterfeit religion, based on theological dogmatism.[115]

At the close of his *Autobiography*, White offered a remarkable *apologia pro vita mea*. He wrote that the purpose of the *History of the Warfare* "was to strengthen not only science but religion." He aimed not only to free science from "trammels which for centuries had been vexatious and cruel," but also to "strengthen religious teachers by enabling them to see some of the evils in the past which, for the sake of religion itself, they out to guard against in the future."[116] He declared that "I have sought to fight the good fight; I have sought to keep the faith,—faith in a Power in the universe good enough to make truth-seeking wise, and strong enough to make truth-telling effective,—faith in the rise of man rather in the fall of man,—faith in the gradual evolution and ultimate prevalence of right reason among men."[117] After so many years of reading and writing, he came to the conclusion that "religion in its true sense—namely, the bringing of humanity into normal relations with that Power, not ourselves, in the universe, which makes for righteousness—is now, as it always has been, a need absolute, pressing, and increasing." Man possesses an inherent need to worship, both public and private. Since this feeling is universal, "it would seem best for every man to cultivate the thoughts, relations, and practices which he finds most accordant with such feelings and most satisfying to such needs." The decline of religious feeling, according to White, is something to mourn. "It will, in my opinion, be a sad day for this, or for any people, when there shall come in them an atrophy of the religious nature."

But in recognizing the inherently religious nature of man, White was quick to mention that, like all other things, religion evolves. The history of religion, he noted, demonstrated its progress. An anthropomorphic god is "fading away among the races controlling the world." The power of "right reason" is gradually casting off the tyranny of theological dogmatism. "To one who closely studies the history of humanity," he contended, "evolution in religion is a certainty." The lesson to be learned from history, according to White, is that "just so far as is possible, the rule of our conduct should be to assist Evolution rather than Revolution." Religious revolution has and will always occur. But as history has shown, such revolutions quickly become bloodbaths. White was optimistic, however, for he saw the more liberal Christianity of his day as purer and better than all that had come before, and therefore it alone would lead to a final reconciliation between science and religion.[118]

CONCLUSION

As a young man, Andrew Dickson White had become disillusioned by what he perceived as the "harsh" doctrines of orthodoxy. His early education personally convinced him that theologically specific education stood in the way of both religious and scientific progress. As a professor, politician, and then university president, he took up the task of both educational and religious reform. With this aim, he founded the nonsectarian Cornell University. Clerical opposition quickly confirmed what he had already believed about theological dogmatism. His opponents blasted him and his university as "infidel," "atheist," and "godless"—but neither White nor Cornell were "godless." At first, he simply ignored such criticism, but soon felt compelled to defend himself and his trustees. He thus

Figure 2.5. Photograph of Andrew D. White entombed in Sage Chapel. Courtesy of Rev. Daniel T. McMullin, Associate Dean for Spirituality and Meaning Making and Director of Cornell United Religious Work.

determined to take the offensive. Adapting material from his historical lectures, he composed a narrative he believed would put his critics to shame. In the course of three decades, he developed this narrative into a weighty two-volume treatise. It was designed to demonstrate not the irreconcilability of science and religion, but how they might be reconciled. His *History of the Warfare* repeatedly called on theologians to recognize that their opposition to science was a serious error—how such opposition was detrimental not only to scientific progress, but more importantly to religious progress. White hoped that when the myths that had been associated with religion were cleared away, the essence of Christianity would emerge, "pure and undefiled." Like Draper, White's ideal religion was free from dogma. He thus embraced Arnold's definition of religion as "a Power in the universe, not ourselves, which makes for righteousness."

Unlike Draper, however, White sought to establish Christianity on a new basis. While Draper emphasized a liberal Protestantism that was at once more rationalistic, White argued that access to divinity could only be found through feeling, intuition, and the moral conscience. Here were their real distinctions. Both drew from a rich Protestant heritage, one emphasizing what has often been called a "religion of the head" and the other a "religion of the heart." White, moreover, portrayed the conflict in providential terms, as a dialectical process in which God continually worked out the evolution of religion. Each "battlefield" ultimately stimulated the growth and progress of religion. Religion was indeed in a process of evolution, and White believed he merely attempted to chart the course of that evolution.

As we shall see, White's historical narrative would be broadly misunderstood. A few religious thinkers grasped White's intentions and praised his work. At the same time, those who lauded it also admitted that it was a devastating blow to orthodox Christianity. When critics asked him how he had separated the "kernel from the husk," he could offer no satisfactory answer. Perhaps more importantly, freethinkers and atheists embraced and appropriated his narrative as a weapon against all religion. Atheist booksellers added it to their lists and recommended it to their readers. White struggled to the end of his life to explain what he had intended in his *History of the Warfare*, publishing his *Autobiography* as an *apologia pro vita mea* for his oeuvre. Despite being entombed in the nondenominational Sage Chapel at Cornell University (or perhaps because of it), White's narrative became, however inadvertently, one of the most effective and influential weapons of unbelief.

ENGLISH PROTESTANTISM AND THE HISTORY OF CONFLICT

ATTENTIVE readers of the first two chapters should have been struck by the repeated trope of the "Reformation" in the intellectual formation and context of Draper and White. This is no accident. It is important to recall that the outlook of most nineteenth-century Englishmen had been "Protestant, progressive, and whig."[1] While public attitudes toward Catholics ranged from mild suspicion to vehement abhorrence, anti-Catholic sentiment was pervasive, infusing almost every facet of Victorian life.[2] Antipathy towards the Roman Catholic Church was indeed a transnational cultural phenomenon. It is thus a mistake to continue emphasizing Draper's anti-Catholicism as some historians of science have, as if such a sentiment would have offended his mostly Protestant audience. That Draper was anti-Catholic is not disputed. What needs to be pointed out, however, is that he was no more or less anti-Catholic than other nineteenth-century Protestants, including White. Simply put, to accuse Draper of anti-Catholic sentiment is to tell us very little indeed. More importantly, such claims ignore the fact that the vast majority of Anglo-American thought was tacitly or explicitly anti-Catholic. As D. G. Paz put it, anti-Catholicism was "an integral part of what it meant to be a Victorian."[3]

At the dawn of the nineteenth century, a number of factors greatly increased Protestant anti-Catholic fear and loathing. During the first half of the century, for instance, overpopulation and adverse social and economic conditions in Ireland forced many Catholic peasants to emigrate to Britain, becoming a flood during the great potato famine of 1847. Wishing to avert civil unrest, the British parliament passed in 1829 the Catholic Emancipation Act, which revoked many civil disabilities imposed on Catholics since the time of the English Reformation. Then, in the 1830s, the Oxford Movement, or Tractarianism, emerged. Led by Oxford dons John Keble (1792–1866), Edward B. Pusey (1800–1882), John Henry Newman (1801–1890), Richard Hurrell Froude (1803–1836), and others, these Tractarians were a group of High Church Anglicans who, through a series of pamphlets known as *Tracts for the Times* (1833–41), sought to enhance and restore the "Catholic" character of Anglicanism. They were quickly accused of being leaders of a secret papist school of divinity in Oxford.[4]

Anti-Catholic feeling was further exacerbated in 1845 when many of the Tractarians left the Church of England and converted to Roman Catholicism, especially after John Henry Newman submitted to the Roman Catholic Church. Protestants were also enraged when British Prime Minister Robert Peel (1788–1850) increased the annual subsidy for the Royal College of St. Patrick at Maynooth, a leading Catholic seminary in Ireland. At the turn of the decade, Pope Pius IX (1792–1878) also began reinstating the Catholic hierarchy in England. In 1850, for instance, Pius IX had restored parishes, dioceses, and a dozen territorial bishoprics, in addition to installing a new Archbishop of Westminster. Dubbed the "Papal Aggression," anti-Catholic sentiment reached a frenzy when Nicholas Wiseman (1802–1865), the first Roman Catholic Archbishop of Westminster since the Reformation, declared in a pastoral letter of "governing" various counties in England. This letter allegedly elicited Queen Victoria to quip, "Am I Queen of England or am I not?" In response to widespread hostility from all segments of society, Parliament issued the Ecclesiastical Titles Act of 1851, which forbade anyone outside the Established Church from using any episcopal title. Pius IX went on to promulgate his infamous encyclical *Quanta cura* and the antimodern *Syllabus of Errors* in 1864, which included language condemning free speech, separation of church and state, free public schools, Bible societies, and Protestantism itself. This condemnation of mostly modern ideas was subsequently reinforced by the First Vatican Council in its declaration of papal infallibility in 1870.[5]

The upsurge in anti-Catholic sentiment can also be measured by the numerous anti-Catholic societies emerging during the nineteenth century, including the British Society for Promoting the Religious Principles of the Reformation (1827), the Protestant Association (1835), the National Club (1845), the Scottish Reformation Society (1850), and the Protestant Alliance (1851), among many others, which flooded the reading market with newspapers, magazines, pamphlets, petitions, and other anti-Catholic material.[6] The general press also consistently described Roman Catholics as "dark," "foul" "cruel," "bigoted," "heartless," "superstitious," and "hypocritical," while casting Protestants as "noble," "brave," "honorable," and "fighters of freedom."[7] At a more popular level, anti-Catholic feeling was disseminated through the Victorian sermon, where ministers often condemned Catholicism as "monstrous," "apostate," and even "satanic."[8] Just as influential were the numerous popular novels that advanced a strident Protestantism.[9]

More important for the purposes of the current chapter, many nineteenth-century Protestant historians argued that "true Christianity" had reemerged at the Reformation. As we have seen in previous pages, the sheer vitality of this genre of historical writing cannot be overemphasized. Anglo-American Protestants of all denominations insisted that without the Reformation, there would be no economic success, no intellectual and scientific growth, no political liberty—in other words, no modernity. By the end of the nineteenth century, praising the Reformation had become one of the leading hallmarks of popular anti-Catholic discourse. Thus, an ingrained anti-Catholic prejudice was part and parcel of nineteenth-century Protestant theological self-understanding. What needs to be emphasized here, however, is that anti-Catholic writers were also defining, defending, and criticizing what they believed to be the very nature of Protestantism itself. The narratives that nineteenth-century Protestant historians, scientists, and philosophers told were, in short, Protestant self-critique. They did not merely attack Roman Catholicism: they defined what Protestantism was.

While anti-Catholicism was an indelible feature of nineteenth-century Protestantism, it is necessary to recognize that it was an inherited tradition—indeed, one that can be traced back to the time of the Reformation itself. In this chapter, I trace the emergence of a Protestant historiography that became an essential feature of the history of science narratives found in writers like Draper and White. Protestant historiography was a remarkably formidable and protean weapon, combining theological criticism, historical evidence, and rational analysis. In Roman Catholicism, Christian faith had been based on tradition understood in the sense of authoritative belief and wisdom handed down carefully from generation to generation. This notion of tradition had been rejected by sixteenth-century Protestant reformers. In its place, the reformers constructed a new vision of history designed to expose the corruption of the Roman Catholic Church and thus justify Protestant secession. Crucially, similar rhetorical strategies were subsequently adopted and exerted between contending Protestant groups. With the onset of the religious wars and the discoveries from voyages across the Atlantic, there was an attempt to solve internal religious conflicts by rationalizing faith and reducing it to essential components. By the early nineteenth century, early histories and philosophies of science functioned as Protestant self-critique, designed to provide Protestant Dissidents or Nonconformists with a set of beliefs through which they could differentiate themselves from both Catholic and mainstream Protestant religion. The historical narratives of Draper and White have been misunderstood by historians of science largely because they have failed to put them within the context of this Protestant heritage, to which we now turn.

THE PROTESTANT REFORMATION AND THE USE OF HISTORY

Let us start at the beginning. If we return to the sixteenth century, we witness the inauguration of a new vision of history. The reputation of the Roman Catholic Church had already been tarnished by the work of the humanists, particularly the Dutch satirist Desiderius Erasmus (1466–1536). Though not particularly fired by a reforming zeal, his *Praise of Folly* (1511) and *Colloquies* (1518) used derogative terms to characterize the scholastics, and thus provided a vocabulary for how Protestants would subsequently describe the history of medieval Christianity. Erasmus and other humanists used the rhetoric of "corruption," "superstition," and "irrationality" to refer to the Roman Catholic Church, and many subsequent Protestants in the sixteenth and seventeenth centuries viewed such sardonic oratory as accurate historical descriptions of the Catholic tradition.[10]

Certainly, in their critique of Rome, Protestant reformers initially proved their point through *sola scriptura*. But in their attempt to defend such Protestant principles, they were compelled to explain them by reference to the doctrines and practices of the early Church, and in so doing needed to trace when and how the Church lost its way. As the struggle between the forces of Reformation and Counter-Reformation intensified, Protestants gradually came to appreciate more and more the study of history and its possible uses in refuting the historical foundations of Rome and the papacy. In short, Protestants had learned to exploit history as a polemical weapon for attacking theological adversaries.

Inspired by the humanist critique and their own theological grievances, early Protestant reformers constructed a new historiographical tradition. The reformers were faced with an urgent need to create a common identity, which involved a repossession of the past: history had to be rewritten to underline the Protestant claim for legitimacy and authority, and to give believers a sense of belonging. Employing notions such as *reformatio* and *renovatio,* they differentiated the "past" from the "present." When they called for a return of *ad fontes* to the original sources of the Christian faith, they envisaged themselves in a dialectical process of recuperating and recapturing a lost era of apostolic purity, thus forging and engendering a chronology composed of "periods." It was not Catholicism but Protestantism that embodied the true church of Christ and his earliest disciples.[11]

This nascent historiography is clearly displayed in the writings of Martin Luther (1483–1546). While Luther was not primarily concerned with explicating a "philosophy of history," his thought was nevertheless steeped in history and its meaning. As early as 1519, for example, in his Leipzig debate with German scholastic theologian Johann Eck (1486–1543), Luther used the historical record to contend that popes, councils, canon law, church fathers, and theologians have constantly contradicted each other. Based on Scripture *and* history, Luther thus denied the divine origin and authority of the papacy. Shortly after the Catholic Church condemned his writings as heretical, Luther composed a series of treatises attacking the Church for corrupting Christianity. In his *Babylonian Captivity of the Church* (1520), he accused the Church of inventing the sacraments to keep the laity in subjugation and thralldom. The allusion to Israel's Babylonian exile should not be missed. The post-exilic Hebrew prophets interpreted the Babylonian exile as a time of divine purification, envisioning a new exodus that would bring redemption and the restoration of a "righteous remnant." For Luther, as Babylon held Israel captive, now the papacy was holding Christianity in bondage.

Luther's understanding of Christian history signified an explicit rejection of all institutional guarantees for historical continuity. Following the Augustinian model, he imagined the world divided by two kingdoms in violent struggle. Since the days of Cain and Abel, history is a narrative of this dual progeny locked in a terrible conflict. But history was also a movement in which God was ever active in using men as His instruments to accomplish His will. The ultimate purpose of recorded history, according to Luther, was to bring man to a knowledge of God through His works. Luther pointed to the corruption of the Church as something that has been repeated throughout history. The pattern of Church history itself is one of apostasy, struggle, and degeneration. But with the principles of *sola scriptura* and *sola fide*, Luther believed he was recovering the original teachings of the apostolic church. The task of reformation or restoration, however, required the destruction of the established Church. In short, Luther had effectively linked Protestants with the faithful remnant of God's elect, trumping Catholic narratives of continuity, all the while presupposing the periodic need to purify the church by separating the "sheep from the goats."[12]

Though not a historian, Luther used history as a polemical weapon against the Catholic Church. This understanding of history became the dominant Protestant view. Subsequent Protestants recapitulated this narrative in their struggles with Catholics, identifying themselves as that faithful remnant destined to restore true Christianity. Luther's colleague and ally Philip Melanchthon (1497–1560), for example, would go on to make Wittenberg the center of the new historiography. Melanchthon composed a universal history, *Carion's Chronicle* (1532), which he originally wrote in collaboration with German mathematician and astronomer Johann Carion (1499–1537). The text became the standard historical textbook in Germany for some two hundred years. This historiographical tradition was further amplified by Matthias Flacius Illyricus (1520–1575), a former student of Melanchthon and leader of the "Magdeburg Centuriators" and chief organizer of the *Ecclesiastica historia integram ecclesiae Christi* (1559–1574), which was a systematic attack on ecclesiastical history. The exhaustive researches of the Magdeburg Centuriators in Germany recovered an immense collection of textual material to demonstrate the apostolic origins of Protestantism. Published in thirteen volumes, each book covered a century of history, and the controlling theme throughout was that history was the scene of constant conflict or warfare between good and evil, between God and the Devil, between the work of Christ and the intrigues of Antichrist. These "second generation" reformers divided ecclesiastical history into epochs, and more vehemently vilified the papacy as corrupting the simple teachings of Christ and the early church.[13]

The Reformation therefore gave rise to a new historical consciousness. While central elements of his history were later ignored, Luther had established an historical precedent. Protestants were obliged from the outset to supplement their study of the New Testament with a close historical survey of the Church. History became the handmaiden of Protestant theology. By rewriting the story of the past, Protestant reformers crafted a narrative and interpretive framework that became normative for much of Protestant Europe. A new history of Christendom was thus devised. For most Protestants, religious outlook and historical understanding became inextricably meshed. With the acute sense of successive "ages" in history, the reformers were convinced that the Reformation had opened a new epoch in the history of Christianity in which the hand of God could work inexorably for good. Especially important for our purposes, and where we can explicitly see continuity in the historical writings of Draper and White, many early Protestants came to believe that the conflict with the Church was necessary for the progress of religion.[14]

By the time the Reformation reached England, there had already been a longstanding history of protest against papal authority on the island, with John Duns Scotus (1265–1308), William of Ockham (1287–1347), John Wycliffe (1330–1384), and William Tyndale (1494–1536) as obvious early examples. Indeed, Luther changed his mind about the use of history after reading the work of English historian Robert Barnes (1495–1540), who surveyed the corruption of the papacy in his *Vitae Pontificum*, or *Lives of the Roman Pontiffs* (1536). Luther had Barnes's book published in Wittenberg, writing in its preface:

> [In] the beginning, not having much expertise in history, I attacked the papacy *a priori* (as is said), that is, from the holy Scriptures. Now I wonderfully rejoice

> that others are doing this *a posteriori*, that is, from history. And I seem to myself clearly to triumph since, with the light becoming clear, I perceive that the histories agree with the Scriptures. For what I have learned and taught from the teachers St Paul and Daniel, that the pope is the Adversary of God and of all, this history proclaims to me, pointing out this very thing with its finger, not revealing it in general (as they say), and not merely the genus or the species, but the very individual.[15]

English reformer Miles Coverdale (1488–1569) also decried the "great poison poured into the church whereby religion sore decayed." The superstitions, idolatry, pomp, and pride of the Church were "alterations, pervertings, and contrary to all old ordinances, having no ground in God's word, and are clean against God."[16] English churchman John Bale (1495–1563) likewise blamed the whole Catholic Church for the evil that befell the faith, describing history as the space of time within which a struggle is waged between two opposite powers: "the true Christian church, which is the meek spouse of the Lamb without spot," and "the proud church of hypocrites, the rose-coloured whore, the paramour of antichrist, and the sinful synagogue of Satan."[17] Bale's friend the Marian exile John Foxe (1516–1587), author of *Actes and Monuments* (1563), or more popularly known as the "Book of Martyrs," adopted the new historiographical scheme and gave the Church of England a particular sense of ascendency. His book of martyrs was given near-scriptural status, often appearing alongside the Bible in churches; it continued to be reprinted as late as the end of the nineteenth century.[18] Other English Protestants, including such men as Matthew Parker (1504–1575), John Jewel (1522–1571), John Whitgift (1530–1604), Richard Hooker (1554–1600), among others, followed the new historiographical scheme, constructing polemical narratives that supported the deep historical roots of their own faith.[19]

As Alexandra Walsham aptly observes, the English Reformation in particular gave rise to empirical inquiry, "to the urgent need for the sanction of history that sent both Protestants and Catholics scurrying into the archives and that propelled the antiquarian endeavors that created the great libraries of books and manuscripts in Cambridge, Oxford, London, and across the Continent, upon which we are now reliant."[20] Adapting Protestant historiography to the English historical context, English reformers claimed that pure apostolic Christianity had been transferred intact to English well before the intrusion of the Church of Rome. From this point of view, English history appeared as an endless struggle with Rome. It is interesting to point out that the language of "youth and age" pervades the polemical literature engendered by the English Reformation. As we have seen with Draper and many of his sources, the notion had become deeply embedded in English Protestant thought that people in different stages of life (infancy, childhood, youth, manhood, old age, and death) had distinct characteristics, temperaments, and instincts that corresponded typologically with particular eras of human history.[21]

Another particularly effective rhetorical strategy, and a direct consequence of the new historiography, was comparative religious history. English Protestants began comparing different religious traditions in order to trace the origins and growth of idolatry in Catholic beliefs and practices. These reformers saw "superstition" as synonymous with Catholicism and paganism, effectively linking the papacy with paganism. By comparing other religions, English Protestants such as Oliver Ormerod (d. 1626), Edward Brerewood (1565–1623), Samuel Purchas (d. 1626), and countless others came to believe it was possible to identify the "popish hierarchy" as the "old pagan hierarchy revived." The theological potency of tracing the "paganization" of Christianity, of accusing Catholicism of "pagano-papism," was indeed profound and would be substantially adopted by Protestant polemicists in England and on the Continent.[22] In the hands of early English Protestant reformers, the charge that the Catholic Church had taken its model from the "priestcraft" of the pagans was a severe indictment. But as we shall see later, Protestants would use this same charge against other contending Protestant traditions.

The advent of Protestantism thus precipitated a significant shift in the perception, investigation, and interpretation of the past. A new historiography had been invented that necessitated a fundamental and irrevocable shift of historical orientations. Reformers had recast history in order to give historical legitimacy to their protest. The new historiography invoked by Protestants remained a powerful polemical weapon for the next generation of reformers. By the seventeenth and early eighteenth centuries, English anti-Catholic propaganda was pervasive. Anti-popery or anti-Catholicism was thus a stock-in-trade of English Protestantism, one that remained central to Protestant identity into the nineteenth century. To be consistently critical of Rome was thus an essential trait of Protestant piety.

REASON, SCIENCE, AND RELIGION IN ENGLAND

It must be emphasized, however, that arguments directed at Roman Catholics also appeared in disputes between contending Protestant groups. Henry VIII (1491–1547) broke from Rome for emphatically political rather than religious reasons. He was succeeded by his young son Edward VI (1537–1553), who also did little to reform ecclesiastical policy. When Edward died, his older half-sister Mary I (1516–1558) came to the throne and reestablished Catholicism through force. When Mary died, the crown passed to her Protestant half-sister, Elizabeth I (1533–1603). Like other English Protestant works printed at the time, both Illyricus's *Ecclesiastica historia* and Foxe's *Actes and Monuments* were dedicated to Elizabeth. She attempted to reach a new religious settlement by promulgating the Act of Supremacy of 1558, the Act of Uniformity of 1559, and the *Thirty-Nine Articles* of 1563.

The so-called Elizabethan settlement had left many dissatisfied, however. When Protestant refugees returned from Geneva and Germany, they brought with them strong Calvinist views that were characterized by their opponents as "Puritan." Dissenting religious groups like the Puritans memorably condemned the so-called Elizabethan settlement as only "halfly reformed." In their *Admonition to Parliament,* a manifesto published in 1572 and written by London clergymen Thomas Cartwright (1535–1603), John Field (1545–1588), and Thomas Wilcox (1549–1608), these Puritans demanded that Elizabeth abolish all "popish" remnants and "restore" the Church of England to its ancient office as found in the New Testament. While Protestant historians portrayed the Catholic Church as corrupt, Dissenting Protestants wielded history to undermine the position of the newly Established Church. The Puritan call for the restoration of the primitive church was thus essentially the same historical argument the original reformers used against Rome.[23]

Failing to pacify Puritan demands, Elizabeth called on her clergy to defend the new religious settlement. Unsurprisingly, Anglican apologists in turn offered counterhistorical arguments defending the Elizabethan religious settlement as a necessary step in the progress of Christianity. Among these apologists, perhaps the most important was the philosopher and theologian Richard Hooker. In his monumental *Laws of Ecclesiastical Polity* (1593–97), Hooker defended the settlement against the religious "enthusiasm" of the Puritans by emphasizing a new hermeneutical approach to Scripture. By "enthusiasm," Hooker meant those radical reformers who sought to reform the English church according to God's word and pattern itself on Calvin's Geneva. More generally, as English philosopher David Hartley (1705–1757) later quipped, it was that "mistaken Persuasion in any Person, that he is a peculiar Favorite with God; and that he receives supernatural Marks thereof."[24] According to Hooker, one must avoid the overzealous Puritan desire to govern the church following the precepts of Scripture alone. While Scripture was infallible, man was not. Scripture no doubt contains all things necessary for salvation, but it must be understood within the sphere of natural human reason. Hooker thus emphasized the role of human reason and even natural philosophy in justifying religious beliefs and practices. Indeed, he believed human reason is consubstantial with divine Reason, and thus made a plea for an orderly and reasonable approach to religion, just as God Himself governed the natural world through laws. Hooker's defense, in short, was a justification of the place of "reason" and the "laws of nature" in religion.[25]

The struggle between contending Protestant groups became more pronounced after the death of Elizabeth and the coronation of James I (1566–1625). While the Puritans demanded the suppression of Catholicism, James continued Elizabeth's policy of toleration. The confessional conflicts unleashed by the fragmentation of Christendom encouraged many to look for an alternative vision of faith. While some Protestants continued to propound God's book of Scripture, others, following Hooker's suggestion, turned to God's book of Nature. We see this most prominently in the work of a host of "English virtuosi" in the seventeenth century, beginning with Francis Bacon (1561–1626). Bacon, whose father, Sir Nicholas Bacon, was one of the chief architects of the Elizabethan religious settlement, was likewise concerned with the growing religious discord in his day. In his *Advertisement touching the Controversies of the Church of England* (1589), for instance, Bacon criticized the Puritans for ignoring the "will of God as revealed through the laws of nature."[26]

Numerous studies have pointed out the religious or theological foundations of Bacon's thought, particularly in his work on natural philosophy.[27] What needs to be emphasized here is that Bacon had deliberately aligned his campaign for the new learning with the larger struggle of the Reformation, effectively linking the prophetic ethos of the new knowledge of the natural world to the prophetic orientation of the Protestant Reformation. In his *Advancement of Learning* (1605), for instance, Bacon described medieval scholastics as the "great undertakers" who had "deformed" the understanding of God's creation. Explaining the revival of learning during the sixteenth century, Bacon pointed to Luther, who out of his struggles "against the bishop of Rome and the degenerate traditions of the church . . . was enforced to awake all antiquity, and to call former times to his succours to make party against the present time: so that the ancient authors, both in divinity and humanity, which had long slept in libraries, began generally to be read and revolved." Bacon went on to assert that religious reformation and the advance of learning were both evidence of God's providence: "when it pleased God to call the Church of Rome to account for their degenerate manners and ceremonies . . . at one and the same time it was ordained by the Divine Providence, that there should attend withal a renovation and new spring of all other knowledge."[28]

He observed that there were many impediments, or "tacit objections," to the new learning. These objections often arose unconsciously, or were simply assumed without serious thought. He thus carefully intimated that whatever fundamental tension existed between the new learning and religious objections, it was in fact a misunderstanding. In this way, he condemned the pride and theological wrangling of the scholastics, not religion. Thus theology, not religion, was the obstacle. Indeed, the new learning, according to Bacon, had positive religious value: "Let no man, upon a weak conceit of sobriety or an ill applied modern," he famously wrote, "think or maintain that a man can search too far or be too well studied in the book of God's word or in the book of God's works."[29]

In the process of defending natural philosophy against such tacit objections, Bacon was writing natural philosophy into Christian history. The natural philosopher in the *Advancement of Learning* is a Christian—more precisely, a Protestant reformer. By embracing the new historiography of the reformers, Bacon brought natural philosophy into the Christian story of salvation. In his *Valerius Terminus*, an unpublished draft of the *Advancement of Learning*, Bacon represented Christ himself as a proponent of the *sola natura*. Reading the book of God's works is "a singular help and preservative against unbelief and error, for saith our Saviour, '*You err, not knowing the Scriptures nor the power of God*'; laying before us two books or volumes to study if we will be secured from error; first the Scriptures revealing the will of God, and then the creatures expressing his power; for that latter book will certify us that nothing which the first teacheth shall be thought impossible."[30] In quoting Christ's reprimand to the Sadducees, Bacon aligned the errors of these teachers of the law with medieval clergymen. The implication is clear: knowledge of Scripture was incomplete unless it was known within the bounds of natural philosophy. Bacon had collapsed the "two books" metaphor, which can be traced all the way back to Augustine of Hippo, into one vast single text. Moreover, by identifying the text of nature with the "power of God," Bacon had given the reformation of natural knowledge an indispensable role within the larger narrative of the Protestant Reformation. The natural philosopher, in short, is a fellow laborer in the divine plan.

Bacon went on to give a more popular rendering of his argument in *The New Atlantis* (1627). In this unfinished "fable," Bacon envisioned a utopian community of scholars and intellectuals interpreting and controlling nature for the benefit of humanity. The tale begins with an unnamed narrator and his crew of lost sailors, somewhere beyond the shores of Peru, "in the midst of the greatest wilderness of waters in the world." That they begin their voyage from a Spanish and, hence, Catholic colony is no mere coincidence, for they are providentially "saved" when winds carry them northward to the undiscovered island of Bensalem, a Christian civilization cut off from the rest of the world for two millennia, and thus from the corrupting influences of the Catholic Church. Once on the island, the sailors learn the history of Bensalem, how its inhabitants were converted not by missionaries but by the miraculous appearance of the Bible enclosed in a "small ark." The sailors also learn of "Salomon's House," which is a society of natural philosophers who study nature in order to better understand God and to relieve man's estate. Bacon's *New Atlantis* abruptly ends with the sailors' conversion and commissioning. The great "high priest" of nature, the Father of Salomon's House, is presented as a new pope. He commissions the sailors to propagate the new gospel of natural philosophy. The new learning, in short, was not a deviation from Christian tradition—it was the covenant fulfilled. It was part of God's providential plan from the very beginning.[31]

The significance of Bacon's depiction of the island as Protestant was not lost on his contemporaries. By writing the new learning into this reimagined Christian history, Bacon made it much easier to conceive a Protestant future dominated by science. In other words, by linking the new learning with the idea of the restoration of Christianity, Bacon gave science a decisive role in the new social order that the Reformation brought to England. If science was doing the true work of the church by opening up the book of God's work, then its practitioners were fellow laborers in the church's ministry. The new learning, in short, came to symbolize a new Protestant ethos.

Indeed, subsequent generations looked to Salomon's House as a model to emulate. Unsurprisingly, the new learning came to be used as a weapon against Roman Catholicism. Even something as seemingly innocuous as a library came to be an instrument of polemic. Bacon had proposed that one of the chief functions of the "Merchants of Light" of Salomon's House was to collect books from all over the world. Thus, when Bacon had sent Sir Thomas Bodley (1545–1613) a copy of his *Advancement of Learning* in 1605, he congratulated him on founding the library of Oxford University, welcoming the foundation of the Bodleian Library as "an Ark to save learning from deluge."[32] The Bodleian Library was thus much more than a university library. Bodley had installed as its first librarian Thomas James (1573–1629), who was, according to Paul Nelles, a "rabid antipapist." James intended and designed the Bodleian Library to be a "store-house of Protestant learning and a bulwark against Roman Catholicism."[33] The library thus served as an instrument of Bacon's vision of the "advancement of learning" or, as Anthony Grafton puts it, "an arsenal of erudition for the Protestant side in the great intellectual war that waged over the Christian past."[34]

Bacon's vision of the "Christian virtuoso," a Christian experimental natural philosopher with the liberty or autonomy to investigate nature solely guided by the light of nature, was also taken up by a host of Protestant writers. The existence of God could be demonstrated, and His attributes determined, by a concerted effort in examining the universe. As noted above, while most of this work offered a religious defense of the new learning, it also aimed to demonstrate, implicitly or explicitly, the "reasonableness" of Protestantism. Paul Kocher, Richard Westfall, and others have demonstrated that Jacobean theologians and writers felt confident in man's ability to use reason and obtain natural knowledge of God's creation, in addition to espousing the belief that Protestant England should be the bulwark of the Reformation.[35]

As the seventeenth century drew to a close, both theologians and natural philosophers concentrated more and more on establishing "rational" foundations for Christian belief. The so-called "English virtuosi," those who followed Bacon's vision, attempted to provide such a foundation by demonstrating how God has revealed himself in nature and how a more "rational" Protestantism provided an atmosphere more conducive to the new learning. Protestantism, it was argued, embodied the principles that would allow for the progress of learning, society, and religion itself. Thus, when polymath Samuel Hartlib (1600–1662) and his circle of collaborators took up Bacon's philosophical program in the establishment of the Royal Society in 1660, they were following a broader narrative that came to represent the English Reformation. The decidedly Protestant character of the society is evident in Bishop of Rochester Thomas Sprat's (1635–1713) official *History of the Royal Society of London, For the Improving of Natural Knowledge,* first published in 1667.[36] In his introduction, Sprat explains that Bacon deserved a special place of honor because he "had the true Imagination of the whole extent of this Enterprise."[37] What Sprat meant by "whole extent" was the drama of Christian history. Sprat envisioned natural philosophy as regaining what was lost at the Fall—and, implicitly, what the Catholic Church failed to achieve. The new experimental philosopher "has always before his eyes the *beauty, contrivance,* and *order* of *Gods Works.*" This was indeed the "first service, that *Adam* perform'd to his *Creator,* when he obey'd him in mustring, and naming, and looking into the *Nature* of all the *Creatures.* This had bin the only *Religion,* if men had continued innocent in *Paradise,* and had not wanted a *Redemption.*"[38]

Sprat argued, in other words, that the science of the Royal Society was an instrument of the Reformation that would restore primitive religion. He believed the experimental philosopher had equal claim to the Reformation and made several pointed comparisons between the corruptions of natural philosophy and the Catholic corruptions of Christianity.[39] He claimed that the "*Church of England* will not only be safe amidst the consequences of a *Rational Age,* but amidst all the improvements of *Knowledge,* and the subversion of old Opinions about *Nature,* and introduction of new ways of Reasoning thereon. This will be evident, when we behold the agreement that is between the present *Design* of the *Royal Society,* and that of our *Church* in its beginning. They both may lay equal claim to the word *Reformation;* the one having compas'd it in *Religion,* and other purposing it in *Philosophy.*"[40] For Sprat and many others during the period, the experimental study of nature was a religious discipline. What needs to be emphasized here, however, is that the justifications for such work were usually couched in anti-Catholic terms. The new experimental philosopher thus became a symbol and catalyst for spreading the Reformation, in the hope of finally defeating intransigent Catholicism.[41]

A MORE "RATIONAL" CHRISTIANITY

It was becoming more and more apparent in England that the congruity of empirical scientific research and Christianity was defended by the more reasonable and moderate Protestant thinkers. Indeed, Sprat asserted that "the universal disposition of this age is bent upon a rational religion." Thus the Church of England, he added, "cannot make war against reason, without undermining our own strength, seeing it is the constant weapon we ought to employ." The enemy, according to Sprat, were those with "implicit faith" (i.e., Catholics) and "enthusiasts" (i.e., Puritans). Accordingly, the Royal Society subscribes to a religiously "latitudinarian" scientific program. Sprat's official *apologia* for the Royal Society made perfectly plain the open alliance between liberal Protestantism and scientific inquiry, insisting that the quality of the humble Christian and the scientific experimenter were one and the same.

Thus, in addition to appealing to the historical record, a select group of Protestants increasingly began appealing to reason. These Protestants promoted a "reasonable" creed against the superstition and irrationality of enthusiasts and Catholics alike. Recourse to reason had thus become a litmus test. Knowledge of the natural and the supernatural came from the use of reason, and for a growing group of thinkers, reason and religion developed a holy alliance.

The political situation almost called for such a development. When James I died in 1625, his son Charles I (1600–1649) assumed control of the English throne. The Caroline era, however, was plagued by revolutions and civil war, plunging England into one of the most tumultuous periods of its history. In protest to policies enacted by Charles I, a Puritan parliament issued the *Westminster Confession of Faith* (1646), establishing a Presbyterian church order. Charles I was eventually executed by a group of parliamentarians in 1649, which was followed by the so-called Interregnum under the military leadership of Oliver Cromwell (1599–1658). Under Cromwell, England experienced a swell of competing religious movements. After Cromwell died, however, parliament restored the monarchy in 1660, crowning Charles II (1630–1685) king. But with no son, Charles II was succeeded by his Catholic brother, James II (1633–1701), the last Roman Catholic monarch of England. Fears of Rome reached new heights with the ascension of James II, and he was quickly deposed in 1688 in the so-called "Glorious Revolution," replaced with staunch Protestants William III (1650–1702) and his wife Mary II (1662–1694). During these years of intermittent religious conflict and political warfare, a number of writers employed anti-Catholic rhetoric to various ends. Some of the most popular works were later collected and published by Anglican bishop of London Edmund Gibson (1669–1748) in his *Preservative against Popery* (1738), which contained over 100 tracts against Catholicism written during the reign of James II.[42]

Incessant Protestant infighting, however, caused many to look elsewhere for a peaceful resolution. There is little doubt that the divisions within Christianity contributed significantly to the disillusionment of many with historical Christian belief. The late seventeenth and eighteenth centuries saw a development in which many of the central beliefs of revealed religion—key doctrines of Christianity such as the Trinity, the divinity of Christ, original sin—were often cast aside in favor of a religion based on what could be known through the use of reason alone. God was seen not as the biblical Father but as the Supreme Being, the impersonal Creator, a God who is a necessary requirement for the maintenance of the laws of nature but is little interested in worship, doctrine, churches, and other religious trappings. This quest for theological peace was the beginning of, as theologian William C. Placher aptly put it, the "domestication of transcendence."[43]

In a time of rampant enthusiasm, civil war, and religious factionalism, the calm use of reason in dealing with religious issues certainly had its appeal. Confused and exhausted, a growing number of English Protestants began advocating the "reasonableness" of religion as the only way toward peace and reconciliation among contending Protestant sects. This attitude toward reason and religion reached its apogee in the teachings and writings of the so-called Cambridge Platonists, a loose-knit group of divines that included men like Benjamin Whichcote (1609–1683), Henry More (1614–1687), Ralph Cudworth (1617–1688), John Smith (1618–1652), Nathaniel Culverwell (1619–1651), and others. As John Gascoigne has shown, the alliance between natural and supernatural knowledge had become a commonplace at Cambridge.[44] But in combining the Baconian project of the new learning with religion, the Cambridge Platonists reenvisioned old beliefs in light of new knowledge, giving new meaning to religion by placing it firmly on more philosophical foundations. According to C. A. Patrides, the Cambridge Platonists boldly rejected the "entire Western theological tradition from St Augustine through the medieval schoolmen to the classic Protestantism of Luther, Calvin, and their variegated followers in the seventeenth century."[45] In its place they advocated that the spirit of man is reason—human reason, the "candle of the LORD," is what it means to be made in the image of God. For the Cambridge Platonists, human reason is the very voice of God. To go against reason, therefore, is to go against God. All came out of Puritan backgrounds that they to some extent rejected. Linked by friendship and residence, the Cambridge Platonists were part of the movement later described as "rational supernaturalism." Spokesmen for moderation in all things, they followed Hooker and others in delineating a *via media*

Anglicanism—between superstition on the one hand, and enthusiasm on the other.

This band of Cambridge men also rejected the exclusive claims of tradition and Scripture and believed that human beings have a natural knowledge of God. In his *An Elegant and Learned Discourse of the Light of Nature* (1652), for instance, Culverwell argued that reason and faith are compatible, for they spring from the same fountain of light and conspire to the same end: "the glory of that being from which they shine." "So that to blaspheme reason," he added, "is to reproach heaven itself, and to dishonour the God of reason, to question the beauty of His image, and, by a strange ingratitude, to slight this great and royal gift of our Creator." True religion, he contended, "never was, nor will be, nor need be, shy of sound reason."[46]

In his *Select Discourses* (1660), which were sermons preached at College Chapel, Smith also celebrated the inward beauty and divine truth found in man. He maintained that man's ability to reason connects him to God. For Smith, God's existence is not only demonstrated from the order of nature, but the rational soul of man. In his preaching, moreover, Smith sought to "call men off from dogmas and barren speculation." To seek divinity merely in "Books and Writings," he declared, was "to seek the living among the dead." Smith was convinced that there is "a near affinity between atheism and superstition." "Atheism could never have so easily crept into the world," he asserted, "had not superstition made way, and opened a back-door for it." Those who are terrified by the "specters and ghastly apparitions," the temples and groves and altars and other "idle toys to place these deities with," are driven to seek refuge in the "strong fort" of irreligion, according to Smith. But the "spirit of true religion," he proclaimed, "is of a more free, noble, ingenuous, and generous nature, arising out of the warm beams of the Divine love which first hatched it and brought it forth."[47]

Whichcote, considered to be the founding father of the Cambridge Platonists, similarly believed that "to go against *Reason*, is to go against *God*." Like the other men of the Cambridge school, he wished to save religion from the dogmas and religious enthusiasm of his day. For Whichcote, God constantly revealed himself in nature. The function of Scripture is thus to confirm the truths already present in the light of nature. "The written word of God," he wrote, "is not the first or only discovery of the duty of man. It doth gather and repeat and reinforce and charge upon us the scattered and neglected principles of God's creation." According to the editor of his *Moral and Religious Aphorisms*, which were published posthumously in 1703, Whichcote promoted the establishment of afternoon Sunday lectures on philosophical topics to counteract the prevalence of religious infighting. God had established two lights to guide mankind, Whichcote said. The light of reason was the light of creation. The light of Scripture was God's revelation. "Let us make use of these two lights; and suffer neither to be put out."[48]

The most eminent and influential members of the Cambridge Platonists were, however, Cudworth and More. In his enormous work entitled *The True Intellectual System of the Universe* (1678), Cudworth defended the existence of God against atheism, materialism, and determinism. At the same time, he considered enthusiasm and superstition just as pernicious as atheism. Cudworth's anti-atheism apologetic belies the fact that the theological imperatives that pervade his writing are also directed at fellow Protestants. In a 1647 sermon preached at the House of Commons, for instance, Cudworth proclaimed that the essence of Christianity is not found in theological formulae or doctrinal pronouncements, but in keeping God's commandments with a righteous heart. The "bookish Christians" make religion little more than "book-craft" and "paper-skill," he contended. According to Cudworth, Christ is not found in "creeds and catechismes and confessions of faith." Christ did not come into the world to "fill our heads with mere Speculations to kindle a fire of wrangling and contentious dispute amongst us," he warned.[49] The implications of the sermon are clear. Theological bickering and infighting was not "true" Christianity. It was never Christ's intention to ensnare or entangle followers with captious niceties or deep speculations. The teachings of Christ were simple, and so should be faith. Indeed, the "book of nature" contains, Cudworth argued, "the whole visible and material universe, printed all over the passive characters and impressions of divine wisdom and goodness." By investigating the external world, the mind "hence presently makes up an idea of God, as the author or architect of this great and boundless machine; A mind infinitely good and wise."[50] In short, Cudworth was concerned to demonstrate not only that God exists, but that there is a right conception of God.[51]

More similarly exemplifies the growing concern for presenting Protestantism as a rational enterprise. Like the other Cambridge Platonists, he made extensive use of contemporary natural philosophy in his many books defending his religion against atheism, enthusiasm, and Roman Catholicism alike. More is indeed an excellent example of not only English Protestant anti-Catholicism but Protestant self-critique. His most well-known work, *An Antidote against Atheism* (1653), was intended to be a "carefull *Draught* of *Naturall Theology* or *Metaphysicks*." He argued that the truth of God's existence is "as clearly demonstrable as any Theorem in *Mathematicks*." The great theater of "External *Phaenomena* of *Universal Nature*," the cycles of day and night and the seasons; the rising and setting of the sun, moon, and stars; the tilt of the earth and the law of gravity; the valleys and mountains, the seas and forests; the form and beauty of vegetables, fruits, and plants; and the bodies of birds, fish, beasts, and men—all, in short, are tokens of the Creator, and therefore leave the atheist without excuse.[52]

Although More's stated aim is to combat atheism, he spends the bulk of *An Antidote against Atheism* on religious enthusiasm. The enthusiast, according to More, boldly declares "the careless ravings of his own tumultuous fancy for undeniable principles of divine knowledge." In fact, like the other Cambridge Platonists, More claimed there was a close connection between atheism and enthusiasm. In his Preface, More makes the connection explicit, writing that "*Atheism* and *Enthusiasm*, though they seem so extremely opposite one to another, yet in many things they do very nearly agree." He portrays both as "diseases" that can and ought to be countered by the exercise of reason. More went on to publish an entire discourse on the nature, causes, kinds, and cure of enthusiasm.[53] One remedy against enthusiasm, in religion and politics, according to More, is to avoid revealed knowledge altogether, whether found in sacred text or doctrines of the church. The supreme revelation is imprinted onto the human mind, a "naturall Sagacity" universally received by all men, and thenceforth serves as the final epistemological authority.[54]

Whatever differences between them, the Cambridge Platonists were united in pursuing the reformation of religion along more rationalistic lines. In prose, sermons, and poems, they declared that religion should be pared down to the essentials. They therefore underplayed dogmatism and opposed superstition, enthusiasm, and fanaticism. Concomitantly, they also placed more emphasis on the inner "spiritual light" than on outward rules of worship and propositional beliefs. The Cambridge Platonists thus condemned both enthusiasts and Catholics alike. The rational theologies of the Cambridge Platonists emphasized the relationship between core religious doctrines, natural law, and a personal divine providence, with evidence for this intrinsic relationship to be found in the ideas already present in the mind, in the natural world, in the associated records or observations of the experimental philosophers, and in history. "Religion" for the Cambridge Platonists referred primarily to a "rational Christianity" of some kind, particularly a rational Protestantism. This type of providential rational natural theology, or natural supernaturalism, became particularly prominent in the religious culture of the Anglican Church in the Restoration period. The important point here is that the Cambridge Platonists wished not to dispense with organized religion but to reform it.

The ideas expressed in the writings of the Cambridge Platonists formed a vital link with subsequent developments in English Protestant thought. Numerous scholars have already noted the close association between the Cambridge Platonists and Latitudinarian divines, who similarly sought to minimize doctrinal discord by emphasizing human reason in understanding revelation. Indeed, the tendencies of the Cambridge Platonists found fuller expression in the rational theology of the Latitudinarians, in the ideas of such men as William Chillingworth (1602–1644), John Wilkins (1614–1672), Simon Patrick (1626–1707), Isaac Barrow (1630–1677), John Tillotson (1630–1694), Edward Stillingfleet (1635–1699), Joseph Glanvill (1636–1680), Gilbert Burnet (1643–1715), among others.[55] The terms "latitude men" or "latitudinarian" were first used in the seventeenth century to describe a group of English clergymen who desired a more moderating position. With the benefit of hindsight, for example, Burnet looked back over the *History of My Own Time* (1724–34) and gave a sympathetic account of the Latitudinarians, writing:

> All these, and those who were formed under them, studied to examine farther into the nature of things than had been done formerly. They declared against superstition on the one hand, and enthusiasm on the other. They loved the constitution of the Church and the liturgy, and could well live under them: but they did not think it unlawful to live under another form. They wished that these things might have been carried with more moderation. And they continued to keep a good correspondence with those who had differed from them in opinion, and allowed a great freedom both in philosophy and in divinity: from whence they were called men of latitude. And upon this men of narrower thoughts and fiercer tempers fastened upon them the name of Latitudinarians.[56]

Well acquainted with Tillotson, Stillingfleet, Wilkins, and others, Burnet himself claimed he "easily went into the notions of the Latitudinarians" as a young man; and as bishop of Salisbury he implemented a latitudinarian agenda, advancing it through pulpit, politics, and polemic.[57] Burnet's massive *History of the Reformation* (1679–1715) is a case in point. He posited a view of history that should now be familiar: a Protestant historiography designed to narrate religious decline and recovery. Burnet agreed with other Protestant historians that soon after Christ ascended, a corrupted clergy quickly emerged, greedy for worldly wealth and power. The Catholic Church had thus perverted the original teachings of Jesus and the Apostles. With this degeneration, the stage was set for the medieval dark ages. The sixteenth-century reformers rescued Christianity from these popish machinations and corruption. But in Burnet's mind, Protestantism would go on to lose sight of its original purity. No longer an invigorating revival of primitive Christianity, it had sunk into a spiritual sloth. Instead of rejoicing in the Christian renewal of human nature, Protestants became obsessed with inessential forms of their faith. The divisions within the new Christendom put a visible stop to the progress of the reformation.[58]

Progress would come, according to Burnet and according to other Latitudinarian divines, with a more rational Protestantism. Chillingworth, for instance, claimed that the Catholic Church held "many things not only above reason, but against it, if anything be against it," whereas Protestants like himself "believe nothing which reason will not convince that I ought to believe it."[59] In his tellingly titled *The Religion of Protestants, a Safe Way to Salvation* (1638), Chillingworth argued for the supremacy of Protestant reason, asserting that Protestantism is not just another religion, but the only reasonable one. But in arguing against the alleged irrationalism of Rome, Chillingworth also rejected the notion of absolute certainty in matters of religion. While he famously argued that "the Bible, I say, the Bible only, is the religion of protestants," he also maintained that:

> I will take no man's liberty of judgment from him, neither shall any man take mine from me. I will think no man the worse man nor the worse Christian—I will love no man the less for differing in opinion from me; and what measure I mete to others I expect from them again. I am fully assured that God does not, and therefore that man out not, to require any more of any man than this—to believe the Scripture to be God's word, to endeavour to find the true sense of it, and to live according to it.[60]

Indeed, writing to a friend about the subject of Arianism, Chillingworth asserted that "the doctrine of Arrius is eyther a Truth, or, at least, no damnable heresy."[61] Interestingly, while he believed the articles of faith to be true, he cautioned that such knowledge should not be understood "as certain as that of sense or science," and that therefore strong adherence to them is a "great error." According to Chillingworth, "faith is not knowledge, no more than three is four, but eminently contained in it, so that he that knows believes, and something more; but he that believes, many times does not know; nay, if he doth barely and merely believe, he doth never know." In short, assent to a truth, Chillingworth demanded, must be proportionate to the evidence that is given.[62] In the absence of doctrinal agreement between Protestants, then, recourse to reason must be the sole criterion for ascertaining religious truth.

This rational Protestant view is spelled out even further in the works of Stillingfleet. As a restoration Latitudinarian, he argued in his *Irenicum* (1659) for "moderation" in religious practices and beliefs. He lamented that "our *Controversies* about *Religion,* have brought at last even *Religion* its self into a *Controversie,* among such whose weaker Judgments have not been able to discern where the plain and unquestionable Way to Heaven hath lain." "Why men should be so strictly tied up to such things," Stillingfleet wrote, "which they may do or let alone, and yet be very good Christians still?" He points out that the most eminent divines of the Reformation period did never conceive any one form "necessary." Moreover, Stillingfleet maintains that Christ's "design was to ease Men their former Burdens, and not to lay on more." He called on Protestants to imitate the "Primitive Church" in that "admirable Temper, Moderation, and Condescension which was used in it."[63]

In his *Origines Sacræ* (1662), Stillingfleet strove not only to confute atheism, but to proclaim the reasonableness of Protestantism over against the superstition, idolatry, and irrationality of Catholicism. He argued that religion is "reasonable," that "All Natural Worship is founded on the Dictates of Nature, all instituted Worship on God's revealed Will."[64] Of course by "Religion," Stillingfleet meant Protestantism. He made this abundantly clear when, two years later, he went on to publish *A Rational Account of the Grounds of Protestant Religion* (1664), where he essentially argued that Catholicism was irrational. Far from mutually exclusive, then, Protestantism and reason are naturally complementary, even identical: only popery, in its superstition, idolatry, and implicit faith, was antagonistic to reason. "Where-ever God requires us to believe anything as *True*," wrote Stillingfleet, "he gives us *evidence* that it is so: where-ever it appears the thing is *inevident*, we may lawfully *Suspend* our *Assent*." Stillingfleet appealed for the establishment of "the Empire of Reason," and, perhaps more remarkably, defined faith as "a *rational* and *discursive Act* of the Mind." Without reason, in other words, "Faith would be an unaccountable thing." "Faith" must be a "rational act," whose assent "can be no stronger" than the "reasonable grounds" upon which it is based.[65]

By the mid-seventeenth century, latitudinarianism was pervasive among Anglican clergymen. Indeed, it was Tillotson, a skillful preacher at Lincolns Inn and later Archbishop of Canterbury, who most effectively communicated the Latitudinarian message of the reasonableness of Protestantism. His sermons, as reported by his biographer, were attended by "a numerous audience brought together from the remotest parts of the metropolis, and by a great concourse of the clergy, who came thither to form their minds."[66] As London's most popular preacher during the second half of the seventeenth century, Tillotson argued for and preached sermons on "The Wisdom of God in the Creation of the World." To see God's wisdom and goodness in creation one must examine his works. But Tillotson himself was not a natural philosopher, so he left it to "those who have studied nature" to "discourse these things more exactly and particularly." Tillotson confined himself to generalities. He went on to briefly survey the latest scientific findings in astronomy, geology, zoology, and human anatomy. According to Tillotson, all of this had a purpose, which was for the service and pleasure of mankind. And while he ended this sermon declaring the folly of the atheist, he also took the opportunity to criticize Christians for not being more interested in nature: "It is a great fault and neglect among Christians, that they are not more taken up with the works of God, and the contemplation of the wisdom which shines forth in them."[67]

Tillotson also followed the Cambridge Platonists in equating atheism with superstition. In discussing the "Advantages of Religion to Societies," he explained that "For some ages before the Reformation, atheism was confined to Italy, and had its chief residence at Rome. All the mention that is of it in the history of those times the papists themselves give us in the lives of their own popes and cardinals, excepting two or three small philosophers that were retainers to that court. So that this atheistical humour among Christians was the spawn of the gross superstitions and corrupt manners of the Romish church and court."[68] Elsewhere he spoke of the dangers of "being saved in the Church of Rome," arguing that Catholic doctrine and practice were not only "superadded" to the pure and primitive message of the early church but are "gross absurdities." He sincerely believed the Church of England was "the best constituted church this day in the world," and that its teachings secured men from both the "wild freaks of enthusiasm" and the "gross follies of superstition."[69] Indeed, in another sermon Tillotson argued that true Christianity had been "miserably depraved and corrupted, in that dismal night of ignorance" during the ninth and tenth centuries, when "many pernicious doctrines and superstitious practices were introduced, to the woeful defacing of the Christian religion, and making it quite another thing from what our Saviour had left it." But according to Tillotson, Protestantism "happily cut off" and "cast away" these "manifold errors and corruptions." "Though our Reformation was as late as Luther," he explained, "our religion is as ancient as Christianity itself."[70]

Furthermore, in a sermon preached at Whitehall in 1679, he exhorted listeners to guard against believing in "false prophets." According to Tillotson, all revelation must be subjected to faculty of reason, and no divine doctrine should be accepted that is contrary to the principles of "natural religion."[71] Tellingly, Tillotson believed that "natural religion" imposed "great duties" on the Christian, such as practicing justice, mercy, and piety. Natural religion was a natural instinct or reason, found both in the social consensus of humanity and the laws of the natural world, in addition to the "inward dictates and motions" of the Spirit of God.[72] Elsewhere he defined natural religion as the "obedience to the natural law, and the performance of such duties as natural light, without any express and supernatural revelation, doth dictate to men." Indeed, to believe that natural law opposed revealed religion is to be confused by the devil! Tillotson repeatedly asserted that revealed religion depended on the natural, that "natural religion is the foundation of all instituted and revealed religion." It was the "great design of the Christian religion" to "restore and reinforce the practice of the natural law."[73] As late as 1692, Tillotson continued to preach that "nothing can in reason be admitted to be a revelation from God which does plainly contradict" the principles of natural religion.[74]

Tillotson saw himself as an instrumental figure in spreading the principles of natural religion first, and then from them the "proof of the Christian religion."[75] In doing so, he advanced and promoted the latitudinarian insistence that Protestantism was a rational religion that was morally beneficial to society. To this end, Tillotson also edited and published Wilkins's (1614–1672) *Of the Principles and Duties of Natural Religion* (1675) and *Sermons* (1677), further contributing to the fame of the view that moderate Protestantism is the most reasonable religion. Wilkins tied the scientific and latitudinarian movements together. He was a founding member of the Royal Society, chairing its first meeting in 1660 and serving as secretary for the next eight years. As bishop of Chester, Wilkins was also a well-known mathematician, natural philosopher, and a prominent supporter of the Baconian project.

Wilkins assisted in the rationalization of theology by further accommodating religious belief to new scientific principles. Considerable attention is devoted in *Principles and Duties* to the nature of evidence. He made a distinction between knowledge or certainty on the one hand, and opinion or probability on the other. Under knowledge, he placed physical, mathematical, and moral principles, whereas religious principles are merely probable. According to Wilkins, religion is "that general Habit of Reverence towards the Divine Nature, whereby we are enabled and inclined to worship and serve God after such a manner as we conceive most agreeable to his Will, so as to procure his Favour and Blessing." But natural religion is that "which Men might know, and should be obliged unto, by the mere Principles of *Reason,* improved by Consideration and Experience, without the help of *Revelation.*" In other words, while Wilkins may have used a variety of arguments for defending the existence of God, he ultimately categorized "reason" and "faith" as separate things.[76]

Indeed, throughout *Principles and Duties,* Wilkins seemed unwilling to admit the necessity of Scripture until the very end. At times he also showed himself quite critical of the biblical text, subjecting it to the same standards of evaluation as any other written source:

> Now the *History of Moses* hath been generally acknowledged to be the most ancient Book in the World, and always esteemed of great Authority, even amongst those Heathens who do not believe it to be divinely inspired: And there is no Man of Learning, but must allow to it (at least) the ordinary credit of other ancient Histories; especially if he consider what ground there is for the Credibility of it, from Theology of the darker Times, which is made up of some imperfect Traditions and Allusions, relating to those particular Stories which are more distinctly set down in the Writings of Moses.[77]

He goes on to say that he will not draw upon the revelation of Scripture at all, and thus avoid having to reconcile it with the natural world. What he will do, however, is give it the same credit as other ancient writing. This allows Wilkins to locate common notions implanted among all religious traditions, which was important in his defense of natural religion. "Such kind of Notions," he wrote, "are general to mankind, and not confin'd to any particular Sect or Nation, or Time." He added that the "κοινὴ ἔννοια, Common Notions, λόγοι σπερματικοὶ, Seminal Principles; and *Lex nata*, by the *Roman Orator*, an innate Law, in opposition to *Lex Scripta*, and in the Apostles Phrase, *the Law written in our hearts*," are shared by most traditions.[78]

Wilkins used the most recent discoveries in the natural sciences to show "how firm and deep a Foundation Religion hath in the Nature and Reason of Mankind." But as a latitude-man he also called on Protestants to bring to their faith "the same Candour and Ingenuity, the same Readiness to be instructed, which he doth to the Study of human Arts and Sciences; that is, a Mind free from violent Prejudices, and a Desire of Contention." Man must be "subdued by those clear Evidences, which offer themselves to every inquisitive Mind, concerning the Truth of the *Principles of Religion* in general, and concerning the *Divine Authority* of the *Holy Scriptures*, and of the *Christian Religion*."[79] Indeed, Tillotson, in the preface of *Principles and Duties*, confidently declared that Wilkins's rational theology could serve as an "antidote against the pernicious Doctrines of the *Antinomians*, and of all other *Libertine-Enthusiasts* whatsoever."[80]

The Latitudinarians consistently considered their rational theology both a defense against atheism and a deliberate attempt at integrating the new science with traditional religious thought. The increasing "mechanization" of nature during the seventeenth century provided new visions of God and, concomitantly, new "physico-theological" treatises. Writers such as Walter Charleton (1619–1707), Robert Boyle (1627–1691), John Ray (1627–1705), Christopher Wren (1632–1723), Robert Hooke (1635–1703), Isaac Newton (1642–1727), William Derham (1657–1735), and many others believed the new natural philosophy could be used in defense of Christianity. However, as we have seen with the Cambridge Platonists and the Latitudinarians, the English virtuosi sought to demonstrate not only how God has revealed himself in nature, but how a "rational" Protestantism provided an atmosphere more conducive to the sciences.[81] Charleton, Boyle, Ray, and other English virtuosi wanted the factions in the church and all English Christians to rise above their doctrinal differences and devote themselves to what they saw as the more important task—namely, establishing a religion that would unite all Protestants. This new reformation of religion, they believed, would come through the study of the works of God. As Boyle had argued in his *Christian Virtuoso* (1690), a commitment to experimental knowledge leads the Christian "to yield an hearty and operative assent to the principles of religion." He described the natural philosopher as the "priest of nature," implicitly undermining the current priesthood, Catholic and Protestant alike.[82] The inventions, discoveries, and observations of the natural philosopher was worship, and thus the empirical investigation of nature could form the basis of a more rational theology. Protestantism, in short, embodied the principles that would allow for the progress of learning, society, and religion itself.

The emphasis on God's works in nature, however, shifted the focus of religion away from sin, grace, and redemption to the works of nature themselves and the efficacy of human reason. Reason overshadowed revelation. The rational theology of the Cambridge Platonists, Latitudinarian divines, and English natural theologians would eventually come to undermine the very reasonableness of orthodox Christianity itself. English virtuosi William Whiston (1667–1752) and Samuel Clarke (1675–1729), for instance, held typical heterodox views. Besides producing his famous physico-theological treatise *A New Theory of the Earth* (1696), Whiston, Lucasian Professor at Cambridge, was also the author of the five-volume *Primitive Christianity Revived* (1711–12), where he blamed Athanasius for corrupting the pure message of Christ. Clarke, who is mostly known for his sermons entitled *A Demonstration of the Being and Attributes of God* (1704), in which he espoused cosmological, teleological, and ontological proofs for the existence of God, also became involved in the Trinitarian controversy with his *The Scripture-Doctrine of the Trinity* (1712), where he promulgated the view that Scripture attested that God alone is self-existent and the Son and Holy Spirit are not. Indeed, the prevalence of Arianism in the Church of England at the end of the seventeenth century was extensive.[83]

English rational theology of the seventeenth century laid the foundations for the deistic critique of revelation in the eighteenth century. In their polemic with Rome and religious enthusiasts, these rational theologians attempted to construct a rational theology capable of gaining the reasoned consent of an impartial examiner. Deism was merely the logical development of this principle. Rather than limiting the true faith to those fundamental doctrines shared by all Christians, the English deists simply broadened the perspective and located the true faith in the "religion of nature"—that is, in those basic rational beliefs shared by all men in all ages. As Peter Gay aptly put it, "Liberal Anglicanism and the dawning deist Enlightenment were connected by a thousand threads."[84]

A MORE "RATIONAL RELIGION"

Thus when the so-called English deists first appeared, they had an abundant selection of Protestant histories and natural theologies that supported their critique of orthodox Christianity. Peter Harrison has suggested that the desire to demonstrate the reasonableness of religion becomes "deism when doubt is cast upon the adequacy of revelation as a medium of religious truth."[85] Indeed, there were many who were all too eager to exclude every element of mystery from revelation. With a more diffusive Christianity emerging, such men as Edward, Lord Herbert of Cherbury (1583–1648), Charles Blount (1654–1693), Matthew Tindal (1656–1733), Thomas Woolston (1669–1733), John Toland (1670–1722), Anthony Collins (1679–1729), Thomas Morgan (d. 1743), Thomas Chubb (1679–1747), Conyers Middleton (1683–1750), and Peter Annet (1693–1769) promoted a noninstitutional and therefore nonpartisan and nondogmatic "natural religion." These deists consistently condemned revealed religion in general and Christianity in particular. They argued that miracles and doctrines such as the divinity of Christ, the Resurrection, and the Trinity were irrational. Moreover, they argued that sacred Scripture was full of legends, self-contradictions, and nonsense. More still, they also criticized what they saw as the immortality of Christianity. They insisted that the church was an oppressive social institution that obtained its power by intimidation. All the deists were influenced by empiricism, mechanical philosophy, and Bacon's insistence that the purpose of knowledge was social and religious progress. Most importantly, the deist critique of church history was culled from Protestant thought. As S. J. Barnett has persuasively argued, the English deists appropriated much of their historical narrative from their Protestant forebears.[86] The border between deism and liberal Protestantism is indeed difficult to establish.

Moreover, as Jeffrey R. Wigelsworth and others have argued, the English deists were neither atheists nor even deists in an exclusive or final sense.[87] Most of the English deists in fact denied that sobriquet. The English deists did, of course, reject the Athanasian Creed and denied the divinity of Christ. They reduced religion to what they regarded as its most foundational, rationally justifiable elements. But so did other Protestants. As Robert Sullivan demonstrates in his focused study on Toland, the liberal "Anglican distrust of any effort to establish Christian commitment on the basis of dogma" opened the doors to a torrent of deism, anti-Trinitarianism, and other unorthodox movements.[88] What is clear, then, is that the deists had taken the views of the Cambridge Platonists, Latitudinarians, natural theologians, and other "broad minded churchmen" to their logical conclusions. They extended the Protestant historical and rational critique against Catholicism to Anglicanism and argued that all hierarchically established churches should be replaced with a noninstitutional "natural religion." The English deists, in short, simply broadened Protestant historiography to include most or all of Christianity. For such thinkers, all hierarchical established religion had been and still was priestcraft, instituted by the clergy for gain, and thus advocated a noninstitutional belief in God.

One seminal figure and successor of these ideas was the English polymath Joseph Priestley who, as we saw earlier, was an inspiration to Draper. Indeed, in

Priestley we see the confluence of a number of movements and ideas of the seventeenth and eighteenth centuries. Perhaps best known today as the discoverer of oxygen, or what he called "dephlogisticated air," in his day Priestly was better known for his pilgrimage from Calvinist orthodoxy to his own idiosyncratic liberal form of Protestant Christianity. In his *Memoirs*, published posthumously by his son in 1809, Priestley relates how he came to reject the "gloom and darkness" of his Calvinist upbringing for Arianism, and then ultimately arrived at the conclusion that "Christ was a man like ourselves." Jesus is fully man and only man. He later wrote that he "saw reason to embrace what is generally called the heterodox side of almost every question."[89]

Nevertheless, Priestley was a good example of the alliance forged between science and religion, especially in the Dissenting tradition to which he belonged. The Unitarians were very close to the deism of the earlier decades. As Basil Willey noted, the Unitarian "differed mainly in origin and organization."[90] The affinity between Anglican latitudinarianism and Protestant Dissent is well known, and as Dennis G. Wigmore-Beddoes has demonstrated, a similar affinity existed between Unitarianism and Broad Churchmen in the nineteenth century.[91] For now, it is important to recognize that Priestley fused the ideas of the Cambridge Platonists, latitudinarianism, and English physico-theology in his version of Protestantism. But unlike many of the figures discussed in this chapter, Priestley was never associated with the Established Church, and as a Dissenter nothing prevented him from becoming more and more liberal. Nevertheless, his aim was always to "defend Christianity, and to free it from those corruptions which prevent its reception with philosophical and thinking persons, whose influence with the vulgar and unthinking is very great."[92]

Thus, in spite of his scientific achievements, religion was at the core of his life, and the propagation of what he believed to be "true Christianity" was the main object of his voluminous works. His religion was a Christianity against the Christians. True Christianity needed to be rescued from the pseudo-Christians, according to Priestley. His characteristic method of defending "Christianity" was to expose and remove its "corruptions," which for him included nearly everything considered by the orthodox to be of its very essence. He pursued this idea from his earliest writings. In his *Catechism, for Children, and Young Persons* (1767), for example, he declared that the religion of Christ was quickly corrupted by the "papists" until "it pleased God to bring about a reformation, which is going on, and, we hope, will go on, till our religion be, in all respects, as pure, and as efficacious to promote real goodness of heart and life, as it was at the first."[93]

Priestley's conception of progress was inextricably connected to his religious convictions. As first and foremost a Dissenting minister of a rational sort, he promoted a history of religion that demonstrated rational Dissent as the true manifestation of Christianity. In his *Address to Protestant Dissenters* (1768), for instance, he argued that the rational Dissenters were the only ones who possessed a practical and rational Christianity. Indeed, the Dissenters continued the work of the reformation from "popery," emancipating Christianity from superstition and the dogma of orthodoxy.[94] With unbounded optimism he believed that man was moving towards educational and moral progress. For Priestley, science was the main agent of that progress. In his preface to *Experiments and Observations on different Kinds of Air* (1774), he explained that the rapid progress of knowledge in his day is how God extirpated "all error and prejudice, and of putting an end to all undue and usurped authority in the business of religion as well as of science."[95] Scientific progress, in other words, allowed for social, moral, and religious progress. Priestley, then, shared with other liberal Protestants the view that the progress the modern world has made over the ancient world in religion, science, government, law, commerce, manners, and even happiness is due to the providence of God.

Like the Protestant reformers, Priestley saw Scripture as the source and norm of the Christian life. And like later Protestant historians, he interpreted Christian history following the New Testament as the history of the corruptions of Christianity. In an *Appeal to the Serious and Candid Professors of Christianity* (1770), he called all Christians to use both the Scriptures and reason in matters of religion. He argued that both lead to disbelief in the doctrines of atonement, original sin, biblical inspiration, the Trinity, and the divinity of Christ. Indeed, he offered a "concise history of opinions concerning Jesus Christ," concluding that the doctrine of the Trinity was the principal corruption of Christianity. Jesus of Nazareth was "a man approved of God—by miracles and wonders and signs, which God did by him."[96]

In his first extensive theological work, *Institutes of Natural and Revealed Religion*, which appeared between 1772 and 1774, Priestley set out the basic principles of natural and revealed religion for the edification of young members of his Birmingham congregation. He declared that reason exemplified true Christianity, and that it unites man to God. Reason is indeed a form of devotion. Anything that opposed it, then, was anti-Christian. Thus he rejected "perverse" forms of Christianity—Papist, Anglican, and Calvinist alike. Any form of orthodoxy that continued to cling to "superstition" or "enthusiasm" should be rejected. But he also scorned the "infidelity" of the deists with their feigned piety. He declared that in origin, both Jewish and Christian revelation were united in the idea of the "unity of God." Over time this great doctrine was radically corrupted—"For upon the very same principles, and in the very same manner, by which dead men came to be worshipped by the ancient idolaters, there was introduced into the Christian church, in the first place, the idolatrous worship of Jesus Christ."[97]

Like other liberal Protestants before him, then, Priestley contended that most of these corruptions were part of a system of "heathenism" or "paganism" that was introduced into Christianity at a later date. The same theme of corruption was fundamental to his most notorious historical work, *The History of the Corruptions of Christianity* (1782). This was an unambiguous attack on orthodox Christianity. In its opening passage, Priestley claimed that "the unity of God is a doctrine on which the greatest stress is laid in the whole system of revelation." He viewed his religious duty as exposing the way this pure monotheism had been corrupted by the intrusion of foreign and "heathen" ideas. Strikingly similar to Draper a generation later, Priestley argued that the history following the New Testament is a history of the corruptions of Christianity, which included what had been considered its fundamental doctrines, such as the Trinity, Immaculate Conception, Original Sin, Predestination, Atonement, and the plenary inspiration of Scripture. These corruptions, he believed, prevented the universal acceptance of Christianity.[98]

According to Priestley, the only way to preserve the "genuine system and spirit of Christianity" is to expose the falsehood of "what has so long passed for Christianity" and demonstrating "what is truly so." The argument should now be familiar. As we have seen, this same "historical method" had been used by the Protestant reformers, the Cambridge Platonists, the Latitudinarian divines, and even the English natural theologians. Priestley, with the authority and zeal of a pastor, exclaimed that "the gross darkness of that *night* which has for many centuries obscured our holy religion, we may clearly see, is past." "The *morning* is opening upon us," he continued, "and we cannot doubt but that light will increase, and extend itself more and more unto the *perfect day*. Happy are they who contribute to diffuse the pure light of this *everlasting gospel*." Christianity had experienced centuries of distortion, and thus must be corrected. Primitive Christianity, in short, was "properly Unitarian." Indeed, the earliest Christians, including Christ himself, were Unitarians. Priestley intended the work to expose the accumulated errors of Christianity and thus bring unity among Protestants.[99]

In his conclusion Priestley pleaded not for a "progressive religion" but for a "progressive reformation of a corrupted religion."[100] Though of course as a Unitarian he rejected the theology of the reformers, he nevertheless appealed to the progressive character of the Reformation. Indeed, the overarching theme of his theological work is the ideal of the restoration of Christianity to its primitive purity. He thus established himself as a Christian reformer, seeking to release the contemporary faith from its own troubled history. In his work Priestley wished to defend Christianity and free it from those corruptions that prevent its reception by the more intelligent classes. He wished to recast Christianity in a form that would win the assent of intellectuals. His central belief is stated most succinctly in *A General History of the Christian Church* (1802), that "whatever is true and right will finally prevail; that is, when sufficient time has been given to the exhibition of it, rational Christianity will, in due time, be the religion of the whole world."[101]

THE COMING "NEW REFORMATION"

Among the Victorians, the kind of belief in progress that Priestley represented was pervasive.[102] German historical thought at the beginning of the nineteenth century also played a decisive role in how Victorians conceptualized progress in history. Crucially, as Thomas A. Howard notes, German historicism had its foundations in a particular kind of theology, that "nearly all nineteenth-century liberal German theologians saw themselves not as debunkers of religion but as faithful torchbearers of the Reformation." Indeed, nineteenth-century German theologians and historians such as J. D. Michaelis (1717–1791), J. S. Semler (1725–1791), Johann Gottfried Herder (1744–1803), J. G. Eichhorn (1752–1827), and W. M. L. de Wette (1780–1849) believed the principles of the Reformation could still act as a catalyzing agent, advancing civilization away from superstition and darkness towards reason and light.[103] During the tercentenary celebrations of the Reformation in Germany, for example, many theologians and historians promoted a stadial view of human history in which the rise of Protestantism ushered in a new understanding of intellectual and religious freedom. Theologian Friedrich D. E. Schleiermacher, whom, we recall, White diligently read, preached a sermon in 1817 from Berlin's Trinity Church in which he claimed that "we live daily in the free enjoyment of the glorious benefits that befall to Christianity on account of the Reformation of the Church."[104] Even German philosopher Georg W. F. Hegel, an important source of inspiration for Draper no doubt, declared that "the great form of the World Spirit . . . is the principle of the North, and, from a religious perspective, Protestantism."[105]

German literature, music, art, education, historiography, science, and religion, thus had a clear impact on Victorian thought.[106] Among dissident intellectuals, such as Robert William Mackay (1803–1882), Francis W. Newman (1805–1897), William R. Greg (1809–1881), George Henry Lewes (1817–1878), George Eliot (1819–1880), Frances Power Cobbe (1822–1904), William E. H. Lecky (1838–1903), Mary Augusta Ward (1851–1920), and others, all were decidedly influenced by German historical scholarship. In fact, English dissident intellectuals often saw themselves as new "reformers," fulfilling the work that began with Luther. For example, in his radical weekly newspaper, *The Leader*, literary critic George Lewes promoted a "pure religion," which he defined simply as "faith in God obeyed in love to man."[107] In subsequent issues, Lewes went on to proclaim his hope in the coming "New Reformation." To restore religion to its proper function, he asserted, "it must be emancipated, and to achieve that blessed end is the object of the new Reformation."[108] "The New Reformation," Lewes wrote, "will start from a fuller development of Luther's great principle," which was, of course, the "liberty of private judgment." The "New Reformation," however, will go beyond Luther by offering "absolute freedom," giving every soul "the sacred privilege of its *own* convictions."[109]

Dissident intellectuals associated with *The Leader* rejected orthodoxy because they found in Scripture and traditional Christianity ideas that were not compatible with the "spirits and wants of the age." But while Christianity seemed to them a dying religion, dissident intellectuals continued to believe that central tenets of English Protestantism contained vital cores of truth. In a long review, for instance, *The Leader* declared that English Protestantism had manifested "free religious inquiry among earnest men," and that leaders of this new "movement against dogmatic theology" were men "at once devout and practical," and who wished to substitute conventional religion for "one purer and more influential."[110]

Indeed, this review was in response to a book by William R. Greg, whom the editors of *The Leader* considered one of the "pioneers of the New Reformation." Greg, a Unitarian layman, argues in his *Creed of Christendom* (1851) that the "Inspiration of the Scriptures is baseless and untenable under any form or modification which leaves it a dogmatic value," that "the Gospels are not textually faithful records of the sayings and actions of Jesus," and that the "Apostles only partially comprehended, and imperfectly transmitted, the teachings of their Great Master." Yet Greg did not wish his work to be regarded as "antagonistic to the Faith of Christ." He believed that "the removal of superincumbent rubbish is indispensable to the discovery and extraction of the buried and intermingled ore." Historical Christianity, according to Greg, "is not the Religion of Jesus." Rather, he envisaged and promoted a "higher view of religion," one that did not dread the discoveries of science nor the progress of religion.[111]

Another "pioneer of the New Reformation," according to the editors of *The Leader*, was Francis W. Newman. Indeed, Greg claimed it was Newman who convinced him of the need of a new reformation in religion.[112] The editors of *The Leader* were so impressed by Newman's work that they published several long and glowing reviews. In its first notice, the weekly declared that there was "no work so capable of making a path for the New Reformation to tread securely on."[113] Newman, the younger brother of Catholic cardinal John Henry Newman, was brought up as an evangelical Anglican but later came to reject scriptural authority after reading "a Unitarian treatise against the doctrine of Eternal Punishment." In 1853 Newman published his provocative spiritual autobiography, *Phases of Faith*, which chronicled his break with Christianity. Despite abandoning Christian orthodoxy, Newman longed for a religion "animated by primitive faith, love and disinterestedness." Indeed, he believed that "religion always had been, and still is, *a state of sentiment toward God*," and thus "far less dependent on articles of a creed."[114]

Newman would later publish short tracts addressing *The Religious Weakness of Protestantism* (1866) and *Thoughts on a Free and Comprehensive Christianity* (1866). Following a line of argument as old as the Cambridge Platonists, Newman argued that English Protestantism "has always been found on the side of resistance." The more intransigent it has been, the "more Atheistic is the aspect of public life." Having read Greg, Newman declared that continued Protestant insistence on the virgin birth, resurrection, and other miracles of Jesus ultimately fails to satisfy "the demands of ordinary scientific reasoning." Protestantism could not possibly stand on its old basis. But Newman was happy to announce that the theology of the "old school is dying out." While he was encouraged by certain "enlightened Protestant divines," Newman wished to see the "whole Protestant theory" demolished. He concluded that "whatever it was once, Christianity is now in all the great concerns of nations a mere ecclesiasticism, powerful for mischief, but helpless and useless for good." Like other dissident intellectuals, Newman hoped to sweep away all the rubbish of faulty logic, irrational belief, and religious superstition and uncover a purer devotion to a purer God.[115]

The novels of Mary Anne Evans, who published under the name George Eliot, also reflect many of the enlightened sensibilities and ideas of the dissident intellectuals. Raised in an evangelical family, she came to abandon her faith after being disappointed with evangelical social ethics and after studying the religious and theological ideas of German higher criticism. In 1842, in an oft-quoted letter to her father, Eliot had concluded that the Bible was a collection of "histories consisting of mingled truth and fiction." By 1850 she had become the assistant editor for John Chapman's (1821–1894) radical *Westminster Review*, publishing her first article in the quarterly the following year wherein she urged readers to recognize "the presence of undeviating law in the material and moral world." Unsurprisingly, she proclaimed that English Protestantism was "in bondage to terms and conceptions which, having had their root in conditions of thought no longer existing, have ceased to possess any vitality, and are for us as spells which have lost their virtue."[116]

Eliot penned those words in a review of Robert William Mackay's *The Progress of the Intellect* (1850). Mackay was one of the earliest writers to introduce German historical criticism to an English audience. Interestingly, Mackay claimed that "religion and science are inseparable." But by "religion," he meant an "Elder Scripture, writ by God's own hand, Scripture authentic, uncorrupt by man." In the intellectual progress of religion, he argued, "the hypothesis of miracle has lost its usefulness." Miracles, according to Mackay, implied "something inconsistent with the order of a perfect government, something overlooked in the original plan requiring an interpolation contradictory to its general tenour." A "perfect and immutable being," Mackay argued, "cannot break his own laws, or be at variance with himself." Thus, notions of miracles ought to be replaced with belief in the perfection and uniformity of natural law. In the final analysis, according to Mackay, historical Christianity had lost all meaning in his day, and therefore ought to be replaced with a new religion.[117]

Another important dissident intellectual was Mary Augusta Ward, niece of poet Matthew Arnold. Ward authored the immensely popular *Robert Elsmere* (1888), a tale of an Anglican minister in the process of losing his faith in historical Christianity. Ward combined many intellectual and moral tendencies of her age, which looked forward to the inevitable triumph of a new and simpler type of religion. As Bernard Lightman observes, Ward was confident that the unsettlement of orthodox Christianity would usher in a new faith, stronger and more attuned to modern thought.[118] In short, it was a Protestantism with doctrine left out. Indeed, Ward believed it was the guiding hand of God himself behind the intellectual and religious crisis of the period.

Ward celebrated the coming of a new faith in her 1889 article entitled the "New Reformation," published in the *Nineteenth Century.* Ward's "New Reformation" is developed in a dialogue between two fictional characters, Ronalds, an obscure orthodox Anglican minister, and Merriman, who has embraced German historical criticism as Elsmere did in Ward's novel. In fact, Merriman is little more than Elsmere over again, repeating Ward's own personal belief that the old orthodoxy must "translate" or transform to the subjective mode of interpreting religious phenomena, that all supernatural miracles must dissolve, dissipate, and resolve into naturalistic explanations. "All round us," Ward declared through Merriman, "I feel the New Reformation preparing, struggling into utterance and being! It is the product, the compromise of two forces, the scientific and the religious." Ward thus celebrates that the growing "liberal forces now rising" among all religious sects "will ultimately coalesce, [and] science will find the religion with which, as it has long since declared, through its wisest mouths, it has no rightful quarrel, and religion will find the science which belongs to it and which needs it." Ward's "New Reformation," in other words, will serve as a substitute to historical Christianity.[119]

Perhaps the most salient example of the "New Reformation" was the historical writings of William Lecky. At an early age he published a survey of the *Religious Tendencies of the Ages* (1860), which aimed at solving "that great problem of theology, the legitimate province of private judgement."[120] Unsurprisingly, he believed that Catholicism had suppressed private judgment, whereas Protestantism had encouraged it. But Lecky himself subscribed most closely to a latitudinarian position. As historian Jeffrey Paul von Arx puts it, for Lecky latitudinarianism "was the latest expression of the traditional Protestant revolt against spiritual authority," the very "culmination of the development of religious thought," and the "basis for social and political unity and progress."[121]

Indeed, the real religious force in England, according to Lecky, had always been latitudinarianism.[122] Protestantism was only temporary. The progress of religion is on an inevitably upward march from the barbarous superstitions of the Dark Ages to the most radically liberal position of rationalists like Lecky. While Lecky did not deny the existence of God, he did argue that the contending religious parties brought Christianity into doubt. The solution was what he believed was the moderation displayed in the latitudinarian position. Latitudinarians offered the "spirit of charity and of tolerance towards those with whom they disagree." Importantly, he insisted that "Protestantism and dogmatism are logically incompatible." Systematic theology had "been the parent of almost all the errors and of a very large proportion of the crimes that have disfigured the history of Christianity." Such a system had corrupted the simple message of Jesus. Indeed, Lecky strongly believed that "primitive Christianity" was the very essence of the latitudinarian position. Thus, according to Lecky, latitudinarianism was not only the fulfillment of the Reformation, but the continuation of primitive Christianity.[123]

These statements anticipated central features of Lecky's more well-known work, the *History of the Rise and Influence of the Spirit of Rationalism in Europe* (1865), which was an extended attack against dogmatic theology, and one that was remarkably similar in outline to Draper's own intellectual history of Europe.[124] For Lecky, a liberal and nondogmatic Protestantism was essential for moral and religious progress. Like so many of his contemporaries, Lecky did not wish to eradicate religion, or "true Christianity" as he put it. A "true and healthy Christianity," he wrote, cultivates "a love of truth for its own sake," and inculcates a "spirit of candour and of tolerance towards those with whom we differ." In other words, the decline of dogmatic theology and clerical influence was not inimical to religion at all—rather, it called on people to return to the "days of the Apostles," and thus is a "measure if not a cause of its advance."[125]

In describing the triumphal advance of reason over superstition, Lecky argued that rationalism had made extraordinary strides in Protestant countries. By "rationalism," then, Lecky meant "Protestant Rationalism." Like other dissident intellectuals, Lecky believed that "religion in its proofs as in its essence is deemed a thing belonging rather to the moral than the intellectual portion of human nature. Faith and not reason is its basis; and this faith is a species of moral perception."[126]

AN UNHOLY ALLIANCE

By the late 1880s, a series of contentious articles appeared in the *Fortnightly Review* discussing "the great change in theological and religious thought" in the century.[127] Renegade Anglican priest Charles Voysey (1828–1912), who was condemned by the Judicial Committee of the Privy Seal as a "heretic," began the series by explaining that the "new reformation" did not mean wholesale destruction, but improvement. Voysey believed the "new reformation" was the continuation and indeed completion of "old reformation," that great and heroic struggle "to break the tyranny of Church authority and establish the right of private judgement at all in matters of religion." Thus the "new reformation" will be an expansion and wider application of "our Protestant forefathers." While the "old reformation" was gradual, the "new reformation" occurred rapidly as a result of biblical criticism, comparative religious studies, and the enormous increase of scientific knowledge. This "increased eagerness for truth and fact," Voysey argued, is part of the reason so many of his contemporaries (including himself) had rejected the dogmas of Christianity. Perceptively, Voysey observed that once the doctrine of the Fall is rejected, "the whole fabric which rests upon it comes to the ground." He writes: "No fall, no redemption; no curse, no Calvary; no hell, no atonement; no atonement, no incarnation; no incarnation, no Christ; no Christ, no Trinity." At the same time, he asserted that he could not explain reason, consciousness, and love "in any other terms than loyalty to my Maker, the author of my being." Voysey believed "in the existence of some One who has given me life, and put into me this sense of loyalty to Himself before all things, who makes me glad when I obey and sad and wretched when I disobey Him." In superseding historical Christianity for the "New Reformation," Voysey sought to base the new form of religion on conscience and heart. The work of the "New Reformation," he argued, will be a return to a more "natural" religion.[128]

While Voysey might appear an extreme example, many other nineteenth-century liberal Anglicans, or Broad Churchmen, also responded to calls for a "New Reformation," and thus shared a common bond with dissident intellectuals. Indeed, as Voysey himself observed, "hundreds of living clergymen and ministers" have rejected central Christian dogmas.[129] Although neither a party nor a faction but a set of individuals, Broad Churchmen shared the belief that the authority of the Bible and the Church must be subjected to historical and scientific criticism.

Influenced by German idealism and romanticism, Broad Churchmen stressed the importance of religious experience, feeling, and intuition over claims of theological dogmatism. While members of the Church of England, they nevertheless believed that various doctrinal positions, which had long been considered essential to the Christian faith, needed to be modified or even abandoned in the light of modern thought. By relaxing doctrinal restrictions, they hoped to reconcile Christianity with the intellectual tendencies of the age.[130]

Moreover, as historian Duncan Forbes has demonstrated, nineteenth-century liberal Anglicans promoted an idea of history that was organic, in which society and religion was in a process of continual growth, decline, and renewal.[131] For the liberal Anglican, history "resembled the life of an individual, or a plant, or the astronomical progress of a day or a year. It had its boyhood and manhood, its intermediate stages and transition periods, its dawn, high-noon and dusk, its spring, summer, autumn and winter, its blossom and seed-time."[132] Draper's own analogy that "man is the archetype of society," and that "individual development is the model of social progress," is too similar to be mere coincidence. In describing the history of civilization, Draper similarly compared the stages of man as an individual (the age of Infancy, Childhood, Youth, Manhood, Old Age, and Death) to the intellectual development of society (the age of Credulity, Inquiry, Faith, Reason, and Decrepitude).[133]

As we have seen, the liberal Protestant tradition can be traced back as far as the seventeenth century, to the writings of the Cambridge Platonists. But in the first half of the nineteenth century, our *dramatis personae* customarily begins with the Oriel Noetics, such as Edward Copleston (1776–1849), Richard Whately (1787–1863), Edward Hawkins (1789–1882), and Thomas Arnold (1795–1842), but in addition to other clergymen such as Henry H. Milman (1791–1868), Julius C. Hare (1795–1855), Connop Thirlwall (1797–1875), Frederick Denison Maurice (1805–1872), and Charles Kingsley (1819–1875). This first generation of Broad Churchmen were all known for their rationalism, liberalism, and calls for reforms in religious belief.[134] As early as the 1830s, for instance, Arnold had been urging a broader, more critical, and less dogmatic approach to theology. Arnold also joined forces with Nonconformists in opposing the Tractarians and their attempt to accentuate the traditionalism of the Church of England. Launching one of the bitterest attacks against the Oxford Movement, Arnold called them the "Oxford Malignants," comparing them to the "Judaizers of the New Testament."[135]

Other liberal Anglicans sought to liberate Christianity from what they believed were outdated doctrines. In 1853, Christian Socialist Maurice published a collection of *Theological Essays* in which he rejected both the traditional substitutionary view of the Atonement and the notion of eternal punishment.[136] Moreover, in a series of letters between him and a "layman" on the question of the Bible and science, Maurice declared that "divinity [i.e., theology] needs reformation," and that he was grateful to "the physical student [i.e., man of science] if in anywise he helps the Reformation forward."[137]

Broad Churchman Kingsley's own interest in the sciences is also well known.

He participated in many scientific circles, including serving as an elected fellow in the Linnean Society and Geological Society of London. He also kept a busy correspondence with a number of scientific men, particularly Charles Darwin and Thomas H. Huxley. Indeed, in 1863 Kingsley even told Maurice that he was busy "working out points of Natural Theology, by the strange light of Huxley, Darwin, and Lyell."[138] As Darwin's other "bulldog," Kingsley believed that "all natural theology must be rewritten" in light of evolutionary theory.[139] To this end, this clergyman-naturalist published a number of books attempting to co-opt Darwin into a new vision for natural theology. In his evolutionary moral fairy tale *Water Babies* (1863), for example, Kingsley encouraged young readers (and adults) to seek God in the divine element underlying all physical nature, the inflexible laws of nature. This natural theology is more explicitly spelled out in his *Madam How and Lady Why* (1869), where he admonished readers to search "God's Book, which is the Universe, and the reading of God's Book, which is Science." But in encouraging readers to study nature, Kingsley also separated questions of "how" and "why" into separate categories, which was just another way of separating science and religion. His most detailed statement of the new natural theology appeared in "The Natural Theology of the Future," which was a public lecture delivered at Sion College in 1871 and used as the "Preface" of his *Westminster Sermons* (1874). Tellingly, Kingsley declared that natural theology must be both "eminently rational and scriptural," as it was among seventeenth-century, doctrinally latitudinarian clergymen. According to Kingsley, the new natural theology must change "as human thought changes and human science develops."[140] It should not be surprising, then, that Kingsley was also vehemently anti-Catholic, as illustrated in his historical novels *Yeast* (1850), *Hypatia* (1853), and *Westward Ho!* (1855).[141]

The next generation of Broad Churchmen were more radical still, as is illustrated in the notorious publication of *Essays and Reviews* (1860). For this monumental manifesto of liberal Anglican thought, Henry Bristow Wilson (1803–1888), country vicar and former Oxford tutor, enlisted seven essayists to "illustrate the advantage derivable to the cause of religious and moral truth, from a free handling, in a becoming spirit, of subjects peculiarly liable to suffer by the repetition of conventional language, and from traditional methods of treatment." Despite its seemingly innocuous preface and inconspicuous title, Frederick Temple (1821–1902), Rowland Williams (1817–1870), Baden Powell (1796–1860), Charles W. Goodwin (1817–1878), Mark Pattison (1813–1884), and Benjamin Jowett (1817–1893) all brought the full impact of German historical-critical scholarship to England, provoking one of the greatest religious controversies of the Victorian age.[142]

The opening essay was by Temple, Headmaster of Rugby and later Archbishop of Canterbury. In his "The Education of the World," which echoes Lessing's own treatise from the previous century, Temple described the advance or progress of knowledge in history. Like others before him, he divided history into stages, corresponding to the life of an individual—childhood, youth, and manhood. Interestingly, he believed society, like the individual, was capable of "perpetual development." The evolutionary progress of Temple's "colossal man" obviously resembles closely Draper's own analogy in his *Intellectual Development*, published three years later. As we now know, when Draper spoke at Oxford in 1860, he also attended Temple's sermon at St. Mary's church, which had made an indelible impression on him.[143] In this sermon on the "present relations of science to religion," Temple argued that "change in science necessitates a change in its relation to religion." As Kingsley did, Temple also urged his audience to "look for the finger of God" in the universal laws of nature. A reverent study of these laws, he proclaimed, "can and will bring us nearer in temper to their Divine Author."[144]

An essay by Powell, Oxford professor of geometry, differentiated between those "essential doctrines" and those "eternal accessories" of Christianity. Noteworthy is his rebuke of English Protestants for resisting the intellectual tendencies of the age. In tackling the question of miracles, Powell claimed that most Protestants allegedly believed "revelation as once for all announced, long since finally closed, permanently recorded, and accessible only in the written Divine word contained in the Scriptures." The discoveries of modern science, and particularly the discovery of the uniformity of nature, undermined miracle stories—indeed made such interruptions of the natural order "inconceivable." Powell believed that "the more general admission at the present day of critical principles in the study of history, as well as the extension of physical knowledge, has done something to diffuse among the better informed class more enlightened notion on this subject."

In his essay Wilson drew from Arnold and other liberal theologians, calling for greater theological freedom and for modernizing Church formularies in hopes of uniting all Christians. He believed that decreasing church attendance was the result of its "extreme and too exclusive Scripturalism." He appealed to the Protestant principle of "private judgment," and argued that Scripture was a progressive, living organism, and therefore open to continual reinterpretation. He even remarked that Christ did not intend to establish a "theology of the intellect," and that "speculative doctrine" only stifled the "true Christian life, both in the individual and in the Church."

The longest essay of the collection was by theologian and Master of Balliol Jowett. He contended that the "unchangeable word of God . . . is changed by each age and each generation in accordance with its passing fancy." A new critical method is needed, one that takes into account all the advances of knowledge. In other words, we need to read and interpret the Bible "like any other book." "Doubt comes in at the window," Jowett wrote, "when Inquiry is denied at the door." Indeed, "the time has come when it is no longer possible to ignore the results of criticism." By this he meant that readers should forgo the traditions of theology, dogma, and the ecumenical councils. The Bible is an ancient document, and not readily applicable to modern times. In this context Jowett questioned the resurrection, reinterpreted the atonement, and espoused the benefits of a comparative study of religions. A more "rational" interpretation would "dry up the crude and dreamy vapours of religious excitement" and offer "new sources of spiritual health." Jowett believed that Christianity was in a "false position when all the tendencies of knowledge are opposed to it." The Bible itself, according to Jowett, revealed "progressive revelation." Like the dissident intellectuals, Jowett proposed a progressive history of religion analogous to the kind of progress witnessed in the individual, from childhood to adulthood. The interpretation of Scripture, he declared, must "conform to all well-ascertained facts of history or of science." He believed that religion and science were not in conflict, as long as "any scientific truth is distinctly ascertained." While the scientific study of the Bible may force us to relinquish belief in miracles, or the personality of God, or abandon other traditional doctrines, the essence of religion remained. By divesting the Bible from theological dogmatism, in short, Jowett hoped that "the truths of Scripture again would have greater reality."

While this composite volume provoked one of the greatest religious controversies of the Victorian age, the essayists were not without defenders. Maurice, for instance, in 1861 wrote to fellow Broad Churchmen Arthur P. Stanley (1815–1881) that the clergy should accept the *Essays and Reviews* as ushering a "new reformation" in the Church.[145] That same year, Stanley himself produced an important defense of the essayists in an article published in *Fraser's Magazine*. While he disagreed with the general tone of the volume, he pointed out that "the principles, even the words, of the Essayists have been known for the last fifty years, through writings popular amongst all English students of the higher branches of theology." He asserted that "science, history, and the principles of our moral nature are formidable antagonists to Theology if she sets herself against them." Like so many others, Stanley made the distinction between religion and theology, and asserted that a "new reformation" was emerging within the Church through the agency of German critical scholarship.[146] Later, he echoed Kingsley's sentiments in arguing that there is an "essentially progressive element in religion," and that the "conceptions of the relations of man to man, and, still more, of man to God, have been incontestably altered with the growth of centuries."

For Stanley, there is a "higher Christianity which neither assailants nor defenders have fully exhausted."[147]

Although we will look more closely at the affinities between liberal Protestantism and scientific naturalism in another chapter, it is worth repeating Paul White's observation that many of the scientific naturalists used the "resources of liberal theology and romantic criticism" to redraw the boundaries of religion.[148] Indeed, in 1861 Huxley and some of the scientific naturalists organized a defense for the support of the authors of *Essays and Reviews*. The "Scientists' Testimonial" declared that "Feeling as we do that the discoveries of Science, and the general progress of thought, have necessitated some modification of the views generally held on Theological matters, we welcome these attempts to establish religious teaching on a firm and broader foundation."[149] Among its signatories were George Busk, William B. Carpenter, Charles Darwin, John Lubbock, Charles Lyell, and William Spottiswoode. Although never published, this collective effort shows the rally of the men of science in support of liberal churchmen. This unholy alliance became more public however when, in the pages of *The Reader* (a short-lived journal published by Huxley, Tyndall, and Spencer), an anonymous author praised Broad Churchmen for embracing historical criticism and recent scientific discoveries. With the *Essays and Reviews,* the author wrote, "the new theology had publicly burst forth."[150]

The "new theology" of the liberal Protestants portrayed Christianity as a living organism, growing and adapting to its ever-changing environment. Indeed, one of the fundamental arguments of liberals was that Christianity was not static, but developmental, adapting itself to the progressive growth of man's capacity to reason. The story of Christianity's historical progress or development was thus crucial to the liberal case for the continued need of the church to reform itself in the present day. But the sort of biblical criticism championed by Broad Churchmen and other liberal Protestants ultimately did more harm than good. The "septem contra christum" in particular, as Ieuan Ellis observes, not only provided the means of separating religion from science, but also gave the implicit perception that they were in conflict.[151] While there were many affinities between liberal Anglicans and radical Protestants like the Unitarians in the nineteenth century, there is also evidence for strong affinities with dissident intellectuals and scientific naturalists as well.

The Protestant Reformation, then, became a potent metaphor for Broad Churchmen, dissident intellectuals, and agnostics alike. All looked back to the history of Protestantism for vindicating their views. They hailed the Reformation as the triumph of private judgment and individual inquiry over organized ecclesiastical tyranny. For many, the "New Reformation" represented the building of a new religion that would recover what had been lost by Christianity when it perverted the pure ideals of its original founder. In 1864, Stanley excitedly wrote to a friend that "we are on the verge of a religious revolution—a revolution more gradual, I trust, and therefore more safe, but not less important, than the Reformation and ending, I hope, not in further divisions, but in further union." "I agree with you," he wrote to another friend the following year, "that the prophet of the second Reformation has not yet appeared. Perhaps he never will. But that a second Reformation is in store for us, and that the various tendencies of the age are preparing the way for it, I cannot doubt, unless Christianity is doomed to suffer a portentous eclipse."[152]

CONCLUSION

By the nineteenth century, liberal Protestants had successfully shifted the locus of religious authority. Rather than the Bible or Christian tradition, reason, empiricism, and conscience played a larger role in liberal Protestants' confirmation of religious truth. A "new" or "second" reformation became the clarion call of those who sought religious and moral guidance through the free pursuit of science. Many nineteenth-century liberal Protestants accepted and even promoted the division between religion and theology as a strategy for defending religion. Liberal Protestant groups such as those associated with the Oriel Noetics and Broad Churchmen asserted that inductive study of the natural world was a means to identifying a divine Creator and that this same method might be applied to the study of sacred Scripture. Priding themselves on their intelligence, rationality, and cleverness, they shared the belief that the authority of the Bible and the church must be subjected to historical and scientific criticism. Influenced by German *wissesenschaftliche Theologie* and romanticism, together they stressed the importance of religious experience, feeling, and intuition over against the claims of theological dogmatism. Religion to them was the living kernel, theology the dying husk that it inevitably outgrew. By encouraging latitude of opinion, they hoped to bring peace between faith and the modern world.

Important theological changes needed to occur to make such a position possible. It was not that large numbers of hitherto faithfully orthodox believers suddenly had a "crisis of faith" and gave up Christianity. Rather, many found it possible to drop particular beliefs they personally found objectionable or unsatisfactory. Liberal theologians experienced acutely the intellectual pressure of new knowledge upon traditional theology: science, historical criticism, and changing notions of morality applied to traditional doctrine caused Nonconformists and Anglicans to modify traditional beliefs. Christianity was becoming less creedal, less dogmatic, less specific, and more vague.

The emergence of such "looser," more diffuse, and less doctrinally and theologically specific versions of Christianity has a long history. While there is no clear trajectory from the Reformation to modernity, we cannot ignore the fact that numerous Protestant thinkers, from the seventeenth century onward, engaged in a periodic and ongoing struggle to relieve Christianity of "superstitious" excrescences. As Alexandra Walsham observes:

> In its early stages, Protestantism deliberately adopted a rhetoric of rationality and enlightenment. In the polemic that poured from the pulpit and press, it overtly presented itself as a movement that would purge the dross of "magic" from the pure metal of the Christian "religion" and prune away the "superstitious" popish and pagan accretions that had sprung up around it. The reformers liberally employed the metaphor of light dispelling intellectual darkness; they spoke of the Gospel as an instrument for liberating the mass of the populace from the yoke of ignorance in which they had been kept by the papacy; and they poured scorn on the crude materialism and credulity that characterized late medieval piety.[153]

As we have seen, such a polemic of "rationality" is readily found in seventeenth-century Protestant writers, constructing a new historiography designed to expose the corruption of the Roman Catholic Church and thus justify Protestant ascendency. Other Protestant writers also deployed a rhetoric of "reason" and new knowledge to undermine Catholic authority. Similar rhetorical strategies were subsequently adopted and exerted between contending Protestant groups as self-critique, particularly by the Cambridge Platonists and Latitudinarian divines against "enthusiasm" and "enthusiasts." By the nineteenth century, early histories and philosophies of science, such as those found in the writings of White but especially Draper, functioned as Protestant self-critique designed to provide Protestants with a set of beliefs through which they could avoid both Catholic superstition and Protestant enthusiasm. Indeed, as we have seen, Draper's rationalistic interpretation of the intellectual development of Europe, and his subsequent rationalistic account of the history of conflict between religion and science, drew deeply from the well of his Protestant heritage.

AMERICAN NEW THEOLOGY AND THE EVOLUTION OF RELIGION

WHILE liberal Protestantism came of age in North America during the nineteenth century, an interchange of ideas has always existed between European and North American Protestant intellectuals. The new science had revealed a world ordered by fixed, absolute, and immutable natural laws, which could be discovered by the powers of human reason alone. God's role in the process came to be gradually but surely more and more restricted to an original act of creation and to the institution of the natural laws. Since all men were presumed to possess the rational powers by which these laws could be apprehended, further testimony seemed superfluous. The Bible thus came to be relegated to the library of mythological literature, and Jesus Christ reduced to the status of a great moral teacher. The natural religion of the deists, in short, had been, by the beginning of the nineteenth century, mostly absorbed into Christian theology.

But in addition to these rationalistic tendencies, which, as we saw earlier, buttressed Draper's worldview and thus his new Protestant historiography, another stream of ideas emerged with the rise of a humanistic morality, with its emphasis on purely human problems. This humanistic spirit was fundamentally subversive to orthodox Christian belief. By the early nineteenth century, American liberal

Protestantism had made important concessions to this humanism. Central to this new humanistic spirit was religious romanticism, intuitionism, and transcendentalism, which are often associated with the spectrum of views of such figures as George Ripley (1802–1880), Orestes Augustus Brownson (1803–1876), Ralph Waldo Emerson (1803–1882), Frederic Henry Hedge (1805–1890), James Freeman Clarke (1810–1888), Henry David Thoreau (1817–1862), and others. As we have seen, the intuitive or immediate awareness of religious truth was crucial to White's own conception of religion. A particularly important feature of this new American religious movement was the belief in the evolution of religion, which emphasized a historical and empirical approach to understanding religion. As we have seen, although White shared with Draper the belief that natural law revealed the glory of God and that the old dogmas of supernaturalism were no longer tenable in light of the advances of science, White also expected that science would yield a progressively nobler and truer form of religion.

Evolutionary theism was a particularly American liberal Protestant view. While Draper's rationalistic theology was rooted in English latitudinarian thought, White's evolutionary theism, as we have already noted, was rooted in the German theological tradition. And as Annette G. Aubert has convincingly argued, German theologians inspired a significant number of nineteenth-century American intellectuals.[1] The Mercersburg Movement, for example, associated with American theologian John Williamson Nevin (1803–1886) and church historian Philip Schaff (1819–1893), received inspiration from German mediating theology, idealism, and romanticism.[2] Most importantly, Schaff, as one of the most prominent church historians in America in the nineteenth century, attempted to address an American culture marked by religious pluralism and sectarianism by offering a dialectic model of history, including the history of revelation. We have already noted that White regarded Schaff as uniting "the scientific with the religious spirit." Writings such as his *Principle of Protestantism* (1845), where Schaff argued that "the Reformation must be regarded as still incomplete," that "it needs yet its concluding act to unite what has fallen asunder," no doubt must have endeared him to White.[3]

While nineteenth-century American Protestant historians like Schaff adopted a scientific approach to history based on "facts," they also adhered to a concept of "organic development." This view was applied to studies of church history and dogma. British romantic poet and philosopher Samuel Taylor Coleridge (1772–1834), who was also profoundly influenced by German romantic movements, expressed this approach succinctly when he argued that "Christianity is not a theory, or a speculation; but a life;—not a philosophy of life, but a life and a living process."[4] This idea, however, had been there in embryo since the early republic. The claim that God's revelations are a successive, developmental, and "living process" was an essential feature of early American liberal Protestant thought. Therefore, to see how all these streams of thought coalesced in the nineteenth century, and thus make White's position possible, we need to begin by looking at the very founding of the American republic.

REASON, SCIENCE, AND RELIGION IN THE EARLY REPUBLIC

English chemist and Unitarian minister Joseph Priestley had composed his *General History of the Christian Church* in America, which he dedicated to American president Thomas Jefferson (1743–1826). After falling victim to a riot in Birmingham that destroyed his home, laboratory, and chapel, in 1794 Priestley immigrated to the fledging republic to pursue his scientific work and continue promoting his version of Christianity. As one of the founders of the new nation, Jefferson believed the greatest gift that God had imparted to man was reason. Like Priestley, Jefferson rejected orthodox conceptions of God as father in favor of such impersonal categories as "creator," "infinite power," "giver of life," and "intelligent and powerful agent." Jefferson's colleague John Adams (1735–1826), equally inspired by the writings of Priestley, similarly embraced a more "rational" conception of religion. Together they held the commonplace view that Roman Catholicism represented "pagano-papism," a corruption of the original, simple message of Christ.[5]

Both Jefferson and Adams remained serious students of Jesus all their adult lives. Of course, they were convinced that Jesus had only been a sublime moral teacher. Indeed, they believed Jesus taught the kind of rational religion they embraced. Like Priestley and the English deists, then, they believed that his original message was later corrupted into superstitions by ignorant men. Their older colleague Benjamin Franklin (1706–1790) similarly wavered in his views of Christianity, though he never denied orthodoxy outright. Franklin's skepticism and ambivalence toward historical Christianity was largely due to being exposed to skeptical writings he had read, which undermined his confidence in Christianity. In his autobiography, published posthumously, Franklin readily confessed that he rejected the orthodoxy of his youth for a more rational religion. He soon came into contact with some tracts against deism, but "it somehow happened, that they operated a quite contrary effect, to that which had been proposed by the writer; for the arguments of the Deists which had been cited in order to be refuted, appeared to me to be much stronger than the refutation itself. In short, I became a complete Deist."[6] While the Bible had made a deep imprint in his life, Franklin was a pioneer of a distinctly American kind of religion, as Thomas Kidd has recently put it: a doctrineless and moralized Christianity. Franklin's witty advocacy of "common sense" philosophy deemed historical Christianity both unintelligible and doubtful. He sought to separate what he considered "the essentials of every religion" from irrational ideas found in specific religions such as Christianity. Franklin was not alone in his views. Indeed, as Kidd explains, "Throughout American history, Franklin's sort of religion has fueled a multitude of related trends, such as anticlericalism . . . opposition to creeds, and religious individualism."[7]

In addition to Priestley, a number of influential American deists promoted the new religion founded on the power of human reason. American Revolutionary War patriot Ethan Allen (1738–1789), for example, published in 1785 his *Reason, the Only Oracle of Man,* in which he argued that reason was the only means by which man could come to know God. He called on readers to discard warped theologies derived from ancient biblical fables. Reason indicates, Allen contended, that the world had no beginning in time, that Moses's account of creation was a fabrication, and that revelation from God "must consist of an assemblage of rational ideas intelligibly communicated." In other words, revelation "could be nothing more or less than a transcript of the law of nature, predicated on reason, and would be no more supernatural, than the reason of man may be supposed to be."[8] The laws of nature, Allen deduced, were the only revelations of God.

Perhaps more than anyone else, Thomas Paine (1737–1809) epitomized American deism. His well-known anticlericalism endeared him to all freethinkers, but his deism also made him the logical if unacknowledged apostle of American liberal Protestantism. In 1794 he published his *The Age of Reason,* which earned him almost universal opprobrium by the orthodox. Paine, who believed that his own mind was his church, sharply criticized organized Christianity. He ruthlessly applied "reason" to the Bible, undermining its revelations, prophecies, and miracle stories. While he was confident of the historical existence of Jesus, he denied that he was God. Importantly, Paine saw the natural sciences as the best evidence of a supreme being. Clearly not an atheist, Paine professed "I believe in one God, and no more; and I hope for happiness beyond this life." Together with many other progressive and rational theologians, Paine believed that the universe itself—or alone—was singular proof of the beneficence of God. "The Almighty Lecturer, by displaying the principles of science in the structure of the universe, has invited man to study and to imitation. It is as if He said to the inhabitants of this globe we call ours: I have made an earth for man to dwell upon, and I have rendered the starry heavens visible, to teach him science and the arts." "True theology," he argued, is natural philosophy, and the "word of God is the creation we behold." But organized religion, according to Paine, was "set up to terrify and enslave mankind, and monopolize power and profit." Indeed, Paine declared that the "Christian system" not only rejected the study of science but persecuted it, citing the condemnation of Galileo as a prime example of the "militate" conflict between the dogmatic theologians and science. Interestingly, Paine also argued that the medieval period was the long "interregnum of science," a "long chain of despotic ignorance" broken only by the Protestant Reformation. Though it was not the intention of Luther, Paine observed, the good of the Reformation revealed that the only true theology was "natural philosophy, embracing the whole circle of science."[9] Paine's "theological reductionism," as Herbert Hovenkamp put it, excised the most important doctrines of historical Christianity—the Trinity, the Atonement, and the infallibility of the Bible.[10]

AMERICAN NATURAL THEOLOGY

When Protestant colonists from England started to settle along the Atlantic coast of North America in the early seventeenth century, they brought with them the English heritage of natural theology. And with this heritage came, as we have seen, a historiography that challenged religious authority. Puritan minister Cotton Mather (1663–1728), for example, in his *American Tears upon the Ruines of Greek Churches* (1701), argued that an "incredible darkness" had fallen over Western Europe following the fall of Rome, where "learning was wholly swallowed up in barbarity." This darkness was brought to an end "by the revival of letters . . . which prepared the whole world for the Reformation of religion too, and for the advances of the sciences ever since."[11] Here again we see the commonplace belief that the Roman Catholic Church impeded the advancement of knowledge. Furthermore, in his formidable *The Christian Philosopher* (1721), Mather attempted to demonstrate that "*Philosophy* is no *Enemy*, but a mighty and wondrous *Incentive* to *Religion*." The "Christian philosopher" considers well the contrivances of light, stars, sun, moon, planets, comets, meteorology, botany, entomology, and other sciences. Mather also called on the Christian philosopher to study the "*Lord of this lower World*"—that is, mankind. According to Mather, man is a "*Machine* of a most astonishing Workmanship and Contrivance!" He was confident that reason was a "*Faculty* formed by God, in the Mind of Man, enabling him to discern certain *Maxims of Truth*, which God himself has established, and to make true *Inferences* from them!" Mather quoted at length Bacon's "confession of faith."[12] Richard Lovelace has shown that Mather considered himself a major player in the great drama of the completion of the Reformation. "Mather's philosophy of history," Lovelace writes, "assumed that the Reformation of the sixteenth century had been only a partial restoration of the New Testament form of the church."[13]

Much like contemporaries across the Atlantic, eighteenth-century American Protestantism stressed reason in discussing religion. Historian of early America Andrew J. Lewis explains that the early republic was indeed an "empire of reason," and popular works of natural theology built this tradition. Natural theology was truly ubiquitous, "found seemingly everywhere in early republic print culture, from offhand remarks about science as the handmaid to religion to explicit treatments of science as illuminating evidences of God's handiwork."[14] While very little has been written on natural theology in the early republic, it is clear that early American Protestants fused Christianity with natural history. They created and consumed works of natural theology that explained and explicated the power, wisdom, and goodness of God, as manifested in the natural world. In other words, they had no sense that they were helping construct an arena for combat between science and religion a century later.

Besides the writings of Mather, American Protestant natural theologians produced an expansive and diffuse cluster of arguments and approaches to studying the natural world. Unitarian minister and Harvard librarian Thaddeus Mason Harris (1768–1842), for instance, published his *Natural History of the Bible* in Boston in 1793, in which he argued that "the Book of Nature and Revelation equally elevate our conceptions and incite our piety." Offering a natural history of plants, animals, and other phenomena mentioned in the Bible, Harris believed that nature and Scripture "mutually illustrate each other," and that "they have an equal claim to our regard, for they are both written by the finger of *the one eternal incomprehensible GOD.*"[15] Harris later published in 1801 *The Beauties of Nature Delineated*, which were excerpts from *Reflections on the Works of God in Nature and Providence for Every Day in the Year* (1773), written by German minister Christoph Christian Sturm (1740–1786).[16] As a minister, Sturm preached both the love of Christ and religious tolerance. Like many eighteenth-century *Aufklärer*, Sturm admired the beauty and splendor of nature and considered it to be a "*Schule für das Herz.*"[17] While Sturm saw the blessings of God through the works of nature, his *Reflections* does not adhere to traditional Christian doctrines. In the 365 essays encouraging readers to seek and find God in nature, from bees to the solar system, Sturm and Harris come dangerously close to making God identical with nature.

Indeed, American natural theology frequently verged on the pantheistic. Natural theological writers such as Charles Christopher Reiche, George Riley, and others encouraged young readers to connect nature with "the thoughts of God," which they believed would rouse more "noble pursuits."[18] At the turn of the century, books such as John Toogood's *The Book of Nature: A Discourse on some of those Instances of the Power, Wisdom, and Goodness of God, which are Within the Reach of Common Observations* (1802) and James Fisher's *A Spring Day; or, Contemplations on Several Occurrences which Naturally Strike the Eye in that Delightful Season* (1813) sought to instill in both young and old an emotional, aesthetic, and religious response to each flower, each animal, and each rock of the natural world.[19] As Harris put it, "Let us adore God in his wonderful works. Let us endeavor more and more to be acquainted with him. Let us reflect on his greatness, admire his power, celebrate his wisdom, and rejoice in his goodness, displayed in every season of the year, and diffused through every part of creation."[20]

The precarious nature of many American natural theological treatises was somewhat removed by the fact that such writings offered a safe place during times of denominational strife. Hence natural theology was supported by American Protestants who saw in studying nature not only a devotional value but a means to avoid doctrinal and denominational conflict. Rather than get into doctrinal disputes, American popularizers of science and natural theologians asserted the wonder, awe, and gratitude of experiencing God's creation.[21] American chemist and science popularizer Benjamin Silliman (1779–1864), for example, sought to study the Bible "in the light of modern science," instructing his students during the 1820s that "geological facts are not only consistent with sacred history, but that their tendency is to illustrate and confirm it."[22] While he no doubt believed that the study of nature leads man to a "knowledge of his Creator," Silliman also spoke of "progressive creation," arguing that the fossil record could be harmonized with the order of events in Genesis provided that one interpreted the six "days" of creation figuratively.[23]

Silliman, religiously quite moderate, rejoiced when his attempt at reconciling science and religion was "acceptable to the wise and good of all religious denominations." Indeed, he had once dined with Joseph Priestley, whom he found "mild, modest, and conciliatory."[24] Moreover, he told Congregational minister and American geologist Edward Hitchcock (1793–1864), his apprentice in the laboratory with whom he shared many similar ideas, that he completely embraced the "new views" among the geologists and that "no mere divine, no mere critic in language, can possibly be an adequate judge of the subject; or deserve unqualified deference, however able in other subjects." A decade later, he continued to criticize "theological gentlemen" who showed no "disposition to listen to reason and evidence on this subject [geology]."[25]

Silliman more clearly revealed his position that scientific practice must separate science and religion in an 1842 address before the Association of American Geologists and Naturalists. He maintained that scientific explanations must be limited to physical causes known to exist in nature. "Our advancement in natural science is not dependent upon our faith," he wrote. "All the problems of physical science are worked out by laborious examination, and strict induction."[26] While a strong minority of conservative Protestants remained cautious or even suspicious of natural theology, most Protestants were all too eager to align their theology with the scientific enterprise. The rational theology of a previous generation had led most nineteenth-century American Protestants to assume that the most important repository of evidence attesting to God's existence was the natural world. Indeed, natural theology became the go-to argument of nineteenth-century liberal Protestants. But the focus was on the God of the philosophers, the "Creator," cosmic "Architect," or "Author" of nature, not the God of Abraham, Isaac, and Jacob.

The second half of the nineteenth century witnessed what historian James Moore has called the "Protestant struggle to come to terms with Darwin." While natural theology remained a thriving enterprise in the second half of the century, Darwin's *Origin of Species* no doubt forced both scientists and theologians to reassess their approaches. The point that needs to be emphasized here is that liberal Protestants had little trouble reconciling evolution with theology. As Jon Roberts has shown, relatively few American Protestant leaders who remained committed to orthodox formulations of Christian theology actually embraced Darwinism. Those Protestant intellectuals who embraced evolution did so as part of a major adjustment to theology. While liberal Protestants took advantage of the occasion to initiate a thorough reworking of the tenets of Christian doctrine, more conservative Protestants rejected the theory of organic evolution altogether, valuing the authority of a common-sense interpretation of the meaning of Scripture over that of fallible science. Those Protestant intellectuals who accepted the notion of evolution, however, found in science a new way to understand the relationship between God and man. Instead of believing that revelation ended with the events recorded in the Bible, progressive theologians put greater emphasis on human powers of reason, a continuing act of creation and, most importantly, the ongoing revelation of divine truth.[27]

For example, "Christian Darwinian" Asa Gray (1810–1888), Harvard botanist and Darwin's staunchest American supporter, attempted to prove that natural selection was not inconsistent with natural theology in his *Darwiniana* (1876). He responded to critics that Darwin's *Origin of Species* was in fact a theistic work, insisting that Darwin did not deny "purpose, intention, or the cooperation of God" in nature.[28] While Gray's post-Darwinian natural theology looked more Lamarckian than Darwinian, he nevertheless maintained that the "philosophical theist" could still see purpose in evolution. Gray later attempted to relate Darwinism to Christian orthodoxy in his lectures on *Natural Science and Religion* (1880). In these lectures Gray argued that "the business of science is with the course of Nature, not with interruptions of it," thus irrevocably separating scientific practice from religious meaning.[29] While often described as a "deeply religious man," Gray was a man of broad and accommodating religious and philosophical opinions. As A. Hunter Dupree recognizes, Gray was untypical of orthodox Christian thinkers.[30] In his attempt to relate Darwinism to Christian orthodoxy, Gray did not believe that the Bible was divine revelation but the human record of that revelation, and that Christianity itself evolves—positions most orthodox believers no doubt found disconcerting.

American physician and geologist Joseph Le Conte (1823–1901) also attempted to accommodate Christianity to evolution. In works such as *Religion and Science* (1877), *Evolution and its Relation to Religious Thought* (1888), and others, Le Conte described the process of evolution as the unfolding of God's plan. For Le Conte, the law of evolution was "the grandest idea of modern science," arguing that it was "thoroughly established, far more certain than . . . the law of gravitation, for it is not a contingent, but a necessary truth." Unsurprisingly, then, for Le Conte God not only worked through the law of evolution but was intimately involved in terrestrial life. Le Conte's immanentist theology promoted the view that God was "resident in Nature . . . in every molecule and atom, and *directly* determining every phenomena and every event." In other words, evolution is the result of forces resident within nature, which ultimately emanate from a divine source. Le Conte also saw "reason" as the highest form of mental activity. By his reason man could understand through natural laws the purposes of God and come to cooperate with them. In partnership with God or nature, then, man could further his own evolution. As a "theistic evolutionist" of a neo-Lamarckian bent, Le Conte believed science is a "rational system of natural theology" that pointed beyond itself to a divine mind that served as the "energy" that was immanent throughout creation. But Le Conte also insisted that science mandated that Christianity be recast into an evolutionary mold and that its old doctrines be tested against the standards of evolutionary theories. "Religious thought," he explained, "like all else, is subject to a law of evolution." He boldly concluded that evolution demanded "a reconstruction of Christian theology." Otherwise, he added, the "church will die."[31]

While many American liberal Protestants were indeed enamored with arguments from design, others went beyond natural theology in their efforts to demonstrate the harmony between science and religious faith. Rather than relying on "external arguments for the being of God," they opted to look for an "inner manifestation" of God in the realm of feeling. A "new theology" had thus emerged that defended an immanent conception of God, redefined the Bible as an expression of evolving religious consciousness, and recast central Christian tenets in terms borrowed from organic evolution.

THE RISE OF AMERICAN NEW THEOLOGY

Rooted in European liberal Protestant ideas, American liberalism began as a distinctive movement in American thought with the emergence of Unitarianism and transcendentalism in the early nineteenth century. Theological Unitarianism, or the belief in the oneness of God, began in seventeenth-century England and was adopted by eighteenth-century American liberal theologians. But the rational religion advocated by the founders of the republic was eventually combined with such romantic movements as transcendentalism.[32]

The transcendentalists had imbibed German idealism, avoided religious institutions, and preferred informal fellowship through intellectual and literary pursuits. They emphasized the presence of divinity within each human being. This divinity powered the self and was best fostered without the external control of religious institutions. Consequently, American transcendentalists encouraged individuals to pursue spiritual truth on their own. While seventeenth- and eighteenth-century liberal Protestants focused on the "inner light" of reason, liberal Protestants of the new theology looked to emotions as the final source of religious authority. Their primary religious focus was on the immanence of God, the notion that God's presence pervaded all things.

Boston Unitarianism was the immediate context of American transcendentalism; many of its leading representatives were once Unitarian ministers in that city. Emerson, for instance, came from a long line of such ministers. Himself an ordained Unitarian pastor before resigning in 1832, Emerson studied natural theology and philosophy at Harvard and ultimately came to the conclusion that the historical-critical approach to the Bible destroyed any remnant he had of orthodoxy. In 1835 he began his career as popular lecturer, gathering around him a group of young dissident Unitarians who came to champion the new movement of transcendentalism.[33]

In his first book, *Nature* (1836), often considered the primer manifesto of American transcendentalism, Emerson argued for the unity between the divine and the natural. Indeed, he asserted that nature is "the present expositor of the divine mind," that the natural world was a manifestation of divinity. The perception of this divinity is possible because man himself is "God in us." The current of "Universal Being" circulates through us—"I am part or parcel of God," Emerson wrote.[34] Significantly, Emerson's deification of humanity is told through the lens of religious history. Revelation is not fixed and thus cannot be limited to any one age. "It is the office of a true teacher," he proclaimed in his controversial "Divinity School Address" at Harvard in 1838, "to show us that God is, not was; that He speaketh, not spake."[35] In this address he also declared that "historical Christianity" had "fallen into the error that corrupts all attempts to communicate religion." Jesus was indeed a great man, a man who "belonged to the true race of prophets." But "traditionary" faith had turned Jesus into a "demigod, as the Orientals or the Greeks would describe Osiris or Apollo." Historical Christianity, then, was a complete failure. Nevertheless, Emerson looked forward when a "new Teacher" shall "show that the Ought, that Duty, is one thing with Science, with Beauty, and with Joy."[36]

For transcendentalists like Emerson, God's revelation was ongoing, enfolding, and therefore continually evolving. Among the numerous periodicals devoted to spreading this transcendentalist message, the most important and popular were *The Dial*, *The Radical*, and *The Index*, which, according to Clarence Gohdes, "sought to find the source of all truth within the nature of man."[37] Boston had developed a long tradition of Unitarianism and transcendentalism, and organizations like the Radical Club of Boston and the Free Religious Association opened their doors to self-styled liberals engaged in conversation they believed would help bring about the dawn of a new religion.

It must be noted that the religious revivals earlier in the century had already refashioned religious culture by focusing on the individual.[38] American romanticism, which was partly the result of the revivals, transitioned the focus from God-centered to human-centered, emphasizing personal agency and individual liberty. A wave of liberalism moved religious thought toward philosophical idealism, with emphasis on immanence, the humanity of Jesus, and the social call of the Gospels. Whatever moderates there were, they were largely swallowed whole into the emerging liberal Protestant tradition. In his study on the rise of biblical criticism in America, Jerry W. Brown describes the tug-of-war between liberals of varying stripe and conservatives of equally varied stripe. Unitarians used critical scholarship to demonstrate the impossibility of more orthodox dogmatic schemes. But while liberal Protestants employed the methods and results of German scholarship to the acute discomfort of their orthodox opponents, they also maintained that they were discovering the plain unvarnished truth of the Gospels. According to Brown, interest in biblical scholarship and criticism began to wane around the time of the Civil War, as liberal Protestants promoted "a strong dependence upon intuition, inspiration, and sympathy to interpret a biblical text made study of original languages, investigations of authorship, and historical reconstructions unimportant."[39]

While American liberal Protestantism is difficult to define with any precision, there are some identifiable and common markers. As we have seen in previous pages, liberal Protestants often emphasized the immanence of God in nature and human nature. The commitment to divine immanence also manifested itself in discussions of encounters between God and human beings. This was particularly important in the way liberal Protestants understood the deity indwelling in the historical process. That is, liberal Protestants embraced a teleological view of historical process, a conviction that historical development disclosed the gradual unfolding of a divine plan for redemption of humanity. Generally more optimistic about humanity, liberal Protestants were also more open to a universal religious sentiment—or, increasingly, of more variegated forms of religious experience. At the same time, liberal Protestants ignored or actively criticized many of the traditional doctrines of Christianity. There was, thus, a tacit accommodation or adjustment to modern thought and culture, an attempt to integrate Christianity within the broader confines of culture.

Almost uniformly, these advocates of liberal ideals associated intolerance of theological claims of authority with hierarchy and, at least by implication, Roman Catholicism. Conjuring up fears of Rome's "theological and ecclesiastical rule," liberals assailed the theology of those less liberal than themselves, asserting that, in its claims of religious authority and exclusive truth, the Catholic Church was intolerant. Of course, not all anti-Catholicism was liberal nor was it even based solely on liberal principles. Nonetheless, much anti-Catholicism, as we have seen, drew upon the liberal conception of individual independence and mental freedom. Many Protestant Americans recognized Roman Catholics as political enemies of the United States and responded to this perceived threat by organizing nativist movements and undermining the civil liberties of "foreigners."

American anti-Catholicism was, in many respects, a European and particularly British import. Many of the colonists brought with them a deep distrust of Rome and its religion. While in the aftermath of the Revolution the ratification of the Constitution and the Bill of Rights largely eliminated official federal persecution of and discrimination against Catholics, suspicion of the Catholic presence in the new nation did not dissipate. Early in the nineteenth century, when Catholic immigrants from nations such as Ireland and Germany began arriving in the United States, a rhetoric of "nativism" emerged.[40] In particular, American Protestant church historians probed the polity of early Christianity for lessons in support of Protestant government and republicanism against the hierarchies and tyrannies of Roman Catholicism.[41] Samuel Miller (1769–1850), professor of church history at the Theological Seminary at Princeton, and author of perhaps the first American history of science, *A Brief Retrospect of the Eighteenth Century* (1803), contended that the "rise of popery" coincided with the "decline of science." According to Miller, the Catholic Church "has ever been the sworn enemy of learning and science," and that "enlightened Protestants" must wage "consecrated warfare" with popery, for he believed it was by its very nature a "system of tyranny over both the minds and bodies of men." Miller later warned parents to avoid sending their sons to Catholic seminaries, asserting that the "system of Romanism is . . . the decided foe to liberal inquiry, whether in literature, in science, or in duty."[42]

In the 1830s, Samuel F. B. Morse (1791–1872), coinventor of the telegraph and Morse code, published his *Foreign Conspiracy Against the Liberties of the United States* (1835), which was a series of letters to the *New York Observer* encouraging a Protestant united front against the perceived encroaching menace of Catholic immigrants. "Surely American Protestants," Morse wrote, "have discernment enough to discover beneath them the cloven foot of this subtle foreign heresy. They will see that Popery is now, what it has ever been, a system of the darkest *political* intrigue and despotism, cloaking itself to avoid attack under the sacred name of religion." "They will be deeply impressed with the truth," he goes on, "that Popery is a *political* as well as a religious system; that in this respect it differs totally from all other sects, from all other forms of religion in the country." Tellingly, Morse also declared that Catholics were required by their religion to passively obey the Church and that they were denied the "right of private judgment." Perhaps most important, however, in his appendix explaining the "opposite tendencies of Popery and Protestantism," Morse argued that while Protestantism is founded on the Bible, the "Bible prescribes no *form* of faith, or doctrine, or of church government, in which all, in the exercise of the natural and revealed right of private judgment, can agree, each sect adopts that form most in accordance with what it believes to be the spirit of the doctrines which the Bible teaches." In other words, the Protestant, unlike the Catholic, is free to "choose according to the dictates of reason and conscience." As such, the Protestant may call to his "aid all the treasure of science." According to Morse, the Protestant "believes that the divine Author of truth in the Bible is also the author of truth in Nature. He knows, that as truth is one, He that created all that forms the vast field of scientific research cannot contradict truth in Scripture by truth in nature; the Protestant is therefore the consistent encourager of all learning, of all investigation. Every discovery in science, he feels, brings to religious truth fresh treasures. Free inquiry, and discussion, all intellectual activity legitimately belong to Protestantism."[43]

Presbyterian clergyman Lyman Beecher (1775–1863), father of Henry Ward Beecher and Harriet Beecher Stowe, similarly claimed that the United States was being overrun with despotic Catholics. In a series of lectures, *A Plea for the West* (1835), the elder Beecher railed against the long history of the "baleful influence" and "spiritual dominion" of the Catholic Church. For five hundred years the papacy, "by the terrors of his spiritual power over the consciences of their dark-minded subjects . . . bound kings in chains and princes in fetters of iron." With the Reformation, however, came the "extension of commerce and the arts, the illumination of science, the power of scepticism, and the advance of liberal opinions and of revolution and reform."[44]

THE EVOLUTION OF RELIGION

As we have seen, soon after the publication of Darwin's *Origin of Species*, many American Protestant intellectuals felt compelled to assess its theological implications. While a strong minority of conservative Protestant leaders concluded that reconciliation was impossible, a significant majority of liberal and even moderate Protestant voices called for the modification of traditional formulations of Christian doctrine to accommodate evolutionary theory. New York Unitarian minister Samuel R. Calthrop (1829–1917), for instance, accepted the new theory of evolution gladly, arguing that, far from destroying religion, evolution deepened and broadened religion, making it more sublime and giving it firmer foundations. In 1873 he published an essay on "Religion and Evolution," where he argued that the laws of nature and the laws of the spirit are one and the same. According to Calthrop, "we must fill Evolution full of God." Indeed, the "Religious Evolutionist is a God-intoxicated man," and "sees God everywhere, in everything, around everything." Calthrop also saw evolution at work in religion, writing that one cannot "fail to discover that Religion is a Growth," and suggests that one can trace the "Religious Consciousness" back to the "Wild Man's ear," or primeval man. Darwin's theory has made this historical perspective, this "Sympathy of Religions," as Calthrop has it, possible.[45] The following year Calthrop spoke more directly on the issue of "Religion and Science," arguing that for ages theological dogmatism has stifled reverent students from studying nature. The Church, according to Calthrop, "devoured" astronomy, geology, physiology, and "every little green shoot of original thought." He then demanded that theology must cease to wield a usurped authority over the sciences. But according to Calthrop, this was a gain for religion. The naturalist, he argued, sees and feels "a far grander display of Creative Power." Science, moreover, will show how religion has progressed and where it is going. "In a world that grows," he wrote, "Religion must also grow, and grow in accordance with the Laws of Growth."[46]

Another leader of the "new theology" movement emphasizing the evolution of religion was clergyman Newman Smyth (1843–1925). In his *The Religious Feeling* (1877), Smyth argued that nature was "wonderfully suggestive of God." Influenced by the insights of German theology, Smyth sought to integrate the theology of Schleiermacher and evolutionary theory. Like Schleiermacher, Smyth termed this manifestation the "religious feeling," which was, "in its simplest form, the feeling of absolute dependence." He thus encouraged readers to turn to the "inner manifestation" of God in the human soul. According to Smyth, "human progress is always the resultant of conflicting forces." But advances in the natural and historical sciences has always led to "progress in theology," which is "purifying and enlarging the conception of God."[47] Smyth amplified these arguments in his *Old Faiths in New Light* (1879), where he made the case that Christians should not only accept the modern historical-critical approach to Scripture but also adapt to the findings of science. "The mere suspicion," he asserted, "that the advanced scholarship and the old faiths are to-day at variance, is itself a fruitful cause of popular indifference and unbelief." This distrust of scholarship, however, is misplaced. The history of doctrine, Smyth avers, shows that faith has often undergone "rearrangement" in "new lights." Drawing on Lessing, Smyth contended that revelation was a "progressive force," with history and even the biblical narrative itself providing numerous turning points in religious reformations.[48] Smyth continued to develop his new theology in the *Orthodox Theology of To-Day* (1881). Having argued that biblical criticism and science were safe, enlightening, and spiritually enriching, he felt that the old orthodoxy was unworthy of continued refutation. It was plain to him that theology is not exempt from a "general law of human progress," and therefore a "progressive church, led by the Spirit of Christ, must always keep its historical confessions under the process of revision and adaptation to new environments of thought."[49]

In his 1883 book *The Freedom of Faith,* which Gary Dorrien has called the "manifesto" of the New Theology movement, Congregationalist clergyman Theodore T. Munger (1830–1910) argued that the new theology rejects the "mechanical" theologies of the past, seeking instead to align itself in a process of development. It also makes broader use of human reason and interprets Scripture in a "more natural way." According to Munger, the Bible, "like the order of history, is a continually unfolding revelation of God." While the "Old Theology" was in conflict with natural law, the "New Theology" "recognizes a new relation to natural science." Munger had embraced evolutionary theory and insisted that Christians should not resist it. According to Munger, the New Theology was akin to a "religion of humanity," and thus questioned "the line often drawn between sacred and the secular."[50]

In the last decades of the nineteenth century, Lyman Abbott (1835–1922), eminent Congregational minister and successor to Beecher's pulpit of the Plymouth Church in Brooklyn, claimed that evolution was "God's way of doing things." In his *The Evolution of Christianity* (1892), Abbott surveyed human history and saw in it a continuous progressive change from animal instinct to moral virtue, moving according to certain spiritual laws and by means of the force of God. For Abbott the evolutionary laws discovered from the study of nature were identical with the laws of the spiritual life, and he came to believe that the latter were only adequately understood when viewed as analogous to the former. This meant that the older theology must be set aside. Everything now must be seen in terms of change, development, and progress. In the Bible we do not see a complete and perfect revelation but "men gradually receiving God's revelation of Himself." The Bible is the "history of the growth of man's consciousness of God." Abbott's belief in historical progress demanded that he reject traditional doctrines of Christianity. "The institutions of Christianity must be elastic," he wrote, "because Christianity itself is a growing religion." More bluntly, he declared that "the doctrine of the Fall and of redemption (as traditionally stated) is inconsistent with the doctrine of evolution." According to Abbott, Christ came not only to exhibit divinity to us "but to evolve the latent divinity which he has implanted in us."[51]

A few years later, in his *The Theology of an Evolutionist* (1897), Abbott conceived of God's relation to the world as immanent. He argued that there are no "occasional or exceptional theophanies, but that all nature and all life is one great theophany; that there are not occasional interventions in the order of life which bear witness to the presence of God, but that life is itself a perpetual witness to His presence."[52] Indeed, Abbott's conception of human history as the progressive evolving of divinity led him to write that "the difference between God and man is a difference not in essential nature" and argued that the Bible taught that "in their essential nature they are the same."[53] Abbott interpreted evolution as "progressive change—a change from a lower to a higher condition; from a simpler to a more complex condition," adding that "God dwells in nature fashioning it according to His will by vital processes from within, not by mechanical processes from without."[54]

This progressive view of religion and especially Christianity is even more pronounced in American Unitarian minister Minot Judson Savage (1841–1918), who was perhaps America's most prominent pro-evolution preacher. Converting from Congregationalism to Unitarianism in 1872 after struggling with the religious implications of evolutionary theory, Savage gave a series of sermons at the Church of Unity in Boston, which he published in 1876 as *The Religion of Evolution*. In the opening pages of his book, Savage declared that the theory of evolution was "now accepted by nearly all the leading scientific and philosophic students of the world" and that it was "rapidly giving its own shape to the thought of civilization. Science, art, human life, religion, and reform are becoming its disciplines; and their tendencies in the near future must be largely determined by it." Significantly, Savage begins by explaining why so many believe there is a "conflict between religion and science." The essence of this conflict, which "has been going on since the dawn of civilization," according to Savage, is one between the "philosopher" and the "theological defenders" of religion. But while each episode of conflict, each battle or contest fought, has injured "certain theories of theology" and "certain sectarian claims," it has nevertheless helped religion grow "grander and more magnificent every time." That is, religion "has been helped, advanced, uplifted, magnified, and made grander by the conquests of science." With every great change in scientific thinking, there is a corresponding change in religion. Savage believed that both science and religion sought truth and therefore both led to God. They were "at one, and need no reconciliation." Savage expressed gratitude towards the scientists of the past few centuries for helping to destroy "the superstitions, the crudities, the falsehoods, the misconceptions of men, concerning religion."[55]

Savage rejected the entire notion, put forward by "some scientists" and "some frightened religionists," that "evolution is essentially atheistic and irreligious." He saw in evolution the very "life and power and movement of God." Savage conceived of an immanent God who was in the very laws of nature. Universal natural laws were "only universal, all-encompassing, tireless, changeless providence."

Evolution therefore demolished the old materialism and offered "the grandest conception of God." Moreover, since Savage's account of evolution contained no mention of the theory of natural selection, he could conceive of the evolutionary process as progressive, and moral, in nature. "And not only do we see progress along certain definite lines of law that suggest the rightness of this life-force of the universe," Savage announced, "but this progress has lifted up into what we call the sphere of morals, and has been along certain other definite lines of what we call righteousness." Rather than leading to unbelief, evolution led directly to Christianity. "I am a Christian *because* I am an evolutionist," Savage declared.[56]

Savage later took these ideas and expanded them greatly in his *Evolution of Christianity*, published in 1892, which was not altogether different from Abbott's work of the same year. Perhaps the most salient point in Savage's book is that while Protestantism ushered a revolution in religious thought, it "tried as hard as Rome has tried" to "stay the progress of human thought." He wrote that the "Protestant Church has opposed itself to modern science, has opposed every step of growth that it has supposed to be inconsistent with its interpretation of the Bible, as persistently as has Rome." In the end, Savage called for a "free Christianity," one with "utter freedom of thought, fearless application of the scientific method."[57]

Savage's conversion to the religion of evolution led him to believe that fundamental creedal revision was imperative. In 1887 he wrote about his *Creed* "outgrowing the old beliefs," and that the world is "undergoing so radical a change that it is not too much to say we have a new earth, a new sun, new stars, a new God, a new humanity, a new religion, a new church, [and] a new outlook for humanity." He goes on to outline, in biographical fashion, the changes in the new theology. "I gave up," he explained, "belief in a cruel, partial, imperfect God . . . in a disastrously ruined and fallen world . . . in the total depravity of man . . . in miracles . . . in a miraculous, divine incarnation . . . in the suffering and death of God . . . [and] in endless hell." In place of these things, Savage discovered "an infinite, perfect, loving God . . . a world . . . of orderly law and progress from the first . . . a humanity . . . having climbed up to the point where we can say, 'Now we are the sons of God.'" For Savage, these changes led to the realization in the "divineness of all men."[58] The following year he discussed in more detail the "readjustments" of theology, or the coming *Religious Reconstruction*, as the title of his 1888 book has it.[59]

In 1889 Savage published *Signs of the Times*, which reflected a melioristic optimism, writing that "people believe more than they used to in the possibilities of human progress." He pointed out that the hope for a better age was basic to Christianity. Religion, Savage averred, always offered humanity "not a scheme of salvation," but "a work of adjustment," reconciling persons to their own divine nature, to their fellow beings, and to their god. Moreover, modern evolutionary theory, as Savage saw it, pointed to the inevitable progress of social development and human growth. He reveled in the progress of humanity, "developing as naturally as the flower, and, mark, you, as divinely as the flower; for the natural to my mind is divine." Evolution, then, provides evidence for "the one purpose of God"—namely, the "development of a soul." For Savage, religion had nothing to fear from the tendencies of modern thought. Indeed, the "faith of reason" was the only sure foundation for "the new city of God." The new liberalism no longer sought "the patronage of the older faith," nor did it need "to feel out for a handclasp from some man who, if he is honest, has no business to give us a handclasp." He declared that the new theologians "stand for a new revelation of God's truth, a new gospel of humanity." "Let us," he concluded, "consecrate ourselves to the service of our fellow-men, to the service of God, and to labor towards the realization of this age-long hope of the world."[60]

For one final example, we may turn to clergyman Charles A. Briggs (1841–1913), an advocate for theological progress who believed that the "theological crisis" of his day was largely the result of conservative reactionaries. According to Briggs, while conservatives and progressive forces are in perpetual conflict, "progress in doctrine and life is a necessary experience of a living church; and that progress will never cease until the church attains its goal in the knowledge of all the truth, in a holiness reflecting the purity and excellence of Jesus Christ, and in a transformed and gloried world." The authority of religion, which was the "essential question at the Reformation," must be established in human reason. Optimistically, Briggs was assured that "the fruits of this theological crisis can only be great, lasting, and good." "It is evident that the evolutions of Christian theology which have brought on the theological crisis," he declared, "are preparing the way for a new Reformation, in which it is probable that all the Christian churches will share."[61]

Briggs argued that God has made himself known in the form of human reason, and therefore "modern thinkers have a divine calling to withdraw men from mere priestcraft, ceremonialism, dead orthodoxy and ecclesiasticism, and concentrate their attention on the essentials of the Christian religion." Indeed, Briggs saw himself as continuing the work of the Reformation in letting the tools of biblical criticism help contemporary Christians recover the words of the Bible itself. "It will ere long become clear to the Christian people," Briggs insisted, "that the Higher Criticism has rendered an inestimable service to this generation and to the generations to come. What has been destroyed has been the fallacies and conceits of theologians; the obstructions have barred the way of literary men from the Bible. Higher Criticism has forced its ways into the Bible itself and brought us face to face with the holy contents, so that we may see and know whether they are divine or not. Higher Criticism has not contravened any division of any Christian council, or any creed of any Church, or any statement of Scripture itself." The "age of Rationalism," Briggs proclaimed, was like the work of a farmer, "cutting off the limbs of trees, and pruning vines and bushes, and rooting out weeds." Similarly, the rational theologian is clearing away the "dead wood, dry and brittle stubble, and noxious weeds" of "dead orthodoxy, every species of effete ecclesiasticism, all merely formal morality, all those dry and brittle fences that constitute denominationalism, and are the barriers of Church Unity."[62] According to Briggs and other academic freedom proponents, intellectuals must have the liberty to teach, study, and research without restriction by others in order to seek knowledge and truth. Academic freedom would bring about new forms of religious truth in biblical criticism, comparative religion, physical science, and social sciences.

Historians have identified a host of sources from which Protestant liberalism ostensibly drew, from romantic and transcendental philosophy, Hegelian teleology, English Broad Churchmen, to American Unitarianism. No doubt there were others. Whatever its sources, the new theology was a mediating Christian movement. It was reformist in spirit and substance, not revolutionary. Fundamentally it is the idea that genuine Christianity is not based upon external authority. It was further defined by its acceptance of modern knowledge, especially historical criticism and modern science. As these theological liberals strove to make Christianity more credible, palatable, and socially relevant, they often either ignored or softened its harsher doctrines. This theological shift in the century from a more theocentric perspective to a more anthropomorphic one affected views of heaven, human nature, and various other historical doctrines of traditional Christianity. The new perspective that emerged in the last decades of the nineteenth century was less harsh, less punitive, and less frightening than the dominant view in earlier periods of church history.

Most importantly, by attempting to bridge the growing rift between modern thought and Christian faith, advocates of new theology distinguished theology from religion. This distinction, as we have seen in Draper and especially White, allowed liberal Protestant intellectuals to maintain that true science and true religion were compatible, while arguing that theological dogmatism had always restricted the development of science. Liberals deemphasized theology in order to establish what they hoped would be a firmer base for Christian faith and to minimize the challenge of their own theological revisionism. Although liberals rejected many orthodox Protestant doctrines, they wanted to avoid an antagonistic posture. They presented theological change as part of the natural development of religion and associated it with the progressive revelation of the divine through the evolution of human nature. They saw themselves as protectors of true Christianity rather than its enemies. By freeing religion from the millstone of theology, liberal Protestants believed that civilization could attain greater intellectual and academic freedom.

THE SECULARIZATION OF THE AMERICAN UNIVERSITY

According to White, the pre-Cornell period in the history of American education was "the regime of petty sectarian colleges."[63] A less dramatic and more accurate description of the premodern era would be that most colleges were religious oriented. But while most colleges of the period had religious foundations, they were also forbidden either by charter or public opinion to indulge in religious tests for faculty or students.[64] The American university had a particularly Dissenting or Nonconformist heritage. While denominationalism gave strength and purpose to the religious life of many American colleges, sectarianism did not last. Harvard, for example, fell away from orthodox Calvinism very early in the nineteenth century when Henry Ware (1764–1845), a Unitarian preacher, was elected to the Hollis Professorship of Divinity in 1805. While the orthodox Calvinist withdrew from Harvard to found Andover Theological Seminary in protest, by the end of the century it, too, had transformed from a bastion of orthodoxy to a center of "New Theology."[65]

Nineteenth-century American seminaries and universities, then, largely followed trends among European institutions, which dramatically departed from traditional approaches to Christian thought and history. Liberals did not consider Christian ideas static but subject to social, political, and economic development. Historical circumstances played a pivotal role in explaining religion. Many American scholars were concluding that Christian dogma increasingly adopted metaphysics and Greek influences over time. Deeply influenced by the optimism and progressive flavor of their age, early American universities promoted theological perspectives that accentuated God's love and minimized human depravity and evil. Most importantly, educational reformers enthusiastically included the natural sciences into their college curricula.[66] By the second half of the century, liberal theology had become a decidedly academic enterprise.

As we have seen, for many nineteenth-century thinkers, theology and revelation could no longer provide a sufficient answer to the question of religion, not only because of the religious disagreements that emerged during the sixteenth century, but also because the physical and historical science seemed to discredit orthodox Christian beliefs. Thus, such questions and their subsequent answers were collected in tracts or studies that followed a "rationalistic" or "naturalistic" approach. As Paine put it, natural religion is the "only religion that has not been invented, and that has in it every evidence of divine originality, is pure and simple deism. It must have been the first, and will probably be the last that man believes."[67] Other founding fathers concurred. Benjamin Franklin, in a letter to

Ezra Stiles (1727–1795), then president of Yale College, expressed himself in the following terms: "you desire to know something of my Religion . . . Here is my Creed. I believe in one God, Creator of the Universe. That he governs it by his Providence. That he ought to be worshipped. That the most acceptable Service we render to him is doing good to his other Children. That the soul of Man is immortal, and will be treated with Justice in another Life respecting its Conduct in this."[68] When it comes to "revealed religion," Franklin is clear about his disdain for sectarian differences and for any claims about the truth of the "supernatural" character of religion. Likewise, Thomas Jefferson attests to the prevalence of "natural religion." In his 1803 letter to Philadelphia scientist, physician, and political leader Benjamin Rush (1726–1813), Jefferson similarly follows Franklin's descriptions.[69]

The writings of British deists and German mediating theology reached the American colonies by the mid-eighteenth century, and their influence was felt strongly at Yale, Harvard, Princeton, and William and Mary. As George Marsden has demonstrated, the progenitors of American educational reform in the nineteenth century were dominated by liberal Protestants, who saw their task as heralding a new religious spirit. Yet their zeal for a "progressive" and "universalist" Christianity stripped the faith of its specificity and power, unwittingly creating the conditions that relegated religion to the margins of intellectual life. Liberal Protestants reduced Christianity to progressive but innocuous notions of religious "inner feeling." This redefinition of religion made it more logical for the university to pursue higher education without need for explicit reference to Christian particularism. It was not long before "nonsectarian" came to be interpreted to mean "secular." As Marsden puts it, this ultimately eased the transition from "methodological secularization" to "ideological secularization."[70]

In the mid-nineteenth century when most academics believed in the unity of science and religion, the advent of natural and historical sciences dealt a severe blow to the so-called "Baconian compromise," the notion that God is the author of both the Book of Scripture and Book of Nature. As we have seen, the separation of natural philosophy and theology only resulted in the collapse of the "two books" into one single text of nature. Indeed, Bacon had already conceded much autonomy to nature, thus setting it up to compete and ultimately supersede revelation. American colleges taught that God's work was evident in nature, and greater knowledge of science thus buttressed a Christian perspective that encompassed natural theology and the moral mission of the college. But this worldview proved unable to assimilate portions of the growing corpus of scientific knowledge, particularly in geology and biology. As a result, the modern American research university attempted to "reconstruct" or "redefine" Christianity, as Julie Reuben put it.[71] The academic community basically responded by jettisoning "dogmatic theology." It thus sought to preserve the unity of knowledge by reinstalling Bacon's compromise with little revision, reconceptualizing science and religion each in its own sphere with its own truth.

Again, the reaction of most educators to this dilemma, according to Reuben, was not to abandon religion as a central concern of higher education, but to redefine it. Most important for our purposes was the effort of substituting a broad and ecumenical notion of religion for sectarian faith while retaining the idea that academic knowledge was rooted in religious ideas. This idea proved particularly attractive to college presidents. Most of these university reformers did not think in terms of a war between science and religion; rather, they sought reconciliation between the two. Prominent university reformers such as Charles W. Eliot of Harvard, Daniel Coit Gilman of Johns Hopkins, David Starr Jordan of Stanford, William Rainey Harper at Chicago, among others, rejected the notion of conflict by altering the position of religion in higher education to make it consistent with modern scientific standards of intellectual inquiry. This approach thus differed little from what White had proposed in his historical narrative.

In this spirit, president Charles W. Eliot (1834–1926) reorganized the Harvard Divinity School in order to transcend dogma through the presence of multiple denominational outlooks. Religion, Eliot believed, might fit with the scientific ethos of the emerging university, but the theology of the churches did not. In his inaugural address as president of Harvard in 1869, Eliot argued that because a university education should manifest an "open mind, trained to careful thinking, instructed in the methods of philosophic investigation, acquainted in a general way with the accumulated thought of past generations, and penetrated with humility," it thus "serves Christ and the church."[72] The nineteenth century, according to Eliot, differed dramatically from previous centuries, and ministers must therefore be educated in a whole new way. The priest "has lost all that magical or necromantic quality which formerly inspired a multitude with awe," he wrote, "and the divine right of the minister is as dead among Protestants in our country as the divine right of kings." What is required of the minister, according to Eliot, is "candor, knowledge, wisdom, and love." He declared that the "Protestant ministry as a whole will not recover their influence with the people of this country, until the accepted dogmas of the church square with the political convictions of the people." "Whether the creeds and confessions of the Protestant sects are to be recast or not by councils or synods," Eliot announced, "no one can tell, and it is not very important to inquire." It is enough to point out that Protestant history itself "may well incline them to accord their ministers some reasonable right of private judgement."[73]

As we have seen, liberal Protestants applied evolutionary ideas to their views of religion, and thus believed that Christianity should progress as culture evolved.

Indeed, educational reformers argued that universities could be instruments of this religious progress. In his tellingly titled *The Religion of the Future* (1909), Eliot observed that there is a constant "struggle between conservatism and liberalism in existing churches," but modern studies in comparative religion and the history of religion incontrovertibly demonstrates that "religion is not a fixed thing, but a fluent thing." "It is, therefore, wholly natural and to be expected that the conceptions of religion prevalent among educated people should change from century to century." The nineteenth century witnessed immense increases in knowledge, scientific inquiry, and truth-seeking, which likewise modified religious beliefs. Because universities facilitated "the increase of knowledge and the spread of the spirit of scientific inquiry," they also had a part to play in the "changes in religious beliefs and practices."

The dramatic advances in the natural and historical sciences convinced Eliot that the new religion of the future will not be based on any external authority, nor on the personifications of the primitive forces of nature, nor on the worship of dead ancestors, teachers, or rulers; it will not be propitiatory, sacrificial, or expiatory, not gloomy, ascetic, or maledictory; it will not perpetuate Hebrew anthropomorphic conceptions of God. Rather, the religion of the future will combine elements of the "Jewish Jehovah, the Christian Universal Father, the modern physicist's omnipresent and exhaustless Energy, and the biological conception of a Vital Force." The God of this "new religion" will be called the "Infinite Spirit" in which "we live, and move, and have our being." God will be conceived as "absolutely immanent in all things." Concomitantly, the new religion of the future will reject "the entire conception of man as a fallen being, hopelessly wicked, and tending downward by nature." Future generations will rather find knowledge of God through knowledge itself. Moreover, the new religion will recognize in every great and lovely human person an individual willpower that is the essence of the personality of God. "In this simple and natural faith," Eliot asserted, "there will be no place for metaphysical complexities or magical rites, much less for obscure dogmas." For his part, Eliot welcomed the coming new Christian brotherhood.[74]

Chicago's William Rainey Harper (1856–1906), a complex person full of contradictions and enigmas, was a pioneering biblical scholar and one of the first academics in America to adopt the historical-critical method and to promote the reading of Scripture in the original languages. His humanistic training at Yale inspired him to deconstruct stories like the biblical flood, ultimately concluding that such passages could not be read as factually accurate, historical records. Whatever crisis of conscience he experienced, by the 1880s, as a professor of Semitic Studies at Yale, he hoped that the harmonization of theology with modern science would render Christianity attractive to those who were otherwise beyond its reach.[75]

In 1891, as the first president of the University of Chicago, these streams of thought converged and underwrote Harper's ideal of an inclusive, nonsectarian university working to advance what he believed was the "Kingdom of God," the perfection of human reason where society would become ever more "Christian" and God-like.[76] According to Conrad Cherry, "in his commanding messianic vision and his ambitious educational scheme Harper was a figurative embodiment of an era when modernist, ecumenical Protestantism sought to determine the values of the whole of American culture through education."[77] Indeed, Harper seems to have been motivated to start a second Reformation by means of scholarship. After all, if God's purposes were revealed through the progressive flow of history, no institution could do more to press history forward, to advance the causes of social policy, scientific advance, and good governance. As historian James Wind writes, Harper's overarching vision was that the United States, through its institutions of higher education, "could be the messianic deliverer of the world."[78]

Harper accepted the New Theology, believing that it made religion more attractive, simpler, and more acceptable than did the old orthodoxy. Like other liberal Protestants, he believed that God's self-revelation is mediated through the flow of history. In the university, the work of God's kingdom was at hand. In his *Religion and the Higher Life* (1905), Harper argued that "religion is not the enemy of art, science, philosophy, or ethics." But like other liberal Protestants, by "religion" Harper did not mean "the church, for the church is of a transitory and variable character." Rather, religion was imperishable, permanent, a condition of the mind. Harper believed the history of religion demonstrates that "there is no dogma." Christianity represented the "highest and most perfect form of religion thus far developed." But again, by "Christianity" he meant Christianity "in the broadest sense." According to Harper, it was this kind of religion that proved most acceptable to men and women of higher thought. It will be simple, reasonable, tolerant, idealistic, ethical, and consoling. Indeed, this was the religion of Jesus himself and of the greatest minds of the nineteenth century.[79]

According to Harper, the "college student passes through an evolution both intellectual and moral" and is "taught to question everything," including "matters of religion." Thus, in the progress of ideas, a university education had to discredit some religious beliefs as an inevitable part of progress. The loss of faith does not mean loss of religion. Indeed, for Harper, it depended on what kind of faith was lost. Leaders of university reform who were sympathetic to the modernist impulse in religion affirmed that many old doctrines should be superseded by modern beliefs. Advocates of modern religion rejected the concept of orthodoxy and believed that religious doctrine should be treated like scientific theories, constantly challenged and tested, revised and enlarged. Harper believed that

Christianity should be a "religion of toleration." This broadmindedness would lead, he believed, to a new religious unity.[80]

Harper believed that the scientific study of religion would free it from theological control. He combined a hope for religious renewal with optimism in the new research university ideal. Harper's insistence on a "scientific" approach to the Bible depended on his vision of the laws of human progression and development. Indeed, Harper was deeply driven by an evolutionary progressive vision of the development of society in which the tools of scientific research and study would lay bare before humanity the laws of human life and civilization. But in order for these to triumph, according to Harper, mankind must transcend the traditions and superstitions that it inherited and continues to spread. Universities were supposed to serve as the "prophets of democracy" that would take the lead in spreading the forces of enlightenment and education across the land. In short, Harper's ecclesiology replaced the Kingdom of God with the kingdom of man in the form of democracy, with the university in place of the church as its privileged torchbearer.[81] There was, in other words, an inherent elitism in Harper's vision for a biblical science. The importance of critical methods is that they are set over and against "uncritical" and traditional approaches characterized by the current state of religious education. For the ultimate rise and triumph of a scientific civilization, these "unscientific" ways of doing things would have to go.

One final example comes from David Starr Jordan (1851–1931), founding president of Stanford University and ostensibly a "prophet of freedom."[82] Raised by parents with a "Puritan conscience," but who were theologically universalists who greatly admired the work of Theodore Parker, William Ellery Channing, and their followers, Jordan early on "acquired a dislike for theological discussion, believing that it dealt mostly with unrealities negligible in the conduct of life."[83] Influenced by the writings of Henry David Thoreau, Ralph Waldo Emerson, Charles Darwin, John Fiske, and Andrew Dickson White, Jordan spoke about religion within the idiom of modernist Protestantism. He believed in the existence of a "divine being." "Man cannot worship himself," he wrote. "Righteousness is not the work of humanity, but of God in humanity, and only this can man worship."[84] But Jordan also championed efforts to reconstruct religion. He believed that the freedom of individualism in the new university was consistent with the needs of religion. He thought that "freedom of thought and action would promote morality and religion, that a deeper, fuller religious life would arise from the growth of the individual." He thus expressed confidence in the open university and its ability to reconstruct religion.[85]

Although Jordan's statements about religion used the language of Protestantism, he had little reverence for traditional Christianity. In his *The Voice of the Scholar* (1903), a work dedicated to White, Jordan argued that modern religion, the religion of the future, would evolve into a faith that "has no creed, no ceremonies necessary to its practice, no sacred legends or mysteries, and nothing of the machinery of spiritual power which characterizes great religions in other countries."[86] In his *College and the Man: An Address to American Youth* (1907), a book intended for American high school boys and girls, and published by the American Unitarian Association, Jordan explained to young readers that "you will learn from your study of Nature's laws more than the books can tell you of the grandeur, the power, the immutability of God."[87] He followed this short book with another, *The Religion of a Sensible American* (1909), also published by the American Unitarian Association, in which he tries to spell out the "religion of a wise man." The sensible American cares nothing for the "ecclesiastical calendar." Historical Christianity, moreover, "has interested itself in war and conquest, in pomp and pageantry, in dominion over men and lands, in temporal rulership as well as spiritual control." But the religion of the sensible American is "not dependent on organization," has no necessary connection with "creed or ceremony, with litany or liturgy, with priest or preacher, with symbol or miracle, with sacrament or baptism, with pious action or with pious refraining." It has "almost from the first, been entangled in a warfare of creeds." But according to Jordan, even the "religion of Jesus has no necessary connection with any Church." He concludes by calling others to "strive toward a religion which shall be not collective alone, but personal; not the religion of a time or state, but of a man; not one of creed nor of ceremony nor of emotion, not primarily a religion of the intellect, but a religion of faith and cheer, of love and action, of trust in the realities of nature and in the reality of the spirit, a faith that the universe is in hands of perfect wisdom and that in our way we may be at one with it, striving toward abounding life and helping our brother organisms as we meet them to struggle toward all good things."[88]

CONCLUSION

The ideas of these educational reformers obviously came into conflict with the beliefs of more traditional believers. As first president of Cornell University, then, White shared the position of other educational reformers at the end of the nineteenth century. As a trained historian, however, White took this more general liberal Protestant position that he shared with these reformers and applied it to the historical relationship between science and religion. That the secularization of the American research university was mostly due to Protestantism supports Jon Roberts's and James Turner's contention that secularization was an outgrowth of Protestant thought rather than a result of science triumphing over religion.[89] While few university professors or presidents wished to declare war on religion, they embraced liberal Protestant ideals that emphasized intellectual independence, open-mindedness, skepticism, and other values that made them impatient with traditional orthodox Christian faith. Liberal Protestants had "domesticated transcendence," to use William C. Placher's phrase.[90] However, by transmuting God's transcendence to immanence, liberal Protestantism contributed to its own demise. As more conservative Protestants of the century warned, the type of reconstruction of Christian faith that the New Theology represented could only mean the destruction of Christianity. Early modern attempts to bring God within the realm of comprehension produced a view of God that made historical Christian belief an enemy of science and human freedom. By equating religion—indeed divinity—with inner feeling or morality, God became less impressive and mysterious and more like humanity itself.

Within the context of the university, a widespread movement emerged to create a science of religion. Freed from dogma and studied according to the canons of scientific inquiry, the higher truths of religion would become evident—or so it was believed. Instead, this entire enterprise proved self-defeating. Biblical criticism, along with sociological and psychological studies of religious phenomena, yielded an attenuation of religious authority. Religiously progressive Protestants explained conflict by contrasting the ideal of progressive scientific inquiry with the authoritative methods of theology. They viewed the conflict not as between scientific truth and religious truth but between the openness of scientific inquiry and the dogmatism of theology. They presented theological change as part of the natural development of religion and associated it with the progressive revelation of the divine through the evolution of human nature. But the science of religion failed to fulfill its promise, and in the process doomed religious belief to marginal status. As Julie Reuben puts it, "the message that emerged from scientific studies of religion was that religion had no intellectual content."[91] Although the study of religion flourished in the American university at the end of the century, these studies increasingly emphasized the poetic rather than the factual "truth" of Christian belief. Rather than unifying religion and science, liberal Protestants relegated them to different areas of life: science to the intellectual and religion to the inspirational.

The Protestant Reformation, which began as a movement to renew and purify Christianity, but which quickly turned into polemics between Protestants and Catholics, and subsequently between contending Protestant sects, had a tacit and perhaps even explicit role in creating the perception that science and religion were in conflict. This progression lends credence to a notion first articulated by German sociologist Max Weber, and further developed by James Turner, Michael Buckley, and more recently by Charles Taylor and Brad Gregory: namely, that modern unbelief betrays roots to the Reformation.[92] What the last two chapters have shown, however, is that it was a particular kind of Protestantism that both subjected Christianity to rational criticism and subordinated it to experiential religion, which eventually gave birth to a conflict narrative that, in turn, enabled the rise of secularism. There was indeed a deep kinship between liberal Protestantism and secularism, the boundaries of which were remarkably porous. As we shall see more clearly in the following chapter, such diffusive Christianity did eventually succumb to alternatives to Christianity. Near the end of the nineteenth century, many religious thinkers continued to embrace the historiography of Protestantism and even its religious language but rejected entirely its orthodox Christian teaching. This is extremely important because one of these thinkers became the chief publisher, promoter, and popularizer of the narratives of Draper and White.

YOUMANS AND THE "PEACEMAKERS"

THE motivations and connections between John W. Draper, Andrew D. White, and other nineteenth-century Protestant commentators on the relation of science and religion have been often blurred, overlooked, or completely misunderstood by many historians of science. As we have seen, Draper was not simply anti-Catholic, nor was White simply responding to sectarian criticism of his beloved university. Both writers shared a liberal Protestant heritage that emphasized a historiography of reasonable religion against orthodox theology. The real distinction between Draper and White was one between different emphases within this liberal Protestant tradition. While Draper embraced a "religion of the head," a rational Protestantism taken mostly from English latitudinarian thought, White embraced a "religion of the heart," found among German and American Protestant romantic, idealistic, and transcendental writers.

The reasons historians of science have missed or confused these distinctions are presumably complex. But in this chapter, I would like to suggest that it was largely a consequence of how these ideas were diffused by popular lecturer and "science editor" of D. Appleton & Co., Edward Livingston Youmans (1821–1887). Youmans had indeed developed a long patronage relationship with Draper and

White, publishing, defending, and even clarifying certain points of their work in the pages of his widely read magazine, *Popular Science Monthly*. He also published, through Appleton, their major works, Draper's *History of the Conflict between Religion and Science* and White's *History of the Warfare of Science with Theology in Christendom*. Youmans thus played a crucial role in disseminating and popularizing their historical narratives.

While Youmans agreed with Draper and White in their descriptions of the historical record, he did not share their hope that Protestantism would bring about a final reconciliation between science and religion. For Youmans, Protestantism was but one step in the evolutionary progress of religion. As we have seen, evolutionary thinking had become almost irresistible at the end of the nineteenth century, and it was applied to almost every conceivable topic, including religion. Youmans was thus characteristic of a number of English and American thinkers in the late nineteenth century who, despite rejecting the creed of orthodox Christianity, continued to retain some sense of religious belief.

This distinction may be made clearer if we observe that besides supporting the writings of Draper and White, Youmans was first and foremost an advocate of scientific naturalism, especially its leading representatives Thomas H. Huxley, John Tyndall, and Herbert Spencer. Scientific naturalism was no doubt the "English version of the cult of science in vogue throughout Europe during the second half of the nineteenth century."[1] But as we have seen, while many in the last decades of the nineteenth century rejected doctrines of historical Christianity, most did not abandon belief in God. Rather, what occurred was a loosening or redefinition of Christianity with the concurrent appearance of alternative forms of religious belief. As historian of religion John Wolffe writes, "running through all varieties of alternative belief, spiritual experimentation, and agonized doubting was an underlying religiosity. Men and women might reject the teaching of a specific Church or even Christianity as a whole, but they remained desperately concerned to find some kind of religion or, at least, 'ultimate concern' to give meaning and coherence to their own lives and to the society and culture in which they lived."[2]

The Victorian period was indeed an age of doubt, but it was also a period of *disbelieving religiously*. As Lance St. John Butler put it, it was a time of "faith expressing itself doubtingly and doubt expressing itself in the language of religion."[3] Rather than inducting the secularization of society, the late Victorian period should be viewed as a time of religious reconstruction or transformation. For similar reasons, then, the scientific naturalists must not be seen as antireligious. A number of scholars have been attentive to the continuing power of religion in late Victorian scientific naturalism.[4] Over forty years ago, for example, Robert M. Young argued that scientific naturalism was itself a form of natural theology. Huxley, Tyndall, Spencer, and others, according to Young, replaced the old theodicy with "a secular one based on biological conceptions and the fundamental assumption of the uniformity of nature."[5] James R. Moore has also argued that the so-called Victorian crisis of faith was part of a "wider struggle to negotiate new doctrines, new beliefs, new vehicles of consent that would be seen to maintain continuity with and fulfil the best aspirations of older creeds." This was the new "naturalistic theodicy" of the "New Reformation."[6]

Perhaps more than any other historian, Bernard Lightman has challenged conventional interpretations of the scientific naturalists. He has argued, for instance, that "there were many vestiges of traditional religious thought embedded in Victorian agnosticism," and has even suggested that "agnosticism originated in a religious context." Indeed, given the overarching theme throughout our study, a closer reading of the leading scientific naturalists reveals that many were indebted to a Dissenting or Nonconformist tradition, which "sought to set forth a serious new, non-clerical religious synthesis." Though they may have rejected Christian orthodoxy, they nonetheless aspired to a "religion pure and undefiled," stripped of the dogma they considered accretions and perversions of Christ's original message. Lightman goes so far as to argue that the agnostics should be seen as "new natural theologians," who had a "sense of the divine in nature" and attempted to "treat science as a religion since it was the study of divine natural law." In short, the scientific naturalists pursued "genuine religious goals and not merely the substitution of something secular for something religious."[7]

How this Victorian coterie attempted to achieve these goals was by reconceptualizing "religion" in terms of a solely inner spirituality. By privatizing religion in such a way, and thus separating religion from theology, they believed themselves to be harmonizers or peacemakers, claiming that conflict between "science and religion" was simply impossible. That is, science and religion, when properly understood, have never been in conflict. Rather, the conflict had always been between *science* and *theology*, or theological dogmatism, not *religion*. Obviously, then, while they rejected orthodox Christianity, there was clearly much continuity between scientific naturalism and the Protestant tradition. In fact, as we shall see, the scientific naturalists befriended and praised liberal theologians who supported the advance of science. In distinguishing theology from religion, agnostics like Huxley, Tyndall, and Spencer joined a significant portion of other liberal Protestant thinkers who likewise believed that faith is an expression of personal or moral sentiment. As we have already observed, radical Dissenters and Nonconformists, dissident intellectuals, and even liberal Anglicans also argued that "true religion" had been disguised and disfigured by the clergy over the centuries. And like these other groups, the scientific naturalists believed that a "New Reformation" would bring mankind one step closer to fulfilling (indeed, completing) the Reformation of the sixteenth century. Traces of these Protestant elements are readily found in their writings.

But while Draper and White could still claim to be Christian in a way that convinced many liberal Protestants of their good intentions, the scientific naturalists, including Youmans, cut themselves from the Christian heritage that informed their upbringing. But again, that did not make them atheists. Indeed, during this period a number of alternatives to Christianity appeared, many believing that orthodoxy was usurped by newer, higher, and more advanced forms of religion. This was the context when, in 1875, members of the Evangelical Alliance warned that "we live in an age which prides itself on freedom of thought and emancipation from the control of authority. In every portion of Christendom men are more disposed than ever to run into extremes of opinion and practice. While, on the one hand, fundamental truths are increasingly neglected or denied, vain attempts, on the other, are made in many quarters to meet this infidelity by the revival of superstition."[8]

According to many Protestants, perhaps the most infamous "superstitious" alternative to Christianity at the time was the so-called "Religion of Humanity" by the Comtean positivists. Ironically but perhaps unsurprisingly, Huxley was also particularly offended by this spectacle of "effete and idolatrous sacerdotalism," as he himself put it. He facetiously described positivism as an "incongruous mixture of bad science with eviscerated papistry."[9] He had also famously quipped that Comte's Religion of Humanity was nothing but "Catholicism minus Christianity."[10] In short, Huxley attacked positivism by exposing its displaced religious meanings.

As a devoted advocate of scientific naturalism in America, Youmans largely agreed with Huxley's assessment. But if positivism was merely displaced Catholicism, the views promoted by Huxley, Tyndall, Spencer, and other scientific naturalists, including Youmans, might also be succinctly described as "Protestantism minus Christianity." Youmans and the scientific naturalists had taken the mediating theology of liberal Protestantism to its logical conclusions, adopting and adapting its polemical and rhetorical strategies, including its historiography, yet rejecting entirely its most fundamental religious tenets. As we shall see, Youmans appropriated the historical narratives of Draper and White and pressed them into service for his own vision of what the next evolutionary stage of religious belief should be. But in promoting his new "religion of the future," Youmans ultimately distorted, wittingly or unwittingly, the subtle distinctions between Draper, White, and their conceptions of the reconciliation between science and religion. Thus, while this chapter serves as an extended digression into how the narratives of Draper and White were disseminated and popularized, it more significantly reveals how Youmans treated his magazine as an open forum for impassioned opinions on science and religion. The following pages explore how Youmans's effort to diffuse scientific knowledge to a wider audience was inextricably connected to certain metaphysical, epistemological and, in general, nonscientific assumptions. In other words, the articles that appeared in his magazine uncovers a blend of scientific ideas and facts to support an emerging worldview that constituted new ideas about God, nature, and man. In short, Youmans was replacing one "religion" with another. More importantly, by examining a small portion of the articles published in the *Popular Science Monthly*, this chapter demonstrates how quickly and easily the historical narratives of Draper and White could be manipulated and appropriated to ends they had never intended.

THE LIFE AND TIMES OF YOUMANS

Edward Livingston Youmans was born on a farm in Coeymans, New York, the eldest son of Vincent, a farmer and blacksmith, and Catherine Scofield, a schoolteacher.[11] Raised in a devout Congregational home, his parents expected him to join the ministry. Those hopes were thwarted when Youmans took up an interest in popularizing the sciences. He assured his family, however, that this interest in educating the masses was just as important as the ministry, telling them that he too would "save souls."[12] According to his friend and biographer, American historian and popularizer of evolutionary science John Fiske (1842–1901), Youmans believed his "ravening, insatiable thirst for knowledge" was a gift from God.[13]

By the late 1830s, however, Youmans began to suffer from ophthalmia, which had almost left him blind. During this time his sister, Eliza Youmans, became his constant companion, and together they formed a lifelong partnership. In 1840 he went to New York City seeking treatment from several oculists. When his eyesight improved, he began a rigorous course of self-education, studying chemistry, mathematics, and history. While in the city he even attended Draper's lectures on chemistry at New York University. He also began planning to write his own history of science, a work on the "history of progress in discovery and invention," as he put it. In this work, he would eschew the "deeds of popes, kings, and emperors" and focus rather on "observers, explorers, experimenters, and philosophers."[14] With his sister, he scoured the libraries and bookstores of the city for sources. In 1847, his sister took him to D. Appleton & Co.'s bookstore on Broadway, New York's largest bookstore at the time. William H. Appleton (1814–1899), the eldest son and successor of American publisher Daniel Appleton (1785–1849), was in the store at the same time and noticed the two siblings browsing. He introduced himself, and after discussing their literary goals, offered to loan them any book they might need. The meeting became a turning point in the lives of all involved. Appleton befriended Youmans and eventually employed

Figure 5.1. Engraved portrait with authentic autograph of Edward L. Youmans, from the *Popular Science Monthly*, vol. 30 (1887).

him as the firm's science editor, leading to one of the most remarkable publishing ventures in American history.[15]

While Youmans's "history of progress" never bore fruition, he did publish several popular science books, all through Appleton. These included his *A Class-Book of Chemistry* (1851), *Alcohol and the Constitution of Man* (1854), *Chemical Atlas; or, The Chemistry of familiar objects* (1856), and *The Handbook of Household Science* (1857). In these works, Youmans narrated the history of science as moving inexorably from the metaphysical to the physical, argued that science should serve humanity in practical ways, and, most importantly, contended that science should work in harmony with the moral forces of religion. In his *Class-Book of Chemistry*, for instance, he explained the value of studying chemistry in religious terms. He argued that chemistry "throws light upon the sublime plan by which the Creator manages the world." The greatest benefit to studying chemistry, he opined, is that it reveals the "infinite wisdom of the Creator."[16] By using language familiar to natural theology, Youmans ensured these works were enthusiastically received by Protestant readers.

In the 1860s, Youmans was given full editorial authority as "literary adviser on all scientific publications" for D. Appleton & Co. It was through his influence that the firm became the first to introduce many of the European scientists to the American public.[17] An early example of this new role was his *Correlation and Conservation of Forces* (1865), which was a collection of papers by William Robert Grove, Hermann von Helmholtz, Julius von Mayer, Michael Faraday, Justus von Liebig, and William Benjamin Carpenter. Youmans provided biographical sketches of each man, along with an extended introduction in which he defended the recent "materializing tendencies of modern science." He assured his readers that such a tendency toward materialism actually demonstrates the "truth of the spiritual world." Significantly, the book was dedicated to Draper, whose own work, Youmans explained, had "widen[ed] the range of thought by unfolding a *broader* philosophy."[18]

Another early editorial effort that requires our attention was *The Culture Demanded by Modern Life* (1867). In this book, Youmans brought together a collection of essays, lectures, and extracts from various writers and scientists in England, Europe, and America. In his introduction he wrote that "man is individually improvable, and therefore collectively progressive." He contended that modern life "refuses to be stationary," and that man had "outgrown the arbitrary institutions of the remoter past." Thus when "institutions refuse to change," conflict or "war is the consequence." He pointed to the Protestant Reformation as an example of the progressive character of man, demonstrating that religious conceptions have developed "beyond the ecclesiastical organizations to which at first they gave rise." But as we have noted, for Youmans Protestantism was simply another step in the continual evolution of man. Religion, like the rest of humanity, is never at rest. "Ideas of government, religion, and society," Youmans contended, "have been profoundly modified, and . . . new revelations of man's powers and possibilities, and nobler expectations of his future, have arisen." He therefore ominously urged that both science and religion be granted freedom to progress, for the future "welfare of society" was at stake.[19]

Appleton supported Youmans in his publishing efforts, funding several trips for him to visit England and Europe to solicit authors to publish with the firm. It was at this time that he first became personally acquainted with the scientific naturalists.[20] In time, he secured the publication of the entire corpus of Spencer, Huxley, and Tyndall, thus ensuring that they would remain permanently fixed in the minds of North American readers.[21]

THE NEXT GREAT TASK OF CIVILIZATION

According to Fiske, Youmans "looked forward to a time when he might enlarge the scope of his labours by publishing a magazine which should deal with scientific subjects in such a way as to educate the people."[22] The success of such periodicals as *Harper's*, *Atlantic Monthly*, *North American Review*, and others had encouraged American book publishers to join the new market of periodical publishing. In 1868, Youmans had convinced Appleton to begin publishing a new magazine. The following year, Youmans was put in charge of editing *Appleton's Journal of Popular Literature, Science, and Art*.[23] In its first issue, Youmans argued that science is not simply the domain of the specialist. Rather, it concerned everyone: "It is something to be used in reading, conversation, and business, at home and in the street, week-days and Sundays, in school, at the lecture, and the political gathering."[24] In subsequent issues, he diligently reported on contemporary scientific topics, especially from across the Atlantic. He even contributed his own articles on various subjects, including short reviews and excerpts on the lives and work of many leading scientists, philosophers, and historians. But much to Youmans's dismay, Appleton insisted on giving the magazine a "literary and artistic character." Unhappy with this arrangement, by 1870, after only eighteen months at its helm, Youmans resigned his editorship.

While *Appleton's Journal* failed to meet his expectations, Appleton continued to support Youmans's publishing ventures. Thus in 1872, he founded the *Popular Science Monthly*, a magazine that he "conducted" until his death, when his brother William Jay Youmans (1838–1901)—who, incidentally, studied under Huxley in 1866—took it over. In its pages, Youmans portrayed himself as "an interpreter of science for the people," a purveyor of the great and good in the world of professional science. In the "Editor's Table" of its opening number, for instance, he stated the purpose of the magazine in unequivocal terms: as "the next great task of civilization." Thus the *Popular Science Monthly* was designed to bridge the gap between scientists and the general public.[25] To this end, Youmans worked diligently to obtain joint publication rights with other periodicals, professional and popular, thus reprinting articles or portions of articles that appeared simultaneously in *The Nineteenth Century, Fortnightly Review, Contemporary Review, Nature, Macmillan's Magazine, Gentlemen's Magazine, Revue des Deux Mondes, Revue Scientifique, The Lancet,* and the *Journal of Mental Science,* to name a few.

More importantly, Youmans made his magazine a repository of ideas of the scientific naturalists. Its chief object, as William Leverette noted, was the "naturalistic point of view."[26] Indeed, those who featured most prominently in its pages were, unsurprisingly, such men as Huxley, Tyndall, and especially Spencer. With his editorial voice, Youmans defended the scientific naturalists against critics, arguing that their positions actually displayed a genuine attempt at harmonizing science and religion.

THE GREAT CONFLICT

While the scientific naturalists appeared most often in the pages of the magazine, Youmans also consistently defended and promoted the work of Draper and White. Youmans, after all, dedicated his *Correlation and Conservation of Forces* to his teacher Draper, and specifically said that his *Intellectual Development* elevated and enlarged the scope of scientific inquiry. Draper's name also appeared in his *Culture Demanded by Modern Life,* alongside Herschel, Whewell, Lyell, Hooker, Faraday, Paget, Spencer, Tyndall, Huxley, and others. Furthermore, in a remarkable obituary notice on Darwin in 1882, Youmans included news of Draper's death in the very same article. While Darwin was probably the most scientific man of the world, he wrote, Draper was the "most eminent man of science in America." Both were "the most distinguished representatives of the same school of progressive scientific thought," and both will forever be "associated with that great revelation of ideas for which all modern science has prepared."[27] Draper had become something of a scientific hero to Youmans, who imbibed his views on physiology and intellectual history and his overall assessment of the relationship between religion and science.

Draper's first appearance in the *Popular Science Monthly* came with Youmans's praise. In discussing the "discoverer of oxygen," he referred to Draper's lecture on Priestley, which was later published in the magazine as well.[28] He also gave Draper a glowing sketch of his life and achievements and claimed that he belonged to a group of scientists who aided the rapid growth of more liberal and elevated ideas in politics and religion.[29] Importantly, he also advertised Draper's "work of great

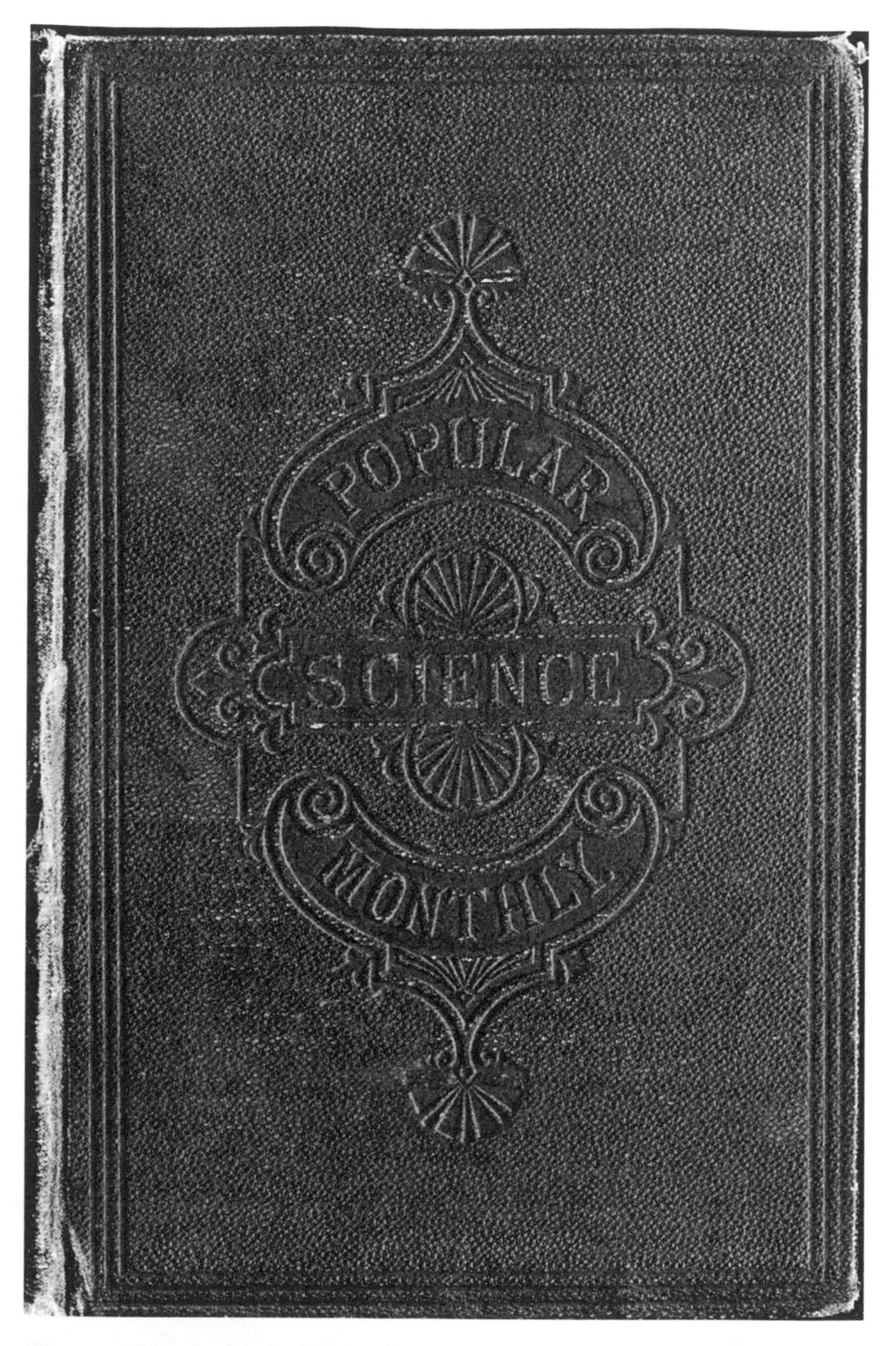

Figure 5.2. A bound volume of the *Popular Science Monthly*, which Edward L Youmans "conducted" from 1872 until his death in 1887.

Figure 5.3. British and American volumes of the International Scientific Series. Note the different national symbols, the Royal Banner of England and the Great Seal of the United States. Photograph courtesy of Howard J. Swatland, Professor Emeritus at the University of Guelph, from private collection.

importance," his *History of the Conflict*.[30] "It is doubtful," he contended, "if there is another man living besides Dr. Draper who has had the peculiar preparation necessary for executing so difficult a task." He had even published Draper's preface to the book, thus disseminating his ideas to those who had neither the time nor the funds to acquire the larger volume.[31]

Launched simultaneously with the *Popular Science Monthly*, Youmans also founded the "International Scientific Series" (ISS), the purpose and scope of which closely matched his magazine venture. According to Bernard Lightman, the ISS was "one of the most famous of all Victorian attempts to codify and popularize scientific knowledge in a systematic fashion to a wide reading public across national boundaries."[32] In 1871 Youmans reported to Spencer that "the progress of liberal thought is remarkable."[33] He endeavored to capitalize on this progress. Youmans envisioned "a series of new works . . . covering the entire field of modern science, to be simultaneously issued on both sides of the Atlantic."[34] To this end, Youmans spent a year traveling throughout Europe to secure authors and publishers for the series. He seemed to believe this was his religious duty, as he told his mother that "I have tried to do well—tried hard to do as well as I knew how in my own way, and I commit myself and my future in hope and trust to that

Divine Power who is the Author of all things, whose ways and works I have tried faithfully to understand—the duty of every rational being."[35]

When he had secured enough authors, Youmans quickly published a full-page "Prospectus of the International Scientific Series" in the first volume of his *Popular Science Monthly*. Its stated purpose was to make available a "series of popular monographs" that embodied "the results of recent inquiry in the most interesting departments of advancing science."[36] While abroad Youmans had formed a "London Advisory Committee" for the ISS, made up, naturally, of Huxley, Tyndall, and Spencer.[37] As a number of scholars have pointed out, the series was a resounding success on both sides of the Atlantic. Altogether, more than twenty volumes were published during Youmans's editorship, such as Tyndall's *Forms of Water in Clouds and Rivers, Ice and Glaciers* (1872), Walter Bagehot's *Physics and Politics* (1872), Spencer's *The Study of Sociology* (1873), Alexander Bain's *Mind and Body* (1873), Henry Maudsley's *Responsibility in Mental Disease* (1874), Oscar Schmidt's *The Doctrine of Descent and Darwinism* (1875), Norman Lockyer's *Studies in Spectrum Analysis* (1878), J. L. A. Quatrefages de Breau's *The Human Species* (1879), and Huxley's *The Crayfish* (1880). One of the best-selling volumes in the series was of course Draper's *History of the Conflict* (1875), selling nearly 20,000 copies. It went through fifty printings in the United States alone, twenty-four in England, and was even translated into at least ten languages: French, German, Italian, Dutch, Spanish, Polish, Japanese, Russian, Turkish, and Portuguese. Indeed, as Roy MacLeod put it, Draper's book gave the series "the greatest boost in sales and notoriety."[38]

Despite its great success, the reception of Draper's book, as we shall see in more detail later, was greatly mixed. Here it is important to note that Youmans was quick to defend Draper's position in the pages of the *Popular Science Monthly*. That there is "no necessary hostility between religion and science," he asserted, "is unquestionably true." But while he had no doubt that religion and science will be ultimately harmonized, he believed that the world had not reached that "blessed consummation." What Draper's book offered, according to Youmans, was an explication of "the causes, the course, and the consequences of this great struggle." Crucially, Youmans argued that Draper had demonstrated that the history of science was a long conflict with "theological authority," between the "agencies of intolerance and of liberalization." He claimed that Draper's book was an answer to increasing tensions between Catholics and Protestants in the late nineteenth century, particularly with the so-called Ultramontane party of English Catholics. His work was thus a textbook by which the "experience of the past is made the basis for an intelligent judgment of the present." It was like a "novel with a well-sustained plot," and therefore must be read carefully before pilloried.[39]

As more critical reviews began appearing for Draper's book, Youmans became more aggressive in his defense. He wrote that while a number of writers had been "enraged" by the book, this was due to misunderstanding and "folly of passion." One common criticism Draper faced was that he had written a "fiction." Many claimed that religion and science had always coexisted in the world. Youmans thought this was a ridiculous claim. He pointed to the "vast body of theological literature, going back for centuries, that is devoted to the work of *reconciling* religion and science." He asserted that entire libraries had been devoted to reconciling Christian doctrine with, for example, astronomy, geology, and biology, among other sciences. "If there has been no conflict," Youmans protested, "there could be no reconciliation, for the attempt to reconcile that which is already harmonious is absurd." That so many have argued that "truth can never be in conflict with itself," and that "religious truth and scientific truth must harmonize," concedes the very fact that there is indeed some conflict.[40]

Interestingly, according to Youmans, this conflict was both "natural and inevitable." The central question, then, is what causes conflict? Here Youmans extended Draper's original analysis, arguing that conflict between religion and science is the result of a conflict within the theological world, and how particular traditions defined "religion." While many of Draper's critics accused him of not defining "religion," theological groups likewise "assume it to be what its members believe, and what those with different beliefs do not possess." Thus the "anxiety in regard to a definition of religion," Youmans wrote, "has not originated with us." He denied that any one tradition possessed a monopoly on religious truth. Youmans argued that Draper had to take on "broad views" in order to demonstrate more clearly those "responsible for theological resistance to the reception of scientific ideas." Youmans, of course, hoped that this broader definition of religion would "avoid all antagonism with science." But historically speaking, "such an interpretation of religion had not been reached, and . . . is very far from being arrived at the present time." Opinions about the origins and nature of life, he consoled himself, were thankfully quite different between "the Unitarian and the Trinitarian—the Universalist and the Perditionist."[41]

As an "interpreter of science for the people," Youmans often promoted educational reform in his magazine, and thus published numerous articles and lectures by White.[42] More importantly, Youmans also published an extended version of White's "Battle-fields of Science" speech in his magazine, which appeared as a series of essays under the new title "The Warfare of Science" in 1876.[43] Youmans published these essays as a single volume later that same year. The constant refrain of White's thesis, as we have seen, was that science offered religion "a far more ennobling conception of the world, and a far truer conception of Him who made and who sustains it." The triumph of science "has been a blessing to religion—ennobling its conceptions and bettering its methods."[44]

As he did with Draper, Youmans defended White's work throughout his magazine. In his Editor's Table, for instance, he commended the attention of his readers to the "import of [White's] clear-cut thesis, and the vigor, learning, and logical force, with which it is sustained." What was "especially valuable" about White's work, Youmans explained, was its "copious notes and references by which it is enriched and fortified." He added that White's work appeared at an opportune time, for "when the hot temper of controversy leads to much random and reckless statement, it is desirable to know, very clearly, what can be proved, and where the proof can be found." Youmans remarked that White has authoritatively delineated "the battle that Science had had to fight from the beginning, and without remission, with ignorance, prejudice, and intolerance, inspired and directed by ecclesiastical influence." Youmans goes on to declare that "the history of Science has been throughout a struggle with the theologians, and that the Bible has been used by devout believers in its infallible inspiration to crush out the results of scientific inquiry." But in the end, he concluded, nothing is gained from "ignoring historic truth, and bigotry and superstition still offer too vigorous a resistance to the advance of rational inquiry to make it desirable that we should quite forget the painful lessons of the past."[45]

Running a decade from 1885 to 1895, Youmans published serially White's "New Chapters in the Warfare of Science," which was again republished by Appleton in a massive two-volume *History of the Warfare*. Youmans died in 1887, two years after the series began, and his brother William Jay concluded the project. In publishing the last "chapters in the warfare of science," William Jay announced that White had shown "how theologians have been forced step by step to yield the domination which they asserted in astronomy, meteorology, medicine, and other fields outside their own province." The greatest consequence of this history, he wrote, "has been that dogmatism and mysticism in preaching and teaching have found fewer and fewer listeners, while the most intellectual ecclesiastics, feeling the same influence, have shrunk from the dogmatic and mystical extravagances of their predecessors."[46]

"SUBSOILING THE PEOPLE" WITH SCIENTIFIC NATURALISM

As we have already noted, compared to the scientific naturalists, Draper and White appeared seldom in the pages of the monthly. Youmans's friendship and partnership with Huxley began with his first visit to England in 1862.[47] He found himself largely in agreement with the scientific naturalists and went to great lengths to imbue the American public with their ideas. Youmans had even

Figure 5.4. Portraits of Thomas H. Huxley (1825–95), John Tyndall (1820–93), and Herbert Spencer (1820–1903).

attended the notorious X Club, where he discussed with Huxley and other members the possibility of "some arrangement for the publication of English scientific works in America."[48] His brother, William Jay, would thereafter become a faithful pupil of Huxley. At the instigation of Youmans, Appleton published Huxley's *Evidence as to Man's Place in Nature* (1863), which had become quite popular in America. Even more popular was his *Lay Sermons, Addresses, and Reviews,* also published by Appleton in 1870. In his first magazine venture, Youmans published an excerpt from Huxley's speech on "Scientific Education" at the Liverpool Philomathic Society, where he argued for "scientific Sunday-schools." In this speech Huxley memorably quipped that "if any of the ecclesiastical persons to whom I have referred object that they find it derogatory to the honor of the God whom they worship to awaken the minds of the young to the infinite wonder and majesty of the works which they proclaim His, and to teach them those laws which must needs be His laws, and therefore, of all things needful for man to know, I can only recommend them to be let blood and put on low diet."[49] Huxley will later find an eager audience in America, who appreciated his directness, honesty, and unflinching spirit.

Youmans went on to publish over fifty articles, lectures, addresses, and speeches by Huxley in the pages of *Popular Science Monthly,* thus making him a household name in America. That Huxley became "very much wanted in the United States," and that there was "great anxiety to see him and hear him speak," is clear when, in the summer of 1876, Huxley came to America for a lecture tour.[50] Youmans announced in the pages of his magazine that Huxley was coming to New York to give three lectures on "The Direct Evidence of Evolution," and even offered readers tickets to the event.[51] In the wake of this lecture, however, Youmans found himself having to vigorously defend Huxley against critics, arguing that he had been "misrepresented, badgered, and vilified, with a recklessness that would have aroused vigorous resistance and sharp counterstrokes in any man of spirit."[52] The key to Youmans's defense was to portray men like Huxley as a friend of progressive religion. He thus published an article in which Huxley praised Priestley's fearless leadership in both science and religion. As Draper had argued earlier, Huxley claimed that those "opinions which have brought most odium upon him [Priestley] have been openly promulgated, without challenge, by persons occupying the highest positions in the state Church." In other words, Priestley's views, according to Huxley, had become "commonplaces of modern Liberalism."[53]

Youmans also published Huxley's famous controversy with English Prime Minster William E. Gladstone (1809–1898), in which Huxley proclaimed that the conflict between science and religion was "purely factitious," a fabrication of "short-sighted religious people" who mistook theology for religion and "equally short-sighted" scientists who forget that science is limited by what is intellectually comprehensible. Huxley believed a turning point in the moral and intellectual life of civilization came with the Hebrew prophets, when they called believers "to do justly, and to love mercy, and to walk humbly with thy God." Anything subtracted from or added to this percept, according to Huxley, mutilates the "perfect ideal of religion." Ultimately, he concluded, "the antagonism of science is not to religion, but to heathen survivals and the bad philosophy under which religion herself is often well-nigh crushed." Gladstone and others like him have, then, wittingly or unwittingly, hindered the great religious work of science.[54]

More importantly, Youmans also published articles by Huxley where he more explicitly supported the progress of theology. For example, he published Huxley's review of a series of sermons by Broad Churchmen Harvey Goodwin (1818–1891), William W. How (1823–1897), and James Moorhouse (1826–1915), preached in 1887 at Manchester Cathedral during the fifty-seventh meeting of the British Association. Huxley praised these sermons as "excellent discourses," which signaled "a new departure in the course adopted by theology toward science," and thus indicated the "possibility of bringing about an honorable *modus vivendi* between the two."[55]

Near the end of his life, Huxley published a number of articles in the *Popular Science Monthly* that displayed less concern with science and more with the relationship between science and religion, particularly how advancements in the sciences have successfully led to progress in theology. In an 1890 article, for example, he argued that science, as operating in the fields of history, philology, and archaeology, had rendered old theological dogmatism untenable.[56] He perceptively argued that it wasn't simply the physical sciences that have come into conflict with theological dogmatism, but historical criticism, and even the new liberal views of religion. In fact, the attempt to reconcile science with "outworn creeds of ecclesiasticism" was the main instigator of conflict between science and religion.[57] In one of his last essays published in the magazine before his death, Huxley praised the "decline of bibliolatry."[58]

It is well known that Huxley's religious views were nurtured in the company of religious Nonconformity. As he grew to be a man, he was remarkably keen to align his cause with the Protestant Reformation.[59] Although reared, as he himself put it, "in the strictest school of evangelical orthodoxy," he later came to reject orthodox Christianity.[60] Nevertheless, he retained much of its evangelical fervor. Indeed, Huxley admitted to a "profound religious tendency capable of fanaticism, but tempered by no less profound theological scepticism."[61] Moreover, when he coined the term "agnostic" in 1869 to describe his own views, Huxley in fact credited his skeptical epistemology to the commonsense philosophy of Protestant philosophers and theologians William Hamilton (1788–1856) and Henry Longueville Mansel (1820–1871). He explained that the term came to him as "suggestively antithetic to the 'gnostic' of church history, who professed to know so much about the very things of which I was ignorant." Later he even contended that his agnosticism was the outcome of following the principles of the Protestant Reformation. "My position is really no more than that of an expositor," he wrote, "and my justification for undertaking it is simply that conviction of the supremacy of private judgment (indeed, of the impossibility of escaping it) which is the foundation of the Protestant Reformation."[62]

Reared in an environment where anti-Catholic sentiment infused nearly every aspect of Victorian life, it is no surprise that Huxley was also vehemently anti-Catholic. He described the Roman Catholic Church as "that damnable perverter of mankind," and believed strongly that one of the greatest merits of the theory of evolution is that it "occupies a position of complete and irreconcilable antagonism to that vigorous and consistent enemy of the highest intellectual, moral, and social life of mankind—the Catholic Church."[63] Moreover, when he visited the Catacombs in Rome in 1885, he wrote to his eldest son that the primitive church was a "simple maiden . . . vastly more attractive than the bedizened old harridan of the modern Papacy, so smothered under the old clothes of Paganism which she has been appropriating for the last fifteen centuries that Jesus of Nazareth would not know her if he met her."[64] He went on to conflate the Roman papacy with paganism and remarkably sided with the Protestant iconoclasts when he said that "the best thing, from an aesthetic point of view, that could be done with Rome would be to destroy everything." Reflecting on the differences between modern Christianity and the primitive church, Huxley stated explicitly that "the Church founded by Jesus has not made its way; has not permeated the world—but did become extinct in the country of its birth."[65]

Like many of the liberal Anglicans, Huxley presented himself both privately and publicly as a religious rebel who continued the work of Protestant reformers. The intellectual progress of his day convinced Huxley that a "New Reformation" was on the horizon. In 1859, Huxley announced in a lecture to working-class men that science was a divinely sanctioned activity that confirmed the order in both the physical and mental world. This supposed cofounder of the "conflict thesis" maintained that "of all the miserable superstitions which have ever tended to vex and enslave mankind, this notion of the antagonism of science and religion is the most mischievous." He affirmed that "true science and true religion are twin-sisters, and the separation of either from the other is sure to provide the death of both." All the "reformations" in the history of religion, Huxley asserted, have been "due essentially to the growth of the scientific spirit, to the ever-increasing confidence of the intellect in itself—and its incessantly repeated refusals to bow down blindly to what it had discovered to be mere idols." According to Huxley, one must distinguish between the "eternal truths of religion" and the "temporary and often disfiguring investiture which has grown round them."[66] That same year Huxley told his friend Frederick D. Dyster (1810–1893) that his popular science lectures were "meant as a protest against Theology and Parsondom in general—both of which are in my mind the natural and irreconcilable enemies of Science." "Few see it," he added, "but I believe we are on the eve of a new Reformation."[67]

Huxley candidly confessed his religious opinions to Broad Churchman Charles Kingsley. He claimed that science embodied the "Christian conception of entire surrender to the will of God." The man of science sits down before "fact as a little child," prepared "to give up every preconceived notion," and follows "humbly wherever and to whatever abysses nature leads." He explained that his redemption came not by some hope of immortality or future reward, but in reading Scottish author Thomas Carlyle (1795–1881), who taught him to "know that a deep sense of religion was compatible with the entire absence of theology." Importantly, he also claimed that he reached his position like the heroes of the Reformation, after exercising his private judgment. "I can only say with Luther," he wrote, "'*Gott helfe mir, Ich kann nichts anders.*'" He concluded that Kingsley and his friends should recognize him and his allies as the "new school of prophets," and that the Church of England could be saved only by the efforts of liberal thinkers like Kingsley himself.[68]

The mention of Carlyle is important. Earlier in the century Carlyle was one of the leading figures who had introduced German romanticism and idealism to the British public. His reading of Johann Wolfgang von Goethe had encouraged him to separate religion and spirituality from contemporary institutional and dogmatic Christianity. For Carlyle, the true realm of religion had become a sense of wonder and humility. Moreover, Carlyle believed the reigning "sham" priesthood should be replaced with the more industrious, honest, courageous, and effective leadership of professionals in the arts and sciences. As Frank Turner wrote, "Carlyle believed the problems of Britain's social and physical well-being should be addressed by leaders whose authority and legitimacy stemmed from talent, veracity, and knowledge of facts."[69] Carlyle's own biographer aptly called him a "Calvinist without the theology."[70]

Huxley and many of the other scientific naturalists saw themselves as this new intellectual clerisy, often portraying the man of science as prophet, priest, and savior of modern society. Like Carlyle, Huxley believed that "religion has her unmistakable throne in those deeps of man's nature which lie around and below the intellect, but not in it." He argued that religion, rightly conceived as belonging to the realm of feeling, could never come into conflict with science. Conflict only arose when theology was confused with religion.[71] The main target of Huxley's aggressive rhetoric, then, was what he variously called "Parsonism," "clericalism," "Ecclesiasticism," or more simply as orthodox theology—not religion. By separating religion from theology, Huxley believed he was affecting the continuing progress of the "New Reformation." "We are in the midst of a gigantic movement," he wrote his wife in 1873, "greater than that which preceded and produced the Reformation, and really only the continuation of that movement."[72] The following year he publicly pledged his allegiance to the "New Reformation" when he observed that the "act which commenced with the Protestant Reformation is nearly played out, and a wider and deeper change than that effected three centuries ago—a reformation, or rather a revolution in thought . . . is waiting to come on." All the issues that motivated the first Protestant Reformation—corruption, dogma, blind obedience to tradition—were again at work in the nineteenth century. This reformation, once complete, would create a new church.[73]

Although Huxley never provided anything like a systematic outline of his religious beliefs, he disclosed in an 1892 letter to biologist George John Romanes (1848–1894) that "I have a great respect for the Nazarenism of Jesus—very little for later 'Christianity.' But the only religion that appeals to me is prophetic Judaism. Add to it something from the best Stoics and something from Spinoza and something from Goethe, and there is a religion for men."[74]

Huxley's belief that science purified religion of false accretions and debased conceptions of God was also stressed by another champion of scientific naturalism—John Tyndall. Indeed, his first contribution to Youmans's *Popular Science Monthly* was an 1872 article on "Science and Religion," which was a reprint of his *Contemporary Review* article "On Prayer," and which would later become the centerpiece of the infamous "prayer gauge debate." Tyndall began by offering a "history of the human mind," repeating the debates concerning the existence of the antipodes, the Galileo affair, the revolution in geology, and the excitement over Darwin's theory of evolution. The consistent error throughout these episodes of conflict, he argued, was theological dogmatism. "In our day, however, the best-informed clergymen are prepared to admit that our views of the Universe, and its Author, are not impaired, but improved, by the abandonment of the Mosaic account of Creation." Religion will survive, he claimed, "after the removal of what had been long considered essential to it." He further observed that in most recent times "religion has been undergoing a process of purification, freeing itself slowly and painfully from the physical errors which the busy and uninformed intellect mingled with the aspiration of the soul, and which ignorance sought to perpetuate."[75]

Youmans published Tyndall's article on science and religion as a way of preparing the ground for his own American lecture tour in 1873. As he later did with Huxley, Youmans advertised Tyndall's expected trip to the United States throughout his magazine, referring to him at one point as the "wise man from the East."[76] Tyndall was also honored by Youmans and other leading American scientific luminaries with a dinner at Delmonico's in 1873 in New York.[77]

Perhaps one of the most controversial articles published in the *Popular Science Monthly* was Tyndall's notorious Belfast address in 1874.[78] This address became the object of much theological opprobrium, transforming Tyndall's public image as a respected and popular lecturer into an aggressive and radical materialist.[79] But Youmans attempted to rebuild that image in the wake of criticism. For example, he published an article by German physician Hermann von Helmholtz in which he praised Tyndall as a foremost scientist and popularizer.[80] Youmans also published an article by Harvard physics professor John Trowbridge, who thundered against those religious critics who had condemned Tyndall's address. "The Church should be eager," Trowbridge declared, "to receive the discoveries of investigators of the strange and wonderful works of the Creator, with confidence that all can and must be reconciled with revealed religion."[81] Youmans himself attempted to repel attacks against Tyndall. He asserted that despite criticism, the address was "received with an unanimity of commendation" among the more intelligent classes. He even contended that theologians who use the "harmonies and adaptations of Nature as proofs of wisdom and design on the part of the Creator" are deeply indebted to scientists like Tyndall, "who have disclosed this order, harmony, and adaptation, by the study of matter." Why, then, condemn the scientist, he asked, "if pushing on his investigation yet further, he claims to discern yet higher potencies and possibilities in this divine material of which the universe is constituted?" Tyndall's address, Youmans concluded, demonstrated that there is "no necessary hostility between religion and science," and that whatever the current conflict, it "will be ultimately harmonized."[82]

Youmans also published several essays by Tyndall where he worked out more fully his own personal philosophy of religion. In an 1875 article, for instance, Tyndall proclaimed that "the liberal and intelligent portion of Christendom must differentiate itself more and more, in word and act, from the fanatical, foolish, and more purely sacerdotal portion." Many had already abandoned the more "grotesque" forms of theology, replacing it with a more "wonderful plasticity of the theistic idea, which enables it to maintain, through many changes, its hold upon superior minds; and which, if it is to last, will eventually enable it to shape itself in accordance with scientific conditions." In an 1879 article, Tyndall was particularly hopeful because he found himself in many ways in complete accord with such men as Unitarian minister James Martineau, transcendentalist Ralph Waldo Emerson, and social critic Thomas Carlyle. But while Tyndall affirmed a personal "communion" with the "Divine," he nevertheless admitted that "its mystery overshadows me." Where others profess to "know," he could only "feel," as Kant felt when he viewed the "starry heavens," or observed "the sense of moral responsibility in man."[83]

Like Huxley, then, Tyndall spoke optimistically of "changes in religious conceptions and practices." In discussing the opening of the British Museum on a Sabbath day, for example, Tyndall explained how religious scholars such as Frenchmen Ernest Renan (1823–1892), German Friedrich Max Müller (1823–1900), and Scottish theologian John Caird (1820–1898) have demonstrated the developmental character of religion. Tyndall argued that throughout history there is a "purpose and growth, wherein the earlier and more imperfect religions constitute the natural and necessary precursors of the later and more perfect ones." These changes or "transmutations" in religion, Tyndall suggested, were "often accompanied by conflict and suffering." We see this, he explained, with the transition from Roman paganism to Christianity, from Jewish Christianity to Gentile Christianity, from Peter to Paul. Each transmutation, therefore, each period of growth, required some struggle, "in which the fittest survive." In short, even the errors, conflicts, and sufferings of bygone times "may have been necessary factors in the education of the world." This background leads Tyndall to his main point on the issue of keeping the Sabbath, which he viewed as a problem of dogmatic theology and not religion. According to Tyndall, Jesus deliberately broke the Sabbath, crowning his "protest against a sterile formalism by the enunciation of a principle which applies to us to-day as much as to the world in the time of Christ"—namely, "The Sabbath was made for man, and not man for the Sabbath." Those theological dogmatists who wished to keep museums closed on the Sabbath ignore Paul's own injunction that one can come to know truths about God by considering his creation.[84]

Like Huxley, Tyndall also saw himself as a "reformer," and was proud of his kinship with William Tyndale (1494–1536), the sixteenth-century martyred translator of the English Bible.[85] Indeed, an early biographer noted that Tyndall had mastered the works of important Protestant apologists such as John Tillotson, Jeremy Taylor, and William Chillingworth.[86] Moreover, the early correspondence between Tyndall and his father, who was an ardent Orangeman of Northern Ireland, reveals that they frequently discussed anti-Catholic books and tracts.[87]

Tyndall also came under the spell of Carlyle's writings. Tyndall later recalled that he first came across some extracts of Carlyle's *Past and Present* as a surveyor in Preston. "I found in it," he wrote, "strokes of descriptive power unequalled in my experience." Carlyle taught him "a morality so righteous, a radicalism so high, reasonable, and humane, as to make it clear to me that without truckling to the ape and tiger of the mob, a man might hold the views of a radical."[88] While Tyndall had no doubt imbibed the religious musings of Carlyle, he also had become particularly fond of the transcendentalism of Ralph Waldo Emerson. He told his old friend English mathematician Thomas A. Hirst (1830–1892) that "in Emerson you behold one of the noblest souls that ever was struck in clay. Every time I rise from his book I find a new vigour in my heart. He teaches one to be so independent that you almost feel disposed to quarrel with himself, just to shew how little you cared about even him."[89] In a fascinating November 1848 letter, Tyndall also told Hirst that although he could not articulate the details of his "religion," he openly admitted his debt to Carlyle and Emerson. Reading their writings, he professed, gave credence to the phrase "born again." These two writers taught him that "religious life is progressive," and that "Religion is not a *persuasion*, it is a *life*."[90]

It was also during the 1840s that Tyndall had essentially abandoned his belief in orthodox Christianity. While he felt unable to believe in the strictures and traditions of established religion, he continued to believe in God. In an 1847 journal entry he wrote, "I cannot for an instant imagine that a good and merciful God would ever make our eternal salvation depend upon such slender links, as conformity with what some are pleased to call the essentials of religion. I was long fettered by these things, but now thank God they are placed upon the same shelf with the swaddling clothes which bound up my infancy."[91]

Thus, while he rejected the tenets of orthodox Christianity, Tyndall did not become an atheist. Historian of science Geoffrey Cantor has recently demonstrated that in these formative years Tyndall sought to "transcend the petty wrangling over theological issues that divided the different denominations."[92] Like Huxley, then, he looked forward to when "methodism, churchism and many other isms . . . [would] sink and a purer[,] lovelier and more practical faith [would emerge]—a faith which Jesus taught and John understood shall bend with benignant influence over our altered world."[93] In 1850 Tyndall pithily summarized his religious views in his journal, writing that "I would lay my hand upon the writings of John in the New Testament, and the religion there taught is not the religion of the understanding but a deeper one." A decade before Huxley, Tyndall similarly proclaimed the "plea of Martin Luther must be mine 'I cannot otherwise—my God assist me!'"[94]

Like Huxley, Tyndall believed that science and religion were not at war as long as they kept to their respective domains. Science, Tyndall maintained, held no powers of veto over "those unquenchable claims of his moral and emotional nature, which the understanding can never satisfy," including those feelings "of Awe, Reverence, Wonder" from which "the Religions of the world" have drawn their strength. Invoking the German idealists, Tyndall looked forward to the day when "Heart" and "Understanding" might "dwell together in unity of spirit and in the bond of peace."[95]

Youmans no doubt provided Huxley and Tyndall a medium to spread the good news of the religion of the future. No one, however, owed so much to Youmans as did Herbert Spencer. Youmans had been Spencer's most enthusiastic "disciple" and "agent" in America.[96] Charles Haar put it aptly when he wrote that "What Huxley was to Darwin in England, Youmans was to Spencer in America."[97] Indeed, the "the hospitable reception of Herbert Spencer's ideas" in America, according to Fiske, was largely due to Youmans's efforts.[98] In 1860, after reading Spencer's prospectus for his "System of Philosophy," Youmans was so impressed that he contacted Spencer directly, offering his help in advancing the project in America. Indeed, when he visited England in 1862 it was primarily for the purpose of meeting Spencer. Thereafter Youmans dedicated the rest of his life to explaining, defending, and disseminating Spencer's doctrine of evolution to American readers.

Like Huxley and Tyndall and so many others during the century, Spencer justified his rejection of orthodoxy on moral grounds. "How absolutely and immeasurably unjust it would be," he wrote, "that for Adam's disobedience . . . all Adam's guiltless descendants should be damned, with the exception of a relatively few who accepted the 'plan of salvation.'"[99] But as one biographer has observed, Spencer's assumptions and outlook belonged to a provincial Dissent heavily influenced by a Calvinistic outlook.[100] His father, William George Spencer (1790–1866), was a religious Dissenter who drifted between Methodism and Quakerism, and Spencer seems to have imbibed from him an instinctive aversion to authoritarianism. But as he makes clear in his own *Autobiography* (1904), Spencer had rejected the God of orthodox Christianity at an early age. And like Tyndall and Huxley, Spencer was fascinated with the writings of Carlyle and Emerson. The latter made such an impression on him that he even visited the poet's home and grave during his own American tour in 1882.[101]

As an evolutionary philosopher, Spencer believed evolution was at work everywhere, from the formation of the solar system to the development of organic life. Unsurprisingly, then, Spencer saw religion as part of this cosmic evolutionary process. Interestingly, Spencer's work contributed to the notion that evolution had a direction and purpose, which was dictated by an Unknowable deity—a notion that was particularly appealing to liberal Protestants.[102] At the beginning of the so-called Darwinian controversies, Spencer declared the "pervading spirit of irreligion" of the day was not caused by science but by theological opposition. Science, according to Spencer, is a religious act, and devotion to science "tacit worship." By developing the concept of cosmic evolution into a system of philosophy, he had constructed an all-embracing worldview that provided its adherents with a sense of meaning in nature. According to Spencer, understanding the history of science not only contributed to our grasp of how the universe works, it also revealed a teleological principle working throughout the history of the natural world.[103]

It seems likely that Spencer picked up these teleological views from his father and other Dissenters and Nonconformists. We learn from English clergymen Thomas Mozley (1806–1893), for instance, that Spencer's father championed some form of evolutionary deism.[104] As we have already seen, this position became popular during the eighteenth century, among thinkers who hoped to reconcile belief in divine providence with the laws of nature. During the early nineteenth century, the laws of nature were combined with the belief in a progressive process of natural development. While God was separated from the cosmos, the natural order was nevertheless benevolently designed by him. This evolutionary deism, or as Charles Taylor put it, "providential deism," was promoted by countless liberal Protestants in the nineteenth century.

Spencer began laying out his own vision of evolutionary deism in his multivolume *System of Synthetic Philosophy* (1862–96), which was intended as a synthesis of all knowledge connected through his concept of evolution. In his first book of the system, *First Principles*, which appeared in 1862, he explained the basic elements of a universal law of evolution. Importantly, he prefaced this discussion with a section on "the Unknowable," in which he attempted to demonstrate the "fundamental harmony" between science and religion. "Of all the antagonisms of belief," he claimed, "the oldest, the widest, the most profound and the most important, is that between Religion and Science." However, instead of perpetuating this antagonism, Spencer proposed to locate the "basis of a complete reconciliation." In fact, the central aim of the book is to understand "how Science and Religion express opposite sides of the same fact." The common element that emerges, according to Spencer, is that "the Power which the Universe manifests to us is utterly inscrutable." Behind all religious experience is some unfathomable mysterious Power. The data of the natural world likewise presents some persistent Force, "an Incomprehensible Omnipresent Power." "A permanent peace will be reached," Spencer argued, when "Science becomes fully convinced that its explanations are proximate and relative; while Religion becomes fully convinced that the mystery it contemplates is ultimate and absolute." Thus the basis of total reconciliation, according to Spencer, is the idea of a mysterious power underlying all phenomena and driving the evolutionary process forward. That power was Spencer's doctrine of the "Unknowable" or "Unknown."[105]

Youmans adhered closely to Spencer's narrative of the history of science, and shared with him the belief that organized religion, particularly orthodox Christianity, had impeded both the progress of science and religion. More importantly, Spencer's doctrine of the great "Unknowable" resonated deeply with Youmans, who believed it moved religion toward a higher conception of God. Unsurprisingly, then, the leading article of the monthly was the first installment of Spencer's "Study of Sociology."[106] Spencer was by far the most frequent contributor to the *Popular Science Monthly*, rarely missing an opportunity to publish some piece in the magazine during his lifetime. He later remarked in his *Autobiography* that "Professor Edward L. Youmans was of all Americans I have known or heard of, the one most able and most willing to help me."[107]

Youmans was very protective of his idol. When some accused Spencer of "atheistical tendencies," he took these critics to task. "It is the common and very foolish trick of religious partisans," he wrote, "to stigmatize those who differ from them in their views of Deity as atheists." There is no "vestige of truth" in the charge, he added, for "Herbert Spencer is not an atheist, and never has been." On the contrary, he has labored to make atheism "baseless and indefensible," and is therefore in "dead antagonism to atheism." In fact, Spencer had rescued theologians from the "consequences of their own logic," which actually leads to unbelief. Spencer, then, was no "denier or antagonist of religion," and there can be no greater or more mischievous mistake than imputing to him and other "scientific writers of this age any hostility to religion as the motive of their labours."[108]

Youmans had also allowed Spencer to publish his own replies to critics in the *Popular Science Monthly*.[109] This was not always the most effective strategy, however. Indeed, perhaps the most celebrated controversy featured in the pages of the magazine was the "Quadrangular Duel" between Spencer, English positivist Frederic Harrison (1831–1923), lawyer and judge James Fitzjames Stephen (1829–1894), and ecclesiastical historian Wilfrid Philip Ward (1856–1916), son of Catholic convert and editor of the *Dublin Review* William George Ward (1812–1882).[110] Each speaker recognized the need for religion manifested in man. But each differed remarkably on what this religion looked like. It is important that we examine this debate in some detail as it was primarily concerned with the "religion of the future," and thus inextricably connected to late nineteenth-century attempts at reconciling science and religion. The debate, in short, was a public relations disaster. It ultimately reflects the failed project of liberal reformulations of religion, and therefore its attempt at reconciling religious belief and modern thought.

In 1884 Spencer initiated the duel when he published his "Religious Retrospect and Prospect" in the *Popular Science Monthly*; the article appeared simultaneously in James Thomas Knowles's (1831–1908) *Nineteenth Century*. On sending the manuscript to Youmans, Spencer noted that he wanted to diffuse its ideas as widely as possible and that the question of the future of religion was a "burning one, and one in respect of which it is desirable to be clearly understood."[111] In this article, Spencer offered an evolutionary account of the "religious consciousness." By looking at its evolutionary history, Spencer believed he could infer the religious ideas and sentiments of the future. Importantly, he contested the notion that science had somehow replaced religion. Science does not destroy religion, but "transfigures it." According to Spencer, science had enlarged the sphere of wonder and "religious sentiment." Primitive man had only a limited understanding of that wonder. The cosmogony of the "savage" is incomparable to the wonder established by the modern astronomer. This deeper insight, wonder, or feeling "is not likely to be decreased but increased by that analysis of knowledge which, while forcing him to agnosticism, yet continually prompts him to imagine some solution of the Great Enigma which he knows can not be solved." But amid all this mystery, Spencer argued, there remains "the one absolute certainty, that he is ever in the presence of an Infinite and Eternal Energy, from which all things proceed."[112] This was of course not the first time Spencer wrote on religion. Indeed, in his *First Principles*, the first volume of his prospective "Synthetic Philosophy," he began with a discussion of religion and religious ideas, and introduced his notion of "The Unknowable," which was, Spencer had argued, the true home of religion and by which he hoped to reconcile science and religion.[113]

Spencer's "Religious Retrospect and Prospect" elicited strong responses. He reported to Youmans that Knowles complained of "a hailstorm of communications about it," and that some clergymen even threatened to drop their subscriptions to the journal.[114] Thus it perhaps surprised Spencer that his old friend Frederic Harrison criticized him so severely a few months later in a scathing attack on his ideas. At the same time, Harrison was, after all, the indefatigable critic of liberal Anglicanism, particularly the authors of the *Essays and Reviews*. As an unbeliever, Harrison praised the liberal divines for their courage and liberality; however, he ultimately challenged their integrity, protesting that they failed to see the legitimate conclusions of their arguments, which was the complete destruction of orthodox Christianity.[115]

Harrison's strategic attack against liberal Protestantism was very similar to his attack against Spencer's doctrine of the Unknowable. According to Harrison, Spencer's conception of the Unknowable was really only a "ghost of religion." "In spite of the capital letters, and the use of theological terms as old as Isaiah or Athanasius," he wrote, "Mr. Spencer's Energy has no analogy with God. It is Eternal, Infinite, and Incomprehensible; but still it is not He, but It." Ostensibly targeting those liberal Protestants who attempted to assimilate Spencer's philosophy, Harrison emphatically declared that "neither goodness, nor wisdom, nor justice, nor consciousness, nor will, nor life, can be ascribed, even by analogy, to this Force." Religion, according to Harrison, "cannot be found in this No-man's-land and know-nothing-creed." Better to bury this religion "than let its ghost walk uneasy in our dreams." True religion must account for the knowable—it cannot simply be "made up of wonder, or of a vague sense of immensity, unsatisfied yearning after infinity." Spencer's Unknowable cannot and will not ever "make any one any better"; it has "no creed, no doctrines, no temples, no priests, no teachers, no rites, no morality, no beauty, no hope, no consolation"; it does not "bring men together in a common belief, or for common purposes, or kindred feeling, it can no more unite men than the procession of the equinoxes can unite them." It cannot comfort a "mother wrung with agony for the loss of her child, or the wife crushed by the death of her children's father, or the helpless and the oppressed, the poor and the needy, men, women, and children, in sorrow, doubt, and want, longing for something to comfort them and to guide them, something to believe in, to hope for, to love, and to worship." The Unknowable, therefore, has "no part or parcel in human life." The Unknowable, in sum, is to "defecate religion to a pure transparency." Spencer's own attempt to "put a little unction into the Unknowable" by describing it in theological terms, Harrison protested, is, in the final analysis, a "philosophical inaccuracy."[116]

Youmans allowed Spencer to respond with his "Retrogressive Religion," where he charged Harrison with attacking an imaginary doctrine, "demolishing a simulacrum and walking off in triumph as though the reality had been demolished." He then assailed Harrison's "alternative doctrine," his "Religion of Humanity," as an "incongruity." Indeed, papal assumptions, he argued, were more modest in comparison to the assumptions of "the founder of the religion of Humanity." A pope may canonize a saint or two, but Comte, Spencer quipped, "undertook the canonization of all those men recorded in history whom he thought specially worthy of worship." The new religion should not be a "rehabilitation of the religion with which mankind commenced, and from which they have been insensibly diverging." Harrison's Religion of Humanity was, therefore, according to Spencer, "retrogressive."[117]

The controversy rolled on into the following year with Harrison's "Agnostic Metaphysics." In this article, Harrison wrote that he had warned Spencer a decade ago that his "Religion of the Unknowable" would find adherents among dubious theologians. He argued that the "Infinite and Eternal Energy," the "Ultimate Cause," the "All-Being," and the "Creative Power," have all been co-opted by the "Christian World," renewing all the mystification of the old theology. Moreover, Harrison inveighed that Spencer knew too much about the Unknowable—"If his Unknowable be unknowable, then it is idle to talk of Infinite and Eternal Energy, sole Reality, All-Being, and Creative Power." This is, at best, "slip-slop" theology and nothing more.[118]

At this point Spencer complained to Youmans that the "Harrison business" had been "a sad loss of time."[119] Nevertheless, Spencer wanted the "Last Words about Agnosticism."[120] But having grown weary, Spencer simply repeated the same arguments. Youmans once again came to Spencer's defense in his Editor's Table. Tellingly, he wanted to print the controversy, he explained, because it "goes to the root of the issue between science and religion, and is really a contest over the vital question whether there is, or is to be in future, any recognition of any such things as truth in the religious sphere of human experience." According to Youmans, Spencer had demonstrated that there is indeed "a verity at the foundations of all religious systems, which will permanently remain when the erroneous beliefs accompanying this verity are utterly swept away by the progress of science." Remarkably, he declared that Spencer was on the "religious side," whereas Harrison was the true "inveterate antagonist." Harrison labored to replace the "religions which have made Divine Power an object of worship" with the "worship of man and the religion of humanity as invented and expounded by Comte." Youmans caustically concluded that "the devotees of this new religious cult may be sincere, but they are none the less absurd."[121]

The third important figure in the controversy, James Fitzjames Stephen, found both creeds rather pretentious. Stephen, who grew up in an evangelical home but later became positively skeptical of religion like his younger brother Leslie Stephen (1832–1904), argued that Spencer's theory of religious development was weak and that his game of words reminded him of "Isaiah's description of the manufacture of idols." "Effort and force and energy," he wrote, "are to Mr. Spencer what the cypress and the oak and the ash were to the artifices described by the prophet. He works his words about this way and that, he accounts with part for ghosts and dreams, and the residue thereof he maketh a god, and saith Aha, I am wise, I have seen the truth." Moreover, Stephen described Spencer's whole theory as "a castle in the air, uninhabitable and destitute of foundations." More pointedly, he declared that Spencer's Unknowable appeared "to have absolutely no meaning at all. It is so abstract that it asserts nothing. It is like a gigantic soap-bubble not burst but blown thinner and thinner till it has become absolutely imperceptible."[122]

But according to Stephen, Harrison fared no better. "Is not Mr. Harrison's own creed," Stephen observed, "open to every objection which he urges against Mr. Spencer[?]" "Humanity with a capital H (Mr. Harrison's God) is neither better nor worse fitted to be a god than the Unknowable with a capital U." We cannot worship an "indefinite number of dead people," and we certainly do not feel "awe and gratitude" to the multitude, "most of whom are utterly unknown to us even by name or reputation." The men of history are, in the final analysis, "dead and done with." Harrison's language of awe and gratitude toward humanity "represents nothing at all, except a yearning after some object of affection, like a childless woman's love for a lapdog." But "mankind, a stupid, ignorant, half beast of a creature, the most distinguished specimens of which have passed their lives in chasing chimeras, and believing and forcing others to believe in fairy tales about them," does not deserve to be worshiped. In the end, "Humanity is and will for ever continue to be to mankind at large just as poor a shadow of a God as the Unknowable." Stephen concludes with the arresting observation that "if this is the prospect before religion, it would surely be simply to say that the prospect before it is that of *extinction*, that men will soon come to see that nothing can be ascertained, or even regarded as moderately probable, about the various questions which are generally described collectively as religious." Interestingly enough, Stephen argued that the only religion capable of doing what both Spencer and Harrison want their respective new religions to do "must be founded on a supernatural basis believed to be true." In the final analysis, theology, Stephen argued, is essential to religion, "and that to destroy the one is to destroy the other."[123]

From yet another perspective, ecclesiastical historian Wilfrid Philip Ward offered a witty condemnation of both Spencer and Harrison in an article published for the conservative monthly, the *National Review*. Ward accused both men of "monomania." He agreed with Harrison's critique of Spencer, calling it "quite unanswerable common sense." Spencer has no right, logical or otherwise, "to have his cake after he has eaten it." An otherwise serious and cautious thinker, Spencer could not see that "if the death-knell of the old Theology be indeed sounded, all reasonable religious worship must die with it." When looking at Harrison's substitute religion, however, Ward was "startled beyond description." Thus, like the starving man who eats a pair of boots, Spencer and Harrison, desperate to satisfy their religious cravings, have each taken a boot. Their religious language is mere dressing. "The truth seems to be," Ward declared, "that these philosophers having conspired together to kill all real religion—the very essence of which is a really existing personal God, known to exist, and accessible to the prayers of His creatures—and having, as they suppose, accomplished their work of destruction and put religion to death, have proceeded to divide its clothes between them."[124]

Despite these strong reactions from opposing viewpoints, there continued to appear liberal religious thinkers who attempted to salvage Spencer's doctrine of the Unknown from criticism. One close observer of the duel agreed that "it was quite true that the day of Christianity and its contemporary religions was past, and quite true that some substitute was required," and thus the most suitable replacement was the doctrine of the Unknowable.[125] A writer for the *Contemporary Review* characteristically called upon the work of Matthew Arnold to mediate between Spencer and Harrison.[126] In the pages of Youmans's own magazine, Canon George H. Curteis (1824–1894) defended Spencer for his "courageous" position. No religious man, he said, should shrink from calling himself a "Christian agnostic." Indeed, by proclaiming his agnosticism, the Christian follows an esteemed pedigree, one which Curteis traced to the Old Testament prophets. Although Spencer was not a "Christian" philosopher, his "guidance is none the less valuable to those who are approaching the same subject from a different side." According to Curteis, Spencer had "purified" the idea of God for the believer, "pruned away all kinds of anthropomorphic accretions," "reminded the country parson of a good many scientific facts," and "schooled them into the reflection that a power present in innumerable worlds hardly needs our flattery, or indeed any kind of service from us at all."[127]

Another defender was Scottish evangelical Henry Drummond (1851–1897), who had used Spencer's philosophy of evolution to demonstrate the "laws of the spiritual world." In his popular *Natural Law in the Spiritual World* (1883), Drummond asserted that "the results of Mr. Herbert Spencer are far from sterile—the application of Biology to Political Economy is already revolutionizing the Science. If the introduction of Natural law into the Social sphere is no violent contradiction, . . . shall its further extension to the Spiritual sphere be counted an extravagance?"[128] Spencer had read Drummond's book at the Athenaeum Club soon after its publication and immediately recommended the work to Youmans, writing that "I found it to be in a considerable measure an endeavor to press me into the support of a qualified theology by showing the harmony between certain views of mind and alleged spiritual laws. It is an interesting example of one of the transitional books which are at present very useful. It occurs to me that while the author proposes to press me into his service, we might advantageously press him into our own service. Just look at the book and see."[129] Youmans took his advice and reviewed the book in his magazine, noting for his readers that Drummond's book was "directed to the problem of the relations of religion and science, and presents a new view from an advanced stand-point, which many regard as helpful and healthful in its influence." He went on to observe that as a book of reconciliations, Drummond's "volume is one of the best of its class."[130]

What needs to be emphasized here is that Harrison's assessment was largely correct. Many "dubious theologians" did indeed come to embrace Spencer's doctrines. As Richard Hofstadter observed, the ground for the positive reception of Spencer was largely paved by an increasingly liberal religious perspective.[131] The prominent liberal clergymen Henry Ward Beecher had written Spencer in 1866 to tell him that "the peculiar conditions of American society has made your writings far more fruitful and quickening here than in Europe."[132] Why America was ripe for Spencer's ideas, Beecher does not say. But as we have seen, liberal Protestantism was quite pervasive in the last decades of nineteenth-century America.

The "quadrangular duel" also reveals something of the difficulties and deep ambivalence among liberal Protestants and agnostics at the end of the century, particularly those who attempted to find some reconciliation between science and religious belief by redefining religion and separating it from theology. While almost all participants in the debate recognized the profoundly religious nature of man, their attempt at constructing a "religion of the future" that would reconcile men of science to Christian faith was a complete failure. Whereas Harrison criticized Spencer and other agnostics over their failure to supply a viable new religion, Stephen and Ward ridiculed both for offering weak and self-defeating substitutes. And while Stephen and Ward came from different worldviews, they both grasped the inherently unstable and unsatisfying consequences of the liberal Protestant attempt at reconciliation. In short, for both the atheist and the religiously conservative or orthodox, Spencer's Unknowable was just as "unreal" as Harrison's Religion of Humanity.

Before turning to Youmans's own peculiar attempt at reconciling religion and science, it may be beneficial to look very briefly at one more important critic of both doctrines of the Unknowable and the Religion of Humanity—namely, prolific English journalist, critic, theologian, and editor Richard Holt Hutton (1826–1897). As editor of the London *Spectator*, Hutton's hostility toward the scientific naturalists may serve as a useful contrast to Youmans's slavish adherence. While sympathetic to the agnostic struggle with doubt, Hutton could not forgive the scientific naturalists for their pretensions and imperious authority. He sarcastically described Huxley, for example, as "pope" because of his tendency to pontificate about his "infallible" scientific expertise.[133] In an article entitled "Agnostic Dreamers," Hutton declared that "nothing is more surprising than the extravagances of Agnostics." While they have destroyed the world of idols, Christian worship, and belief in the immortality of the soul, the agnostics "immediately proceed to substitute for these idols mere dolls of their own fashioning and dressing—dolls which they make no secret of having deliberately fashioned and dressed up for the occasion."[134]

Hutton also sought to undermine the "positivist dream" of a Religion of Humanity. He accused Harrison and the positivists of stealing "the emotional part from the great religions of the world, without encumbering themselves with the intellectual ground of that 'great and strenuous emotion' into which they wish to absorb the existing metaphysics of faith."[135] He reported to readers that on one occasion Harrison delivered a long address on the "memory of the departed," calling members to give "pious contemplation of whose posthumous energy he seeks to kindle the gratitude of the living to something more of 'depth, and breadth, and glow.'" Hutton mocked the "ritualism" of the positivists, arguing that such words were "unreal." To speak of the departed as a "choir invisible," he wrote, "is very much like speaking of the choir invisible of decayed violins, unstrung pianos, and broken-down organs. The music these instruments once gave might be repeated, if there were some delicate phonograph to register it; but no choir, visible or invisible, could use instruments long since mingled with the dust." It is not at all clear what Harrison intended when he called on his audience to appreciate the departed with greater "depth, and breadth, and glow." "We are puzzled," Hutton admitted, "as to the source of Mr. Harrison's hopes." His enthusiasm and hope for the dead, Hutton contended, are "rhetorical extravagances of language, and nothing else." Harrison and the positivists are, therefore, fashioning "artificial emotion into existence," installing dressed-up dolls "which are supposed to stimulate piety in foreign churches."[136] Hutton even questioned Harrison's alleged early Christian faith, as it appeared to him that Harrison had not the "quaintest idea of what the heart of the Christian religion (or, indeed, of any religion except his own) is."[137]

Ultimately, Hutton followed Stephen and Ward in criticizing liberal theologians for using "unreal words" and establishing "phantom creeds."[138] He characterized the "new theology" of the liberal clergy as a "humanist theology," arguing that such liberal theologians wanted to "exchange the story of the Gospels for a string of noble moral traditions and legendary facts; to beware of the 'passionate certitude' of the Apostles as a dangerous source of corruption; and to regard the theology of the Church as an unhappy crystallisation of faith into dogma."[139] Hutton profoundly disagreed with the apostles of the "New Reformation," those who wished to "'translate' our accounts of the older religious teaching, whether in relation to Judaism or Christianity, or any other religion, to the mind of to-day, as to magnify the human and subjective element in the original story of the incipient and rising faith." He acknowledged that this was an attempt to dissolve and dissipate religious conflict with modern science, but also presciently recognized that any such attempt, when fully complete, ultimately explained God away.[140]

Hutton was an important and insightful critic of theological liberalism. He clearly relished exposing the credulity of "honest doubters," and to this end paid much critical attention to Huxley, Spencer, Tyndall, and other alleged "reconcilers" of science and religion. He was impatient with the spurious authority of the scientific naturalists and saw their attempt at reconstructing Christianity and religion in general as a "sort of make-believe adoration." While historians of science have largely ignored him, Hutton offers us an incisive critique of liberal discussions over the "future of religion" at the end of the nineteenth century. Such discussions, as we have seen, were inextricably linked to attempts at reconciling science and religion. The "quadrangular duel" therefore reflects and anticipates the failure of the liberal Protestant effort to reconcile the breach between modern thought and ancient faith.

THE FREE RELIGION OF THE PEACEMAKERS

Shortly after launching his magazine and book series, Youmans delivered an address at the famous Cooper Union, entitled "The Religious Work of Science." This venue was, of course, the same place where both Draper and White first presented early versions of their narratives. In his speech, Youmans announced that "Science has long been regarded and is still widely believed to be the antagonist of religion." But the time has come, he proclaimed, "when it will be accepted as its most powerful ally and best friend." By "science," he meant knowledge of the order of nature, "the laws by which that order is governed." By "religion" he meant the "feeling entertained toward that Infinite Being, Power, or Cause, by whatever name called, of which all things are the manifestation, and which is regarded and worshipped as the Creator and Ruler of the Universe." According to Youmans, science and religion are not mutually exclusive. "It is the office of science to explore the works of God; of religion to deal with the sentiments and emotions which go out toward the Divine Author of these works." Men of science, he explained, devote their lives to exploring the works of God. They thus labor at discovering "divine" truths in nature. Such work is, therefore, "religious work."

As religious work, then, science has attained its "grandest achievement" in revealing new religious growth. "It has recreated," Youmans said, "the universe in thought, and, by elevating and expanding man's conceptions of the sphere of harmony and law, has exalted our reverential feelings toward the Infinite Power by which it is ordered and sustained." Whenever conflict emerged, then, it occurred not between science and religion but between theology and science. Men of science faced many obstacles; the most obstructive were pernicious "theologians who claimed to be authorized expounders of the divine policy." These theologians insisted on a God who breaks and interrupts the natural order. But the

"Almighty" has been vindicated by men of science—they have shown that the wisdom of God is "witnessed not in the violations but in the perfection of his works."

Youmans argued, in short, that "orthodoxy," or dogmatism, must come to an end if science and religion are to grow and advance. He then praised the "advancing liberality among the evangelicals" in accepting evolutionary theory. In the final analysis, he believed that the doctrine of evolution offered "a grander conception of the cosmical order and a deeper insight into its wonderful workings than had ever before been attained, it is the sublimest tribute that the human mind has ever made to the glory of the Divine Power to which it must be ascribed." Ultimately, the conflict between science and theology is "purely imaginary." The old theology had been replaced by science, which provided a "view that is more eminently religious."[141]

Importantly, the occasion for the address was the annual meeting of the "Free Religious Association."[142] In 1825 the American Unitarian Association was established to promote an Arian Christology, a rationalist interpretation of Scripture, and an optimistic view of human nature. After the Civil War, however, controversy and schism emerged within the denomination, as some members decided that Unitarianism still harbored a residual orthodoxy. Founded in 1867 in Boston by those disenchanted with clerical hierarchies and orthodoxy of any kind, the Free Religious Association (F.R.A.) sought to promote the principles of free thought and moral philosophy without any reference to institutional Christianity. A key factor, therefore, in preparing the platform of the F.R.A. were those liberal theologians whom we have already met, men such as William Ellery Channing, Theodore Parker, Henry Ward Beecher, Theodore Munger, and others. Such liberals had become impassioned converts to scientific method and devotees of evolution, which accorded them additional reasons for their eternal confidence in human progress. The purpose of the F.R.A. was to advocate a rational religion without priesthood; a morality without theology; a god without dogmatism; a religion of liberty with no limits of thought; a religion of reason, action, equality, and love. The mere existence of the F.R.A. was testimony that its supporters had renounced the Christian religion as inadequate and were convinced that it ought to be supplanted by a new religion—a free religion. Composed of a diverse assortment of radical Unitarians, Universalists, spiritualists, transcendentalists, scientific theists, and other disaffected religious minorities, the F.R.A. advanced a new "religion of humanity."

According to its first constitution, the F.R.A. aspired to "promote the interests of pure religion, to encourage the scientific study of theology, and to increase fellowship in the spirit."[143] Its first president, Unitarian clergyman Octavius B. Frothingham (1822–1895), asserted that "all men are religious, or have the possibility of becoming religious, and are designed to be religious, and to find, in being religious, the completeness, the grace, and the beauty of their character."[144] He later explained that the association aimed "to remove all dividing lines and to unite all religious men in bonds of pure spirituality, each one being responsible for his own opinion alone."[145] The catalysts for the movement were periodicals such as the *Radical* and the *Index*, which provided a forum for discussing what direction the religion of the future should take.[146]

But as historian Leigh Eric Schmidt has correctly observed, members of the F.R.A. were "hyper-Protestant and post-Christian."[147] Indeed, Francis Ellingwood Abbot (1836–1903), first editor of the *Index*, claimed a close association to the Protestant tradition, arguing that "the history of Protestantism is the history of the growth of Free Religion."[148] In another issue, Abbot called for the "impeachment of Christianity," declaring orthodox Christianity a great organized superstition "perpetuating in moderns times the false belief, degrading fears, and benumbing influences of the Dark Ages."[149] This characterization, a Protestantism minus Christianity, also encapsulated Youmans's own religious position. Indeed, by the time he delivered his Cooper Union address on the religious work of science, Youmans had already become one of the many vice-presidents of the F.R.A.

It may be suggested that in addition to scientific naturalism, Youmans intended his own magazine and book series to popularize the ideas and aims of the F.R.A. When space allowed, he often advertised the meetings of the organization in the pages of his magazine.[150] On the occasion of its twelfth annual meeting in 1879, for example, he described the movement as the "extreme reaction against the restrictive spirit of ecclesiasticism," Catholic and Protestant alike. The association, he explained, sought to "promote the practical interests of pure religion, to increase fellowship in the spirit, and to encourage the scientific study of man's religious nature and history."[151] As one of its vice-presidents, Youmans sought to promote such conciliatory views toward religion throughout his *Popular Science Monthly*. In its pages, he offered a spectrum of liberal theological thought that he believed was heralding the reconciliation between science and religion.

Youmans indeed found the "progress" in liberal theology exhilarating. Theologians, he said, were beginning to "recognize that theology is also progressive, and that, so far from being an enemy, Science is a helping friend of true religion."[152] He published writers who shared these sentiments. Broad Churchman and amateur scientist Charles Kingsley, whom Youmans referred to as "vigorous and brilliant," appeared occasionally within the pages of his magazine, with articles encouraging young men to study the natural sciences for the glory of God.[153] Anglican curate and botanist George Henslow (1835–1925) also appeared in the magazine, where he declared that evolution was proof of the wisdom and beneficence of the Almighty.[154] English Unitarian and philosopher of science William Stanley Jevon (1835–1882) also argued in Youmans's magazine that evolution should not "alter our theological views."[155] Youmans was also inspired by Broad Churchman Arthur P. Stanley's funeral sermon for Charles Lyell, in which he spoke of a "reconciliation of a higher kind, or rather not a reconciliation, but an acknowledgment of the affinity and identity which exist between the spirit of science and the spirit of the Bible." Stanley argued that "increased insight into nature and origin of biblical records" demonstrated that the creation account in Genesis could not be a literal description of the beginning of the world.[156] Interestingly, Youmans wondered how Stanley's position was any different from Draper's "bold prophecy" that the Pentateuch can no longer be interpreted as authentic. Despite the inconsistency of the theologians, Youmans was encouraged that a "better day is dawning," when churchmen of the highest ranks displayed a "spirit of liberality which more than sympathizes with the life of those great and good men" of science.[157] This progress in theology and liberal feeling demonstrated, Youmans believed, that the "old conflict between religion and science will either die away, or lose so much of its rancorous spirit that it may be coolly and rationally considered."[158]

During the 1870s numerous books appeared by theologians and scientists denying any conflict between religion and science. Youmans gave many of these works glowing recommendations in his magazine. For example, when American geologist Alexander Winchell (1824–1891) published his book on the *Doctrine of Evolution* (1874), which argued that the belief in a providential and personal God is "to no extent imperilled by the admission of the reality of any form of evolution," Youmans took note and announced to his readers that they should seek a copy.[159] When Harvard botanist Asa Gray's (1810–1888) *Darwiniana* (1876) appeared soon thereafter, Youmans published a long literary notice urging upon his readers a "dispassionate perusal of this volume." Gray demonstrated, according to Youmans, that the conflict between natural selection and theological belief is "not necessary, and that an enlightened interpretation of religious doctrine must bring it into harmony with the advanced scientific conclusions."[160]

Indeed, Youmans wanted to make sure his readers knew the so-called "peacemakers." While this was a disparate group of thinkers, Youmans strongly believed their views were leading to a freer, more liberal religion. When Winchell published his *Reconciliation of Science and Religion* (1877), Youmans boasted about its "original and ingenious" argument, and that he had convincingly demonstrated that "man's religious nature has had an unfolding, like the other elements of being."[161] That religion is progressive, Youmans had no doubt. Another "peacemaker" was Unitarian minister James Thompson Bixby (1843–1921), who argued in the magazine that science and religion were mutually indebted to each other.[162]

According to Bixby, science has been a "real helper in the progress of religion," purifying it, uprooting old ideas, and dismantling defunct systems of thought. But religion has also helped science. "What could science accomplish," Bixby declared, "without the emotions of enthusiasm and devotion, the instructive feeling of truth and beauty, the love of Nature for its own sake?" The Greeks and Romans, despite their great intellectual acuteness, never fully developed a scientific understanding of nature. Christianity introduced the "grand thought" of a great human family, a "universal brotherhood." Youmans also reviewed Bixby's *Similarities of Physical and Religious Knowledge* (1876), which he said was "written in a liberal spirit, with much discrimination and judicial fairness, and which aims to get down to the radical harmonies of religion and science."[163]

Yet another "peacemaker" who appeared in Youmans's magazine was acclaimed Canadian geologist and principal of McGill College in Montreal John William Dawson (1820–1899). In his article, Dawson argued that "True religion, which consists in practical love to God and to our fellow-men, can have no conflict with true science." While known for his Presbyterian piety, Dawson seemingly agreed with Draper and White in arguing that "Organizations styling themselves 'the Church,' and whose warrant from the Bible is often of the slenderest, have denounced and opposed new scientific truths, and persecuted their upholders." Indeed, there are several remarkable parallels between Dawson and the so-called cofounders of the conflict narrative. He also argued that "theology was not religion," and that the Protestant Reformation provided a necessary impetus to the rapid growth of modern science. Remarkably, Dawson believed the conflicts between modern science and old theology were "necessary incidents in human progress, which comes only by conflict."[164]

One of the most intriguing "peacemakers" to appear in the pages of the *Popular Science Monthly* was the anonymous author of a series of articles that first appeared in *Macmillan's Magazine*, under the title "Natural Religion." This was Cambridge Regius Professor of Modern History, John Robert Seeley (1834–1895). Youmans had tellingly retitled Seeley's articles as the "Deeper Harmonies of Science and Religion."[165] Seeley's intellectual development is well known. Raised in a devout evangelical home, he came under the influence of the "Germano-Coleridgean" school while studying at Christ's College. Seeley eventually embraced liberal Anglicanism, drawing much inspiration from Broad Church theologians. When Macmillan published anonymously his *Ecce Homo* in 1865, a study on the historical figure of Jesus, Seeley simply followed the Victorian vogue for studies on Christ's human character. Like many other liberal Anglicans, Seeley had a strong distaste for theological wrangling. He thus criticized the High Church party for being overly concerned with policing doctrinal conformity. According to Seeley, the Broad Church was not hindered by such "conservative prejudices." Indeed, the Broad Church has allied "itself with progress, and travels in company with education." The educational policies of the High Church are in fact a "direct way to discredit Christianity, and, in a Christian country, to discredit Christianity is to discredit morality."[166]

In the articles that appeared in *Popular Science Monthly*, Seeley explained that the "Christian Church has from the beginning spoken with a certain contempt of learning." But this was a particular kind of "learning." Men of science resemble Christians in this respect, for they too have great contempt for a kind of learning or philosophy. According to Seeley, "the old religious school, and that new school whose convictions we see now gradually acquiring the character of a religion, agree in combining a passionate love for what they believe true knowledge, with a contempt for so-called learning and philosophy." Thus, while the present strife between Christianity and science seems great, he explained, there are increasingly greater points of agreement. For one, both the religious view and the scientific view "protest earnestly against human wisdom." That is, "both wait for a message which is to come to them from without." Religion waits for God to speak; science for Nature to give an answer. Both agree in "denouncing that pride of human intellect which supposes it knows every thing." What is central to an agreement between science and religion, Seeley seemed to argue, is our conception of God. When men of science speak of "God as a myth or an hypothesis, and declares the existence of God to be doubtful," Seeley observed, they are "speaking of a particular conception of God, of God conceived as benevolent, as outside Nature, as personal, as the cause of phenomena." But, according to Seeley, these attributes of benevolence, personality, and will do not exhaust the idea of God, and this is something almost all theologians agree with. According to Seeley, the man of science, whether he acknowledges it or not, is a "theist." "I say that man believes in a God," he wrote, "who feels himself in the presence of a Power apart from and immeasurably above his own, a Power in the contemplation of which he is absorbed, in the knowledge of which he finds safety and happiness. And such now is Nature to the scientific man." Seeley maintained that "no class of men since the world began have ever more truly believed in a God or more ardently, or with more conviction, worshiped him."[167] When Seeley's *Natural Religion* appeared in 1882, Youmans told readers it "deserves the most serious consideration," for it offered hope for all those "concerned about the religious progress of mankind."[168]

At the end of the decade, Youmans boasted about the growing "religious recognition of nature," which affirmed in his mind the belief that "true religion" could never come into conflict with "true science."[169] As we have seen, the rhetoric of the "religion of the future" became an increasingly popular theme by the end of the century, particularly among liberal Protestants. Because they believed religion had suffered from its association with outmoded theological doctrines, they hoped that scientific study would ultimately modernize religious beliefs. Among the liberal Protestant clergy who attempted to create a new theology along these lines, Henry Ward Beecher was particularly important for Youmans. Indeed, it was Beecher who encouraged Youmans to "Stir them up—subsoil the people with Spencer, Huxley, and Tyndall. I have got them all, and go in for them all." In what has now become a commonplace among the more religiously liberal thinkers of the age, Beecher explained to Youmans that "If the trellis of old philosophies is rotten and falling down, take it away and let us have a better. We can train the vines of the faith on the new one just as well."[170] Beecher had also attended Youmans's address at the Cooper Union in 1873, later telling him that his presentation was like a "poultice."[171]

Youmans published an essay by Beecher in the *Popular Science Monthly* in which he lamented that the clergy "stand off toward the bottom of the list among students of human nature." Beecher urged them to study "human nature," both soul and body, by means of the scientific method. He rhetorically asked, "after eighteen hundred years of preaching of the faith under the inspiration of the living Spirit of God, how far has Christianity gone in the amelioration of the condition of the race?" Not far enough, according to Beecher. "I say that the torpors, the vast retrocessions, the long lethargic periods, and the wide degeneration of Christianity into a kind of ritualistic mummery and conventional usages," he argued, "show very plainly that the past history of preaching Christianity is not to be our model. We must find a better mode of administration." That "better mode" is science. According to Beecher, the study of human nature is now being pursued in every system of liberal education, and if the church does not amend its theological systems to conform to the new facts, they will be left behind. He stressed that "the providence of God is rolling forward a spirit of investigation that Christian ministers must meet and join." Christians can no longer "afford to shut your eyes to the truths of human nature." How does the Christian minister study human nature scientifically? As one of Spencer's disciples in America, Beecher claimed a good starting point begins with Spencer's works. "There is much in him," Beecher declared, "that I believe will be found sovereign and noble in the final account of truth, when our knowledge of it is rounded up."[172]

A decade later, Beecher published an article evaluating the "Progress of Thought in the Church" for the *North American Review*. Like Draper and White, Beecher blamed theological dogmatism for the rise of skepticism among the American people. He noted, however, "that a great change, progressive and prophetic, is passing over the public mind, in matters of religious truth." Rather than fighting these changes, and thus leading to more skepticism, Beecher proposed that religious believers integrate them into a new religious movement. Thus, while the old theology had much to fear from scientific disclosures, he argued, "religion has much to hope." According to Beecher, "the future is not in danger from the revelations of science. Science is truth; Truth loves truth. Changes must come and old things must pass away, but no tree sheds its leaf until it has rolled up a bud at its axil for the next summer."[173]

In his *Popular Science Monthly*, Youmans extolled Beecher's article. "It is an independent, a powerful, and a most significant discussion, which we recommend everybody to read," he wrote. Youmans was especially impressed by Beecher's "hearty recognition of science as an agency for the purification of religion." According to Youmans Beecher showed that the gulf between religion and theology is "wide and deep, and that religion must unload theology or sink with it."[174] Later, when Beecher began preaching a series of sermons on evolution, Youmans closely followed, once again publicly congratulating him on "his determination to bring his pulpit into harmony with those revelations of science that are reshaping the thought of the age."[175]

CONCLUSION

The main concern of this chapter has been to chart the ways in which Youmans promulgated and defended the ideas of Draper, White, and the scientific naturalists, in addition to general liberal Protestant thinkers. He dedicated himself to supporting scientific naturalism, a philosophy proclaiming that only processes governed by law should be used to explain the world. Within this scheme, all elements of the supernatural were discounted. Indeed, the *Popular Science Monthly* was not only an aggregate but a catalyst for everything remotely associated with scientific naturalism. What is more, Youmans went beyond the pages of his magazine, reaching a global audience in his International Scientific Series. In both his magazine and other publishing efforts, Youmans tirelessly defended the philosophy of scientific naturalism.

At the same time, Youmans was also a staunch defender of the New Theology and all who espoused some element of it, however tangential. Indeed, he followed very closely the writings of liberal Protestants. Like other liberal Protestants, Youmans depicted orthodoxy as dogmatic and obstructionist. As we have seen, there is a long religious pedigree among liberal Protestant historians and thinkers who used narratives of corruption to demonstrate that the cause of doubt or skepticism was dogmatism, not religion. For Youmans, however, the crisis of doubt was unwarranted, for "religion" did not depend on "theology." This distinction was central not only to liberal Protestants, but also to Draper and White and other scientific naturalists. It allowed both liberal theologians and scientific naturalists to attack accepted, traditional church doctrine without fear of being labeled an atheist or materialist.

In this sense, Youmans is perhaps best seen as a mediator between scientific naturalism and the New Theology. While both positions are clearly different, as one is essentially outside Christianity and the other inside Protestantism, such distinctions, as we have seen, were blurred by Youmans. For both Youmans and liberal Protestants, the separation of religion from accepted creeds of orthodoxy provided the means of reconciling the discord between new knowledge and ancient faith. As we have seen from preceding chapters, this separation was indeed a liberal Protestant trope, a polemical narrative that first emerged against Roman Catholics and gradually applied to other Protestants as self-critique. In the hands of late nineteenth-century scientific naturalists, it became a weapon against all forms of dogmatism, particularly intransigent Protestant biblical literalism, which was now viewed as the chief obstacle to scientific and religious freedom. The same theologians who were "adversaries of science" were also adversaries of the progress of religious truth. Youmans and the other scientific naturalists could therefore criticize the dogmatism they felt inhibited intellectual progress without undercutting "religion."

It is ironic that both liberal theologians and scientific naturalists saw themselves as protectors of religion rather than its enemies. It should now be clear that Youmans viewed the historic conflict between science and orthodoxy as the means of advancing his own vision of the religion of the future. Like many of his contemporaries, he believed that the new science must inform and reform theology. Science will give new shape to religious belief. The history of human progress, according to Youmans, demonstrated that intellectual and religious progress was part of the same evolutionary process. Science was not in conflict with religion but rather with theological dogmatism. The conflict, therefore, pointed to a theology in crisis. Youmans and the scientific naturalists were not offering a new interpretation but adapting and further disseminating one that already had a long pedigree among Protestant intellectuals.

The growing belief that Christianity, after so many years of corruption, pagan accretion, and theological conflict, was now finally reaching a conciliation through the work of science became the clarion call of liberal Protestantism. But while he clearly appropriated and defended the new Protestant historiographies of Draper and White, Youmans did not share their hope that Protestantism would bring about a final reconciliation between science and religion. For Youmans, the times required a "new theology," a freer religion than even what the most liberal Protestants offered. His relentless support of Huxley, Tyndall, and especially Spencer was an essential component of a movement late in the nineteenth century that continued to adhere to the language of Protestantism but without any loyalty to its theological or doctrinal traditions. As an enthusiast for the narratives of Draper and White, Youmans pressed them into service in advancing a post-Protestant new religion. Thus, when Youmans began his work of publishing and popularizing science in the 1870s, he was already an adherent of "Protestant minus Christianity." Youmans and the scientific naturalists offered similar narratives of the history of science and religion, and all shared the common belief that organized religion, particularly orthodox Christianity, had long been a hindrance to the progress of science. At the same time, as one of the leaders of the F.R.A., Youmans reveals himself to be an extremely complex figure. Like many of the men he published, he devised his own vision of the reconciliation of science and religion. He believed that orthodox Christianity had impeded not only the progress of science but also the progress of religion toward a "higher" conception of God. The reconciliation Youmans described was the end of "orthodoxy," a term by which he meant Protestant as well as historic Christianity broadly defined. He interpreted and appropriated the narratives of Draper and White as demonstrating the progress of religion, and that in his own time traditional Christianity would finally come to an end, to be replaced by a kind of noncreedal religion preached by members of the F.R.A. In short, Youmans's motivations for publishing Draper and White illuminate an altogether neglected aspect of his publishing efforts. While we have examined only a small portion of articles published in his *Popular Science Monthly*, they reveal the significant role Youmans played in publishing, disseminating, and popularizing the historical narratives of Draper and White.

But as we shall now see, the public response to these narratives varied greatly. This reformulation of religion was ultimately self-defeating, as Draper, White, and to a large extent other liberal theologians lost control of their historical narratives to a new generation, further removed from the evangelical upbringing in which they themselves had been reared. Their rhetorical strategy, in other words, seriously backfired. There were many reasons for this failure. For one, they could not provide satisfactory answers to critics, not only to the more religiously conservative, but also to the religiously skeptical, who could not see the need for religion to persist if science answered all questions. Whereas the conservatives confronted them about their vague and ambiguous formulations of religion, freethinkers, secularists, and nonbelievers accused them of losing their nerve. At the turn of the century, a new generation emerged who saw the religiously liberal attempt to reformulate the idea of God and religion as gratuitous, superfluous, and therefore unnecessary.

READING DRAPER AND WHITE

A Failed Reconciliation

THE previous chapters give some indication of the vast changes Christianity experienced during the nineteenth century. Advances in the natural and historical sciences, intentional or not, seemed to many a direct assault on traditional Christian belief. Debates about the character of Christian faith raged both inside and outside the church during the century, and out of these debates emerged new ways of thinking about the nature of faith, the historical Jesus, the character and authority of Scripture, the truth of revelation, and the future of religion.

While the nineteenth century was undoubtedly an intensely Christian era, it was also a time of much doubt and disillusionment. As scholar of religion Linda Woodhead has put it, the nineteenth century witnessed the "reinvention" of Christianity.[1] Those who rejected traditional Christian belief, but who claimed to remain theists, often adhered to some form of liberal theology. This new or "reinvented" Christianity was part and parcel of nineteenth-century liberal Protestantism. For the religious liberal, there was an "acute sense of the need for a reformation of Christianity," an attempt to accommodate Christianity to the modern era.[2] Recognizing that advances in the sciences and historical-critical scholarship had supposedly contradicted established religious ideas, many attempted to ameliorate the emerging malaise by readjusting or reconstructing the meaning of religion. As we have seen, nineteenth-century liberal Protestantism generally responded to higher criticism and scientific naturalism by *transforming* rather than *abandoning* the faith. By the last decades of the century, the New Theology or "new religion" movement had found numerous supporters on both sides of the Atlantic.

What this "new" or "freer" religion looked like, however, was deeply contested, as we saw in the previous chapter regarding the "Quadrangular Duel" between Herbert Spencer, Frederic Harrison, James Fitzjames Stephen, and Wilfrid Philip Ward. However the "new religion" was conceived, many men and women in the nineteenth century believed that the reconciliation of science and religion depended on it. One important strategy used by liberal Protestants and religionists at the end of the century was turning "theology" into a pejorative. By contrasting the ideal of a free, progressive scientific inquiry against the authoritative, reactionary methods of theology, religious liberals imagined dogma, not faith, as the true obstacle of modern thought. Conflict occurred, they believed, not between scientific truth and religious truth, but between contesting theological traditions. If religion would only rid itself of dogmatism and ecclesiastical authority, science and religion would be in harmony. The distinction between, and separation of, religion and theology was thus incredibly important—indeed, everything hinged on it. Many liberal Protestants believed the separation of religion from theology was the best approach to bridging the schism between modern thought and ancient faith, and thus for bringing about reconciliation between science and religion. The separation of religion from theology of course antedates the late nineteenth century. As we discussed earlier, a number of seventeenth- and eighteenth-century thinkers used this distinction in their effort to construct a "rational" or "natural" religion. But the distinction was tainted by a strong anticlerical polemic that largely remained in effect into the next century, when mainstream Protestants were beginning to adopt more modernist ideas. By the late nineteenth century, however, this anti-Catholic polemic had transformed into a Protestant self-critique, and subsequently into the "Protestantism minus Christianity" narrative we have seen in the work of Edward L. Youmans and the scientific naturalists.

As editor Richard H. Hutton observed in 1867, however, while "nothing is commoner than to hear men speak in the present day of their disgust for Theology, and their love for Religion," such conclusions represented a "grave twist in the attitude of modern thought on these subjects, going far beyond, and probably even springing from something far beyond, any misuse of words." A fundamental misconception of words had occurred, according to Hutton. "Religion" cannot stand alone and apart from "Theology." Indeed, religion is merely the derivative that springs out of theology, and thus is considered "comparatively good, bad, or indifferent, in proportion as the 'science of God,' on which it has been nourished, is comparatively true or false." Thus, according to Hutton, it makes little sense to talk about religion as something separate from theology. But the more pressing issue was "a misdirection of thought," the belief that "all true thought of God is inferential, and derived from a study of ourselves."[3]

The controversy surrounding discussions over "free religion" or the New Theology is reflected in the public response to Draper and White. A close reading of a selection of periodical reviews and personal correspondence reveals three key aspects of this reception. First, the more religiously liberal welcomed their narratives as genuine attempts at reconciling religion and science. Many liberal Protestants implicitly and explicitly accepted their narratives because they had already independently fostered such narratives. They thus agreed with Draper and White that traditional theology or orthodoxy had proved wholly unable to engage science in fruitful conversation and, therefore, that Christianity needed to be thoroughly modernized.

Second, the more religiously conservative or orthodox, while recognizing their work as attempts at reconciling religion and science, nevertheless vilified Draper and White as instigators of conflict. The orthodox criticized religious liberalism as something entirely different from Christianity. Indeed, the religion of liberalism, they contended, left no room for the Christian affirmation of the transcendence of God. Divinity had thus been reduced to a vague feeling of affirming "wonder," "awe," or "presence," which was nothing more than the deification of human emotions. The effect of religious liberalism, whether among theologians or men of science, is to sanction the reigning culture.

Finally, a younger generation, near the close of the century and further removed from the kind of religious upbringing Draper and White received, appropriated their historical narratives to demonstrate that all religion had been and always will be in conflict with science. As we shall see, the liberal Protestant reformulation of religion was a risky strategy that ultimately failed. Draper and White had lost control of their narratives to a new generation of religious skeptics and unbelievers who saw such liberal attempts at reformulating the idea of God and religion as unnecessary. Science, and particularly evolution, could be invoked to explain all phenomena, from the history of the formation of the universe to the origin and development of the humanity and its social constructs.

In a remarkable twist of irony, both orthodox believers and radical unbelievers found themselves in accord: if theology was found unbelievable or unnecessary, so must be religion. Draper and White, then, were attacked by orthodox and unbelievers alike. Whereas the orthodox confronted them about their vague formulations of religion, unbelievers accused them of losing their nerve. Their narratives were appropriated by secularists and atheists and promulgated at the end of the century without any of their nuance and subtleties. In the end, their narratives were taken out of context, reconstructed into positions they did not hold, and therefore ultimately transformed into something more equivalent to the modern understanding of the "conflict thesis."

RESPONDING TO CONFLICT

Liberals

It must be stated at the outset that Draper's scientific work was overwhelmingly well received. As a professor of chemistry and medicine at New York University, his early studies in chemistry and physiology earned him such accolades as "original," "ingenious," "masterly," "profound," and even "reverent."[4] When he turned from science to historical writing, Draper also received numerous flattering letters from well-known figures. Scottish publisher Robert Chambers, for example, thought his *Intellectual Development* was a "brilliant work." Irish physicist and scientific naturalist John Tyndall believed it had "courage as well as ability." American historian George Bancroft "immediately read every word of it," adding that "it is my hope that you will lead and encourage men to the better experience of reflective judgment and clearer perception of the universality of law."[5] American "cosmic philosopher" John Fiske told his fiancée that while he thought Draper was "lame" as a classical scholar, he completely agreed with his assessment that medieval Christianity had been a "gigantic swindle" and a manifold of "iniquities."[6]

Significantly, a number of American religiously liberal magazines viewed Draper's *Intellectual Development* as an entirely "Protestant" project. *Harper's New Monthly Magazine* gushed that no argument "so magnificent has been essayed since Milton undertook to 'assert eternal Providence, and justify the ways of God to man.'" A reader in the literary *North American Review* interpreted Draper's philosophy of history as the "outgoing of the will of the immutable Creator; and Christianity is not the growth of the human intellect, but the gift of God." The Unitarian *Christian Examiner* found the work clear, earnest, and even reverent in tone, which put Draper in "favorable comparison with the pedantic dogmatism of Buckle, or the frequent harsh austerity of Comte."[7]

Across the Atlantic, the *Athenaeum* celebrated Draper's *Intellectual Development* as "the most instructive and complete of all which have yet been written with a similar ambition." Unsurprisingly, the leading Broad Church Anglican and driving force behind the *Essays and Review*, Henry B. Wilson, writing for the radical *Westminster Review*, called it a "great work" and a "valuable contribution to the study of the philosophy of history." Draper demonstrated, Wilson wrote, "how human life in society, as well as in the individual, is subject to the dominion of

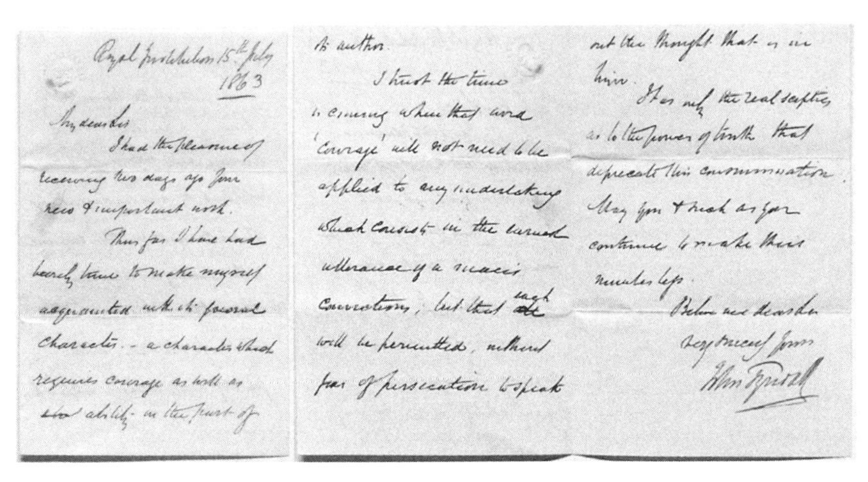

Royal Institution 15th July
1863

My dear Sir

I had the pleasure of receiving two days ago your new & important work.

Thus far I have had barely time to make myself acquainted with its general character — a character which requires courage as well as ability on the part of its author.

I trust the time is coming when that word 'courage' will not need to be applied to any undertaking which consists in the earnest utterance of a man's convictions, but that each will be permitted, without fear of persecution to speak out the thought that is in him.

It is not the real sceptic as to the power of truth that deprecates this consummation. May you & such as you continue to make them numberless.

Believe me dear Sir
very truly yours
John Tyndall

Figure 6.1. Upon receiving John W. Draper's *Intellectual Development of Europe* (1863), John Tyndall exclaimed it had "courage as well as ability." Tyndall would later use Draper's book as a guide in composing his notorious "Belfast Address" in 1874. Courtesy of the Library of Congress, Manuscript Division, John William Draper Family Papers, Container 7.

law." In another issue, English jurist Sheldon Amos praised it as "encyelopædic," [*sic*] "graceful," "eloquent," and "indisputably true." Amos agreed with Draper that the influence of Constantine upon the course of Christian history changed it forever, ultimately leading to its "paganisation." He also noted that "we are apt to overlook what Christianity was once, and what in its own essence it is still." Constantine thus "prostituted" the simple and pristine message of Christ "in order to reinvigorate the putrid mass of the Roman Empire."[8]

When *History of the Conflict* appeared in 1874, Draper continued to receive numerous letters of support. Bancroft, for example, had informed him that he sent a copy of his book to "Prince [Otto von] Bismarck, with a letter touching on all the points." "No man in America," he wrote, "could have come near which you have accomplished. You bring to your work a knowledge of the natural sciences and of the history of man in all his relation."[9] Early in 1865 Draper had sent copies of his *Thoughts on the Future Civil Policy of America* and *History of the American Civil War* to Tyndall. By the end of the year, Draper told Tyndall that he was giving up physics and chemistry for history. "Physics no one reads," he reported, "history every one reads." Tyndall responded that "science is a serious loser by your choice." However, he did notice in Draper's historical writings the "spirit of science," and that his books "thus advance the good cause though in an indirect way."[10]

Moreover, when Tyndall heard of Draper's new book, he also wrote to tell him that "no intelligence that I have lately received is more agreeable to me than that which informed me of your intention to give us a 'History of the Conflict

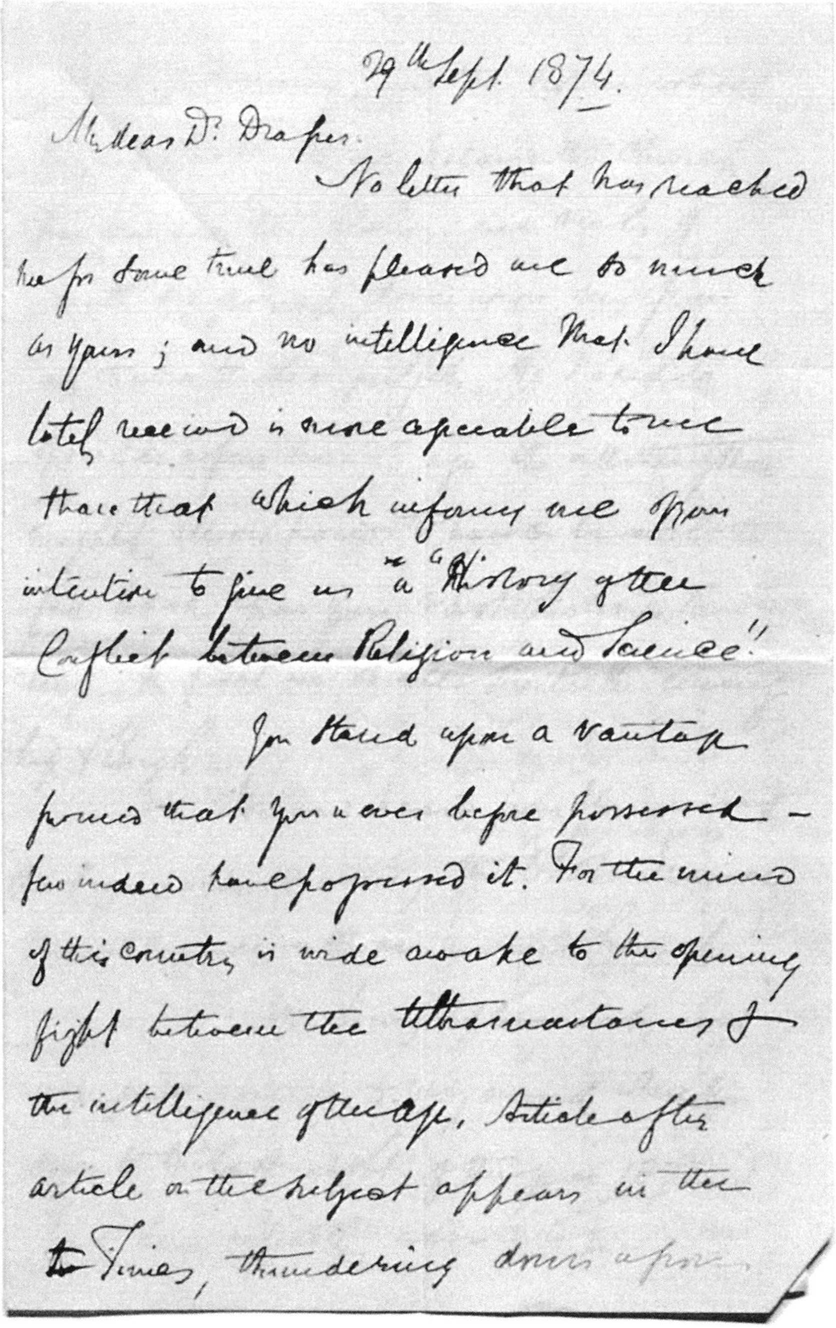

29th Sept. 1874.

My dear Dr. Draper.

No letter that has reached me for some time has pleased me so much as yours; and no intelligence that I have lately received is more agreeable to me than that which informs me of your intention to give us "a History of the Conflict between Religion and Science."

You stand upon a vantage ground that you never before possessed — few indeed have possessed it. For the mind of this country is wide awake to the opening fight between the ultramontanes & the intelligence of the age. Article after article on the subject appears in the Times, thundering down upon

Figure 6.2. When he heard from Edward L. Youmans that John W. Draper was completing a book on the *History of Conflict between Religion and Science* (1874), John Tyndall inwardly shouted "hurrah!" Courtesy of the Library of Congress, Manuscript Division, John William Draper Family Papers, Container 7.

between Religion and Science.'" Indeed, in his "Apology for the Belfast Address," Tyndall recommended readers to seek out "Dr. Draper's important work entitled 'History of the Conflict between Religion and Science.'"[11]

As with his *Intellectual Development*, Draper's *History of the Conflict* received many favorable reviews from liberal readers. For example, a reviewer for *Appleton's Journal*, incidentally the same organ that published the writings of both Draper and White, praised it for its "peculiar force" and "lasting influence," observing that the conflict between religion and science had become the topic of discussion "of every thoughtful man." The heart of the matter, the reviewer explained, was the "religious question"—namely, its changes, survival, and future. Even a local newspaper like the *New Orleans Bulletin* understood that Draper was not just tracing the history of science as much as the "rise and progress of religion." A reviewer for the New School Presbyterian *New York Evangelist* aptly summarized the view of many liberal Protestant readers, describing Draper's book as the "most powerful argument within my knowledge in defense of Protestantism and modern civilization."[12]

Still, if the liberal press were quick to praise Draper for his religiously progressive views, many also attempted to moderate or correct his terminology, particularly his imprecise use of the term "religion" in his title. This also reflects the important shift from an older, rationalistic interpretation of religion to a more romantic one, found particularly among British and American liberal Protestant intellectuals later in the century. A reviewer for the *New Orleans Monthly*, for example, recognized that Draper was "not opposed to Christianity," but did question his choice of terms. Strictly speaking, the conflict is between "the church and science,—the church being regarded as an organization, local or more general, embodying a system of religious belief." The *Universalist Quarterly and General Review* agreed that Draper needed to amend the title of his book from "Religion" to "Church"—or more precisely still, the "Roman Church." Nevertheless, the reviewer believed that the question of the relationship between religion and science was "the question which is now agitating the world of thought," and was grateful for Draper addressing the topic in such a clear manner. Similarly, Unitarian clergyman Thomas Hill maintained in an article for the *Unitarian Review and Religious Magazine* that the "title of Dr. Draper's book is a mistake." He thought the book did demonstrate, however, that conflict between science and Christianity first arose "when the church had been corrupted into a political engine, and a money-grasping corporation." According to Hill, this corrupted church persecuted science *and* religion. "For every martyr of science," he wrote, "history can show a thousand martyrs of religion, slain by the ecclesiastical powers of Rome." The *Westminster Review* also thought Draper's latest work should have been titled "History of the Conflict between Dogma and Science," and not religion and science. It is in "every case those formulas which have been invented for ecclesiastical purposes, rather than the religious sentiment, pure and simple, with which science is in antagonism." Thus, despite their critique of his phraseology, they all agreed with Draper that the conflict began when the Christian Church attained secular power.[13]

Other reviews reflect the incongruity and often strong disagreements between contending liberal factions. Reviews in Henry Ward Beecher's *Christian Union* magazine, for instance, reveal more carefully measured praise. Judging by its title alone, one reviewer remarked, it was a total failure. Its central weakness was its "inability to distinguish between actual religious feeling and the dogmas which are superimposed upon that feeling rather than resulting from its development." But another commentator argued that Draper was no atheist, "as some over-candid religionists are fond of asserting." Draper's God, however, was nevertheless an "impersonal God." Such theism, the reviewer remarked, "can have no influence upon the life; it can inspire no fear, no hope; it makes no action different from what it would have been to the avowed atheist." Yet another commentator faulted Draper for not distinguishing between "the Latin Church and Christianity." But he nevertheless agreed with Draper that the early church committed a "gigantic blunder" when "she allied herself with the civil powers." One final commentator ironically lamented Draper's bias, averring that he needlessly offended "his Christian readers" and, more importantly, destroyed "his influence as a historian." In truth, science and religion are "by nature complementary, not antagonistic." It is only the "dogmatists in either camp that seek for quarrels or impugn each other's motives."[14]

Conservatives

While many of the more conservative periodicals agreed that Draper was attempting to reconcile religion and science, they insisted he ultimately failed. Despite his irenic intentions, they argued, his work could only lead to further antagonism and even unbelief. Presbyterian theologian Henry B. Smith, for instance, in the Old School Calvinist *American Presbyterian and Theological Review*, sharply (but incorrectly) criticized Draper's *Intellectual Development* as "identical with that which Comte, Buckle, and Mill have been elaborating for the last quarter or third of a century." While Draper was clearly not among the skeptical materialists, the drift of his argument, Smith asserted, will "encourage those speculations," and ultimately enthrone "physical laws as supreme." Further, he marveled at "how a believer in God and a divine government, and in man's immortal destiny, can advocate such a view." The work was, therefore, a "mistake and a failure." Similarly, a reviewer for the conservative *Biblical Repertory and Princeton Review* recognized that while Draper proclaimed a "personal God" and "Sovereign Constructor of the Universe," these proclamations were vague and undefined. In fact, Draper had undermined these pious remarks when he argued that divine intervention was "derogatory to the thorough and absolute sovereignty of God." A universe wholly independent of the support, guidance, and control of God made religious belief superfluous. Draper's philosophy, then, could only lead to a "dogma of infidelity and atheism." By reciting the "exaggerations, corruptions, and superstitions that have obtained credence, through the misleading and perversion of man's moral and religious nature," the reviewer explained, Draper's work could only bring about contempt for "all belief in religion and supernatural agencies."[15]

But like the more liberal periodicals, the critical reviews were often just as conflicting. This is partly the result of Draper's apparent anti-Catholic rhetoric, which most Protestants viewed by default if not by admission as "pro-Protestant," irrespective of where they leaned theologically. The *Nonconformist*, for instance, recognized that Draper's sympathies lay with Protestant Christianity and that he sought the "reconciliation of religion and science." But Draper ultimately disappointed, and thus the reviewer bemoaned that his work could have been an "*Eirenikon* by all who believe, both that the Bible gives a true and authoritative revelation from God, and that the book of nature cannot lie." The *Methodist Quarterly Review*, which had once praised highly Draper's scientific work, believed his *History of the Conflict* was "overdrawn." The reviewer accused Draper of depicting men of science as a "glorious army of martyrs" and "ferocious theologians after them with a Bible in one hand and a fiery fagot in the other." The *Southern Review* even quipped that Draper's title was "so adroitly chosen, that it is worthy of the cunning of the Old Serpent." Other conservative periodicals blasted Draper's attempted reconciliation as a charade. A reviewer for *Scribner's Monthly*, for example, reproached Draper's irenic intentions as "a thinly disguised attack, not only upon Christian revelation, but all religion outside of the creed." The *Presbyterian Quarterly and Princeton Review* could not find any clear definitions of either "science" or "religion" in Draper's book. Indeed, the reviewer found it difficult to discern whether Draper believed in God or the immortality of the soul at all.[16]

But the most critical reviews came, perhaps unsurprisingly, from Catholic periodicals. Orestes A. Brownson (1803–1876), a New Englander who danced between Old Calvinist Presbyterianism, universalism, transcendentalism, and agnostic socialism before finally converting to Roman Catholicism, believed the positive reception of Draper's work was a sure sign that society had become profoundly naïve. While Draper acknowledged a belief in a personal God, the admission, Brownson argued, "seems to be only a verbal concession, made to the prejudices of those who have some lingering belief in Christianity." Cast in a "purely materialistic mould," and written to show that "all philosophy, all religion, all morality, and all history are to be physiologically explained, that is, by fixed, inflexible, and irreversible natural laws," Draper only offered a pretended reconciliation. Thus, despite his profession of a "Supreme Being," the "living and ever-present God, Creator, and upholder of the universe, finds no recognition in his physiological system." His philosophy, in the final analysis, was nothing but "pure materialism and atheism."[17]

An interesting feature of Brownson's review is that he accused Protestantism of creating men like Draper. He explained that Protestantism has always been "the religion of the intellect." But this has come at a cost: "Philosophy, science, Biblical criticism, and exegesis, the growth of liberal ideas, and the development of the sentiments and affections of the heart, have made an end of Protestantism." In other words, from its inception, Protestantism had gradually moved away from its religious roots and developed into a "vague philanthropy, a watery sentimentality, or a blind fanaticism, sometimes called Methodism, sometimes Evangelicalism." Protestantism, in other words, was a tangle of contradictions. It both celebrated reason and feared its free exercise; it empowered individuals as interpreters of their own Bibles, but it established authoritarian covenant communities; it could foster a stifling conservatism, yet it carried the logic of radical dissent like a dormant virus. The Protestant appeal to private judgment splintered the old church and created new ones—a process that occurred repeatedly in Protestant America. According to Brownson, the Catholic Church is the "only living religion that does, or can, command the homage of science, reason, free thought, *and* the uncorrupted affections of the heart."[18]

Brownson later returned to Draper in another intensely critical review for his own magazine, *Brownson's Quarterly Review*. Like many other conservative readers, Brownson could not find a clear definition of either religion or science in Draper's book. He argued that a conflict between "religion and science" was impossible, for they were "two parts of one dialectic whole." He claimed that Draper had conflated all religions, from gross fetishism to debased polytheism to absurd fable and obscene rite, making Christianity responsible for it all. This was and still is, he argued, the rhetorical strategy of "Protestant Christianity." According to Brownson, then, Draper was firmly rooted in the Protestant heritage.[19]

The *Catholic World* also roundly condemned Draper's narrative. Its reviewer claimed it was a "farrago of falsehoods, with an occasional ray of truth, all held together by the slender thread of a spurious philosophy." Draper, the reviewer claimed, willfully misrepresented Catholic doctrines and confounded Catholicism with a number of vague and incongruous beliefs. His history was an unbroken tissue of "fatuous drivel." According to the reviewer, Draper is indeed the true instigator of conflict, not the Catholic Church. He may claim the reality of divine governance and the immortality of the soul, but "his whole book," the reviewer explained, "is a cumbersome and disjointed argument in favor of necessity, as opposed to free agency; of law, as opposed to Providence." Like Brownson, the reviewer believed Draper was repeating and multiplying the "old, time-worn, oft-refuted, and ridiculous stories which stain the pages of long-forgotten Protestant controversialists." The real concern over Draper's book was the "unhealthy condition of the public mind which can hail its appearance with welcome." Its positive reception thus exemplified the "diseased mind" of the age.[20]

Unbelievers

Many of the concerns of conservative readers were prescient, as other readers reveal just how easily Draper's narrative could be appropriated for antireligious purposes. A certain Joseph Treat had sent Draper a meandering letter describing the need for a revolution in science. Interestingly, he argued that he could "demonstrate" that all science was based upon an incredible falsehood—namely, Christianity. Great men of science, from Copernicus to Spencer, according to Treat, were tainted by a "Christian character" that prevented them from truly understanding nature. The first step in liberating science from Christian bondage was the study of history. This is the reason he admired Draper's "grand work" of history. Treat seemed to believe that the history of science was the gradual emancipation of science from religion. He asserted that the Bible was the "sole cause of conflict between science and religion," and that Draper's history served well his cause to correct science and eradicate Christianity.[21]

Another good example of this kind of response was made by T. D. Hall, a member of the National Liberal League (later renamed the American Secular Union), which was a group of radicals who broke away from the Free Religious Association.[22] Hall published a pamphlet entitled "Can Christianity be made to Harmonize with Science?" He declared that the historical "facts" in Draper's book are not new. They have been known since the Protestant Reformation. Draper has simply gathered these facts with such skill, and stated them so clearly, that he has "brought them to bear with such overwhelming force." Draper's only failure, according to Hall, was that he "seemed to lack the courage and candor to state the *whole truth*, regardless of consequences"—namely, that science and Christianity are completely and irrevocably incompatible. One cannot abandon core Christian principles and expect it to survive. Doctrines such as the Fall, Atonement, and the Resurrection are the "very life blood of Christianity," and to abandon them is to "cause it shortly to cease to exist as an organized system of religious belief." Declaring essentially that Draper had lost his nerve, Hall maintained that "nothing is ever gained by stopping short halfway, and endeavoring to thrust out of sight a part of the necessary effects of causes which we know to be operating."

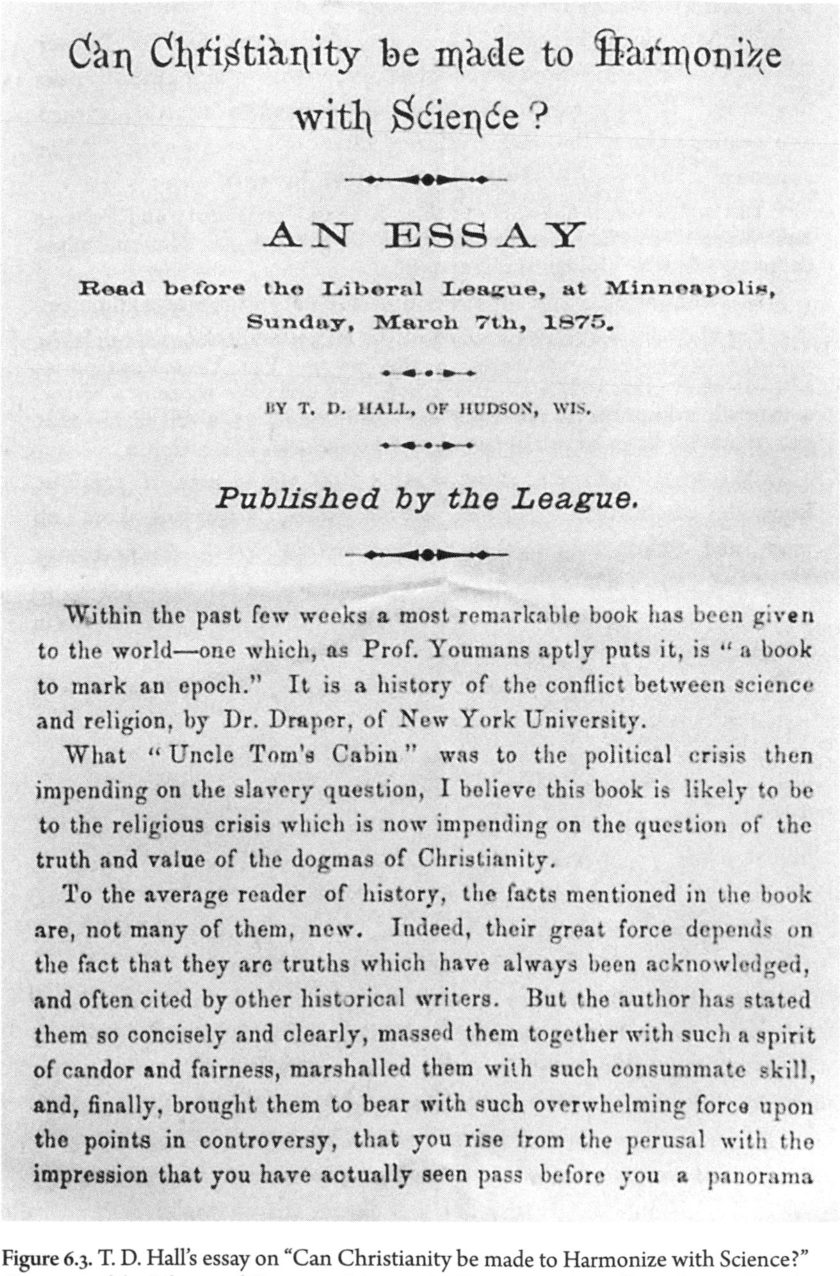

Can Christianity be made to Harmonize with Science?

AN ESSAY

Read before the Liberal League, at Minneapolis, Sunday, March 7th, 1875.

BY T. D. HALL, OF HUDSON, WIS.

Published by the League.

Within the past few weeks a most remarkable book has been given to the world—one which, as Prof. Youmans aptly puts it, is "a book to mark an epoch." It is a history of the conflict between science and religion, by Dr. Draper, of New York University.

What "Uncle Tom's Cabin" was to the political crisis then impending on the slavery question, I believe this book is likely to be to the religious crisis which is now impending on the question of the truth and value of the dogmas of Christianity.

To the average reader of history, the facts mentioned in the book are, not many of them, new. Indeed, their great force depends on the fact that they are truths which have always been acknowledged, and often cited by other historical writers. But the author has stated them so concisely and clearly, massed them together with such a spirit of candor and fairness, marshalled them with such consummate skill, and, finally, brought them to bear with such overwhelming force upon the points in controversy, that you rise from the perusal with the impression that you have actually seen pass before you a panorama

Figure 6.3. T. D. Hall's essay on "Can Christianity be made to Harmonize with Science?" Courtesy of the Library of Congress, Manuscript Division, John William Draper Family Papers, Container 14.

One thing was clear: before the onward march of science, Christianity will and must disappear.[23]

Treat and Hall were both subscribers to the *Truth Seeker*, an American freethought magazine edited by D. M. Bennett (1818–1882) and published by the National Liberal League. Founded in 1876, the National Liberal League sought to promote secularism in the United States and vigorously oppose church influence in the public sphere. The masthead of the *Truth Seeker* embraced "as in one brotherhood" liberals, free religionists, rationalists, spiritualists, universalists, unitarians, friends, infidels, and freethinkers—"in short, all who dare to think for themselves." The second half of the nineteenth century was a golden age of such radical movements, a time when freethinkers on both sides of the Atlantic were celebrating the emancipation of man from religion. Movement leaders tended to reside in urban centers, but thousands of adherents were scattered throughout the rural countryside, linked by a number of circulated freethought tracts, pamphlets, and magazines. Some of its most prominent voices at the end of the century were personalities such as George Jacob Holyoake (1817–1906), D. M. Bennett, Charles Bradlaugh (1833–1891), Robert G. Ingersoll (1833–1899), Samuel P. Putnam (1838–1896), Alfred W. Benn (1843–1915), Joseph M. Wheeler (1850–1898), John M. Robertson (1856–1933), George E. MacDonald (1857–1939), and perhaps most important of all, Joseph M. McCabe (1867–1955).[24]

It is important to note that it was paradigmatic that most of these secularists had liberal Protestant upbringing. As Leigh Eric Schmidt and more recently Christopher Grasso observe, there was an uncanny symbiotic relationship between American Protestantism and secularity. The outspoken American "agnostic" savant Robert Ingersoll, for instance, was the son of a Presbyterian minister who inclined towards liberalism. It was easy for Ingersoll, then, to express a basic faith in science that eventually overwhelmed all religious belief. While freethinker Samuel P. Putnam was raised in the Congregational church, he was gradually influenced by the liberalizing religious movements of Channing, Bushnell, and Emerson that pushed within and then beyond Protestantism, which ultimately led him to abandon theism altogether. Without any authoritarian clerical body, many American Protestants precariously drifted between liberalism to freedom of conscience to outright infidelity. As we have noted earlier, many Americans wrestled with the questions and the answers religion offered them. Protestantism and unbelief always shared a matrix of rival, ally, and enemy. Secularism and liberal Protestantism were perennially intersecting movements.[25]

The point that needs to be emphasized here is that skeptics on both sides of the Atlantic appropriated Draper's historical narrative in their calls for the secularization of society. In 1885, Charles Albert Watts (1858–1946), son of famous secularist publisher Charles Watts (1836–1906), began publishing a short monthly guide devoted to advertising the latest "liberal and advanced publications"—namely, the *Watts's Literary Guide*. The journal became one of the most successful publishing ventures of the secularist movement, and later was renamed the *New Humanist*, which remained in print as of 2019. During his editorship, Charles Albert Watts treated the *Guide* as a "publisher's circular," offering anyone interested in atheism, freethought, or secularism "a complete record of the best liberal publications in this country." Charles Albert filled its pages with summaries of the works of Buckle, Lecky, Spencer, Huxley, Darwin, and others—including Draper.[26]

Soon after launching the *Guide*, Charles Albert Watts organized the Propagandist Press Committee "to assist in the production and circulation of Rationalist publications." These efforts attracted a larger number of subscribers than previously thought possible for a secular organization, and also increased the number of reviews and publications advertised in the *Guide*. Near the end of the century, Charles Albert had formed the Rationalist Press Association and the publishing house Watts & Co. The Rationalist Press Association sought to "stimulate freedom of thought and inquiry in reference to ethics, theology, philosophy, and kindred subjects," to "promote a secular system of education," and to "maintain and assert the same right of propaganda for opinions and ideas which conflict with existing or traditional creeds and beliefs as is now legally exercisable in favor of such creeds and beliefs." To this end, the association published and reprinted books and pamphlets on science, history of religion, biblical criticism, and biographies of rationalists. The Rationalist Press Association, in short, aggressively and deliberately sought to give the perception that modern life was inherently inhospitable to religion.[27]

Writers for the Rationalist Press Association and other secular publishing firms consistently turned to Draper and his historical narrative in support of their own agenda. In his short essay on the *Dynamics of Religion* (1897), for example, prolific secular journalist John M. Robertson listed Draper as an "infidel."[28] Robertson continued to cite Draper's historical narrative as a general history of freethought, but later explained in his *A Short History of Freethought: Ancient and Modern* (1906) that the "survival of theism" in Draper was part of psychological and social pressure, and that this demonstrated the "elements of essentially emotional and traditionary supernaturalism." Draper is an example of how "men engaged in rationalistic and even in anti-theological argument" could continue to cling to religion.[29] Moreover, in his massive two-volume *History of Freethought in the Nineteenth Century* (1929), he argued that Draper's narrative was a "powerful contribution to popular rationalist culture." This work, Robertson explained, was a substantially "evolutionary view of social and mental progress." While he admitted that Draper wrote as a "theist," Robertson argued that Draper was "insistently naturalistic in his whole survey," and therefore acted as a partial solvent to his own theological position. His book became a "freethinker's book, for freethinkers, and it is a curious circumstance that this very aggressive treatise has gone in the ordinary way of bookselling during sixty odd years, without any noticeable polemical notoriety."[30]

Other secularists had reached similar conclusions. Joseph M. Wheeler, whose *Frauds and Follies of the Fathers* (1882) appropriated Protestant historiography to demonstrate the irrationality and credulity of the Church Fathers, recorded Draper as a "freethinker" in his monumental *Biographical Dictionary of Freethinkers* (1889). Samuel P. Putnam, who had notoriously described religion as a curse, disease, and lie, adopted Draper as an authority throughout his *400 Years of Freethought* (1894). George E. MacDonald, in his *Fifty Years of Freethought: Being the Story of the Truth Seeker* (1929), mentioned how the writings of Draper played a central role in D. M. Bennett's rationalistic campaign.[31]

But perhaps the most important secularist to have appropriated Draper's historical narrative was Joseph McCabe. A Roman Catholic monk who abandoned his religious beliefs around 1895, McCabe was a prolific author, writing over two hundred books on science, history, and religion. A vehement advocate of atheism, McCabe frequently forecast the doom of Christianity in light of modern science. He mostly published with Watts & Co. in London, and particularly noteworthy was his massive *Biographical Dictionary of Modern Rationalists* (1920). Like other secularists, McCabe included Draper in his tribute. While he was careful to note that Draper was "a Theist, and believed in personal immortality," his work nevertheless became one of the "classics of Rationalist literature."[32]

McCabe also found a home in Girard, Kansas, with the "Henry Ford of Literature" Emanuel Haldeman-Julius (1889–1951). An atheist and socialist newspaper publisher, Haldeman-Julius began printing a cheap paper-covered "Little Blue Books" series in 1919. He sold an estimated 500 million copies of these pocket-sized books to working-class and middle-class Americans. As "publisher for the masses," Haldeman-Julius intended the Little Blue Books to be a "University in Print," and included a great mixture of classical literature, novels, how-to manuals, and essays on sexuality, politics, philosophy, history, religion, and science. Haldeman-Julius also played a large, albeit almost forgotten, role in the life and literary careers of Will Durant, Bertrand Russell, and Clarence Darrow.[33]

McCabe was by far the most prolific writer for the Little Blue Books series. One of the most successful books in the series was, interestingly enough, his *The Conflict Between Science and Religion* (1927).[34] In this work, McCabe essentially repeated Draper's narrative. But unlike Draper, McCabe gleefully cheered on

Figure 6.4. Emanuel Haldeman-Julius (who attended the Scopes evolution trial in 1925) at his home near Girard, Kansas, with Clarence Darrow. Courtesy of Haldeman-Julius Collection, Leonard H. Axe Library, Pittsburg State University, Kansas.

the decay of religion all over the earth. Historians of the twenty-first century, he argued, will look back with amusement at those men of science and theologians of his own century who protested that there was no conflict between religion and science. "He will read the priests protesting," he wrote, "that there is no conflict between true science and religion, and the professors plaintively chanting that there is no conflict between science and true religion." But according to McCabe, all future historians will recognize that "science has, ever since its birth, been in conflict with religion." The Christian religion, McCabe contended, was the "most deadly opponent" of scientific progress.

More importantly, the vast majority of McCabe's Little Blue Book was a diatribe against "progressive religion." Ironically, he repeated the same arguments of conservative and orthodox opponents of Draper. McCabe called liberal Protestantism the "veriest piece of bunk that Modernism ever invented." According to McCabe, those liberal theologians who reinterpreted traditional religious belief, wittingly or unwittingly, attacked the very foundations of Christianity. The modernists, McCabe wrote, "are Christians who believe that Paul and the Christian Church have been wrong in nearly everything until science began to enlighten the world." But to reject its central doctrines, he argued, is to reject the whole of Christianity. Even those "extreme modernists," such as E. Ray Lankester, Oliver Lodge, Henry F. Osborne, Mihajlo I. Pupin, Robert A. Millikan, and others, who have taken a conciliatory approach, are wrong.[35] These men were members of the scientific community who wrote during the antievolution controversy of the 1920s. Some of them contributed to the pamphlet campaign of the American Institute of Sacred Literature, which published a series of leaflets asserting the harmony of science and religion. But as Edward B. Davis argues, these pamphlets "represented a variety of modernist theological positions reconciling scientific knowledge with religious faith." Indeed, theologian Shailer Mathews, editor of the series, rejected traditional Christianity. "Mathews wanted," Davis writes, "a new Christian faith to replace the old, the religion of Jesus without the Jesus of religion."[36]

For McCabe, however, any conciliatory approach demonstrated a misunderstanding of the true nature of religion. Science is a unified endeavor, whereas religion has never been unified. If one seeks the reconciliation of science and religion, "we shall have to take three hundred different collections of religious beliefs and apply science to them." But this was impossible, McCabe argued. "The land which lies between straight Fundamentalism and straight Modernism," he quipped, "is the Land of Bunk." In the final analysis, the "Christian rationalist," whether he realizes it or not, is still in conflict with science. Those who have succumbed to scientific ways of thought have divested God of all personality, reducing traditional conceptions to abstractions of Power, World-Energy, Cosmic

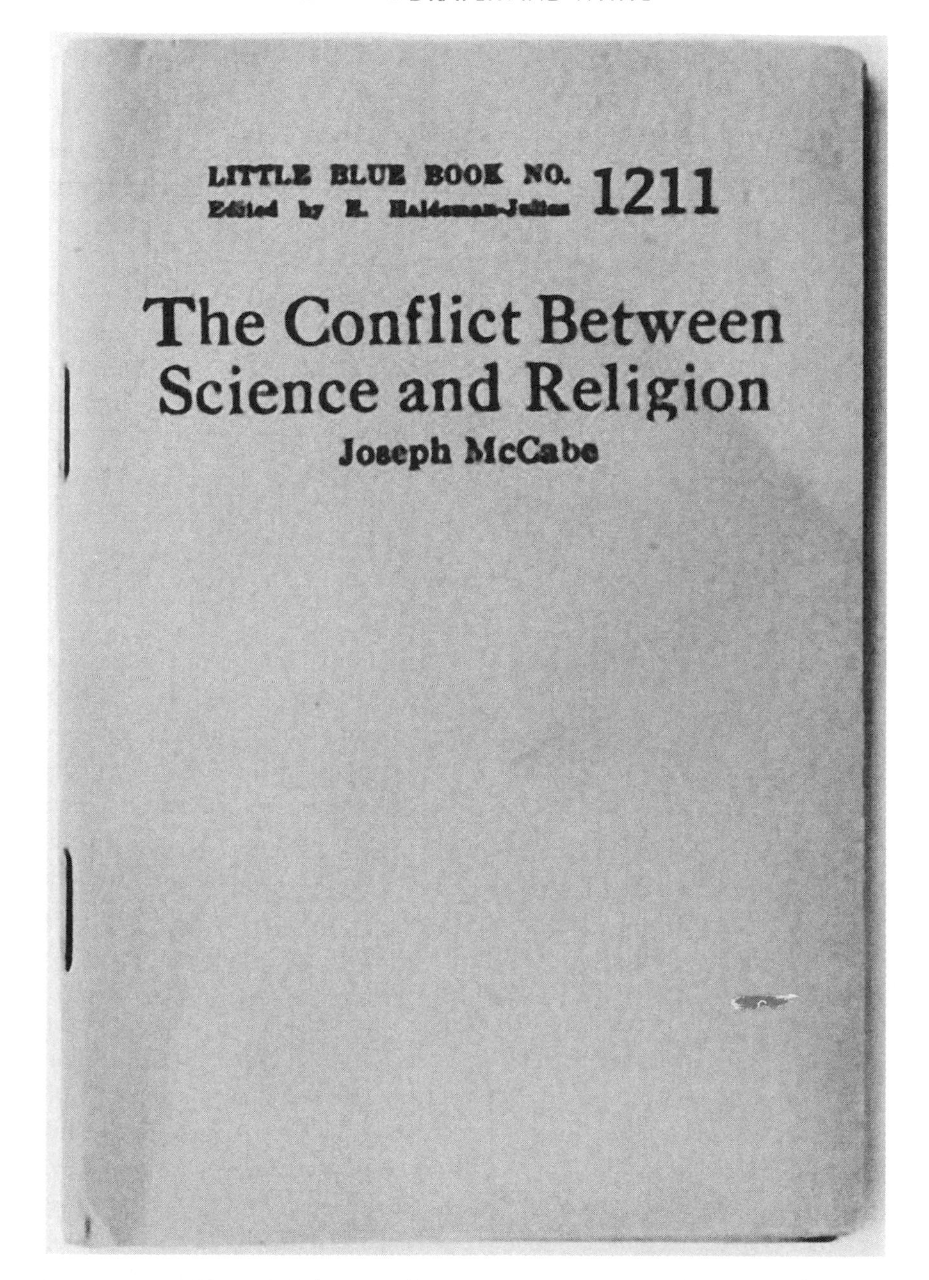

Figure 6.5. Joseph McCabe's (1867–1955) *The Conflict Between Science and Religion* (1927), published in Emanuel Haldeman-Julius's "Little Blue Book" series.

Force, Soul of the Universe, Vital Principle, Urge, Creative Principle, Absolute, and so on. However this deracinated religion was formulated, it is nothing but mere wish fulfillment, and, therefore, resided in the "land of bunk."

REACTING TO WARFARE

Liberals

When we turn to the various responses to White's historical narrative, we must remind ourselves that it differed little from that of Draper, despite White's distinction between religion and theology. That is, while Draper's narrative was clearly influenced by Protestant rationalism, and White's by Protestant romanticism and idealism, they often pointed to the same examples of theological obstructionism. Although White seemingly took a more measured position in his book titles, this distinction between theology and religion was little more than an afterthought. His readers did not fail to notice and accuse him of such duplicitous rhetorical tactics.

Unsurprisingly, liberal readers quickly embraced White's narrative as an effort at reconciling modern life and religion. The London *Popular Science Review* hoped it "will help allay some of the fears which religious people feel in regard to science, which can never come into conflict with genuine religion." Similarly, the *Dublin University Magazine*, usually a more conservative voice among periodicals, nevertheless commended it as an "admirable work," written in a "most tolerant, conciliatory, and rational spirit." White was an example of a true "Christian philosopher." The *Athenaeum*, while criticizing White's understanding of medieval thought, still called on "free men to repeat again and again in the ears of their children the tales of this book." The *Westminster Review* was perhaps the most generous. Its review agreed that the "history of science is not a tranquil narrative of happy revelations, but a troubled history of battle and murder." Although White at times carries the metaphor too far, in the main his work is a "remarkably full and perspicuous history of some very terrible events—events which have an intense interest in relation to the history of progress."[37]

When White published his two-volume magnum opus in 1896, American religious liberals continued to highly praise it. The *Outlook*, the rebranded *Christian Union*, commended White for differentiating between "religion and dogma," and compared his work favorably against Draper's earlier work. White's book "should stand beside the 'History of Doctrine' in every theological seminary," the reviewer declared. His history demonstrated that "the woeful error of these ages has been the merging of the interests of spiritual religion in those of intellectual theology. If the warfare of science has destroyed the body, it has been for the salvation of the spirit." Science, having freed religion from dogmatism, will now "go hand in hand with her." The Congregational weekly *Independent*, to which Henry Ward Beecher was a regular contributor and once served as editor, explained that the book was "an elaborate argument under many heads on the repressive influence theology is said to have exerted on the intellectual progress of Christendom." Its central aim was to "convict theology as a vicious tendency to interfere with scientific freedom to the injury of both science and religion." In attempting to clarify White's argument, the reviewer maintained that the "warfare" was "not between theology and science, but between opposite tendencies in the human mind itself." White demonstrated that it is not theology that comes into conflict with science, but rather an "obstructive and conservative phase of intellectual progress."[38]

Founding president of Stanford University, David Starr Jordan, likewise perceived the enemy of scientific progress as man's innate conservatism. In his presidential address of 1895, Jordan argued that religion was a "condition of the mind and heart—an attitude, not a formula." A pure and undefiled religion does not consist of "creeds" or claims of "orthodoxy."[39] Elsewhere he explained that the conflict had always existed in the human mind. Beginning in the mind, it then develops into organized bodies, and in turn becomes a struggle between freedom and democracy. "The whole conflict is a struggle in the mind of man. It exists in human psychology before it is wrought out in human history. It is the struggle of realities against tradition and suggestion. The progress of civilization would still have been just such a struggle, had religion or theology or churches or worship never existed."[40] Thus when he reviewed White's book for the transcendentalist *Dial*, Jordan repeated these arguments. He praised White's book as "one of the great works of our century," and acknowledged his emphasis on the "struggle of dogmatism." But again, Jordan felt that dogma was nothing more than the "desire of organized conservatism to limit action."[41]

Adherents of the New Theology applauded White's supposedly conciliatory efforts. Theologian Theodore T. Munger, for instance, was an admiring correspondent, often inviting White to his New Haven home and the Unitarian Club.[42] American industrialist and philanthropist Andrew Carnegie wrote that he had followed White's work closely, declaring that the world was rapidly evolving a new and better religion.[43] One Elbridge D. Jackson praised White's view of sacred literature, particularly the Bible, as "eminently precious, not as a record of outward fact, but as a mirror of the evolving heart, mind, and soul of man." Jackson told White that he now had a "better opinion of the Bible," seeing it as "the expression of the very highest flights the human mind and heart have reached in seeking after God."[44] James Gorton, a pastor of a Universalist church, hoped to use White's book in his Sunday school classes.[45]

White's own colleagues from Cornell University, however, noted the fine line he walked. Charles Kendall Adams (1835–1902), for example, White's successor at Cornell, published a review in the popular American magazine *Forum*, proclaiming the book as exceptional and important. At the same time, Adams observed, White's reverent spirit towards "religion" will be lost to many readers.

Indeed, Jacob Gould Schurman (1854–1942), third president of Cornell, could not understand White's desire to preserve some remnant of Christianity. Schurman explained that White's book was neither a history of science, nor a history of dogma, but a history of collisions between the sciences and the "dogma laid down in the creeds of Christendom." He praised it as a "self-attesting encyclopædia," and believed it should be consulted by scientist, historian, and theologian alike. But according to Schurman, "Knowledge is a continuous becoming; it has never attained—it is always on the way." Consequently, religion evolves as society evolves, "the most assured dogmas of to-day may need modification and adaption to the larger vision and deeper insight of to-morrow." Religion, he contended, "would still exist were theology and theologians annihilated." Schurman thus could not understand White's desire to save some outmoded, now largely obsolete version of Christianity.[46]

Conservatives

While many of the religiously liberal emphasized White's "profoundly religious nature," others were unconvinced. American Catholic priest Augustine Francis Hewit (1820–1897), for example, in a series of articles argued that "all science is from God, and is a rethinking of his thoughts." He maintained that "the book of nature and the book of revelation are both alike from God," and thus science and religion were "intrinsically in harmony." Where there is discord or struggle, it arises from mistakes and misunderstandings on one side or the other. As far as White's narrative was concerned, Hewit anticipated the modern revisionist historiographer's emphasis on complexity, making distinctions between conservatism and innovation, personal basis and political circumstance. Hewit went on to call Protestants "inconsistent supernaturalists, who undertake the vain labor of uniting a defense of Christianity with a rejection of Catholicism." The rising "agnostic tribe" drew heavily from these "inconsistent supernaturalists," attacking religion in general as "the product of a long, dark age, on which the light of science is just beginning to dawn." Such was the testimony, he concluded, of Protestantism.[47]

While conservatives often recognized White's perceived irenic attitude, they nevertheless warned that the book would ultimately undermine religion—and Christianity in particular. A reviewer for the politically whiggish but not radical *Manchester Guardian*, for instance, thought White was an opportunist, and that his work demonstrated "little insight into the greatness of the Middle Ages." Rather than dealing with the complexities of history and theological thought, the book was merely a "storehouse of curiosities of superstition," and thus "not a great contribution to the history of thought." The conservative *Speaker* was both disappointed and discouraged, blasting White's book as "a most unscientific work." If "religion and science" was a misnomer, so was science and theology. "Science has from the moment of its birth been at war with ignorance, or prejudice, or folly, but surely not with theology, which must be said, in any tolerable sense of the term, to belong to the great army of the sciences." The Tory *Quarterly Review* recognized White's approach as "religious and reverent," but argued that the same objections raised against Draper could also be raised against him.[48]

Many American readers did not take White's supposed conciliatory approach very seriously. Indeed, they believed it was a ruse. The *Chicago Tribune* published several articles condemning White's understanding of the miracles in the Bible, including pieces by clergyman Alexander Patterson and oriental scholar Elizabeth A. Reed.[49] A reviewer for the *American Historical Review* recognized that White designed his book to "prevent unnecessary damage to Christianity," but noted that his understanding of Christianity was entirely ambiguous. In some of his more conciliatory phrases, White seemed to practice the "very *sancta simplicitas*" he ridiculed. When White appeared to concede that the theologian and the man of science both shared a sincere love of truth, the verdict of his account always seemed to show otherwise. Similarly, a reviewer for the *Nation* could not understand White's desire to preserve religion. "Mr. White seems to us to make a mistake in thinking that he is called upon to offer any suggestions as to the reconciliation to be effected now between science and religion." Others attempted to offer such reconciliations only to be persecuted. That day has come and gone, however. According to the reviewer, "it is science which is established now, and if there is to be a reconciliation, it is religious truth which must justify itself." A writer for the Boston *Arena* concurred, arguing that if White had demonstrated that the Bible had been "the greatest block in the way of progress," why continue forcing children to read it in public schools?[50]

The evangelicals at Princeton College saw White's book as a tool for advancing liberal Protestantism. According to President James McCosh (1811–1894), for instance, White's earlier *Warfare of Science* is an "agglomeration (very indiscriminate and uncritical) of facts to show that religious men have opposed science, and been defeated."[51] Later, in an unsigned review in the *Presbyterian and Reformed Review*, B. B. Warfield (1851–1921) also declared that all that White proved is that "some religious men have been slow to follow the advance of science; and the same could easily be shown of some scientific men: nay, it was largely not because they were men of religion, but because they held too tenaciously to the current ideas of men of science, that these men lagged behind the vanguard of advance."[52]

White also received numerous letters from perceptive readers, and these are perhaps more instructive for our purposes. For instance, Albert Britt, managing editor of the New York weekly *Public Opinion*, pointedly asked White, "are the masses of men yet sufficiently developed to determine and follow out a course of right conduct without the impelling power of a belief in, or a fear of, a personal God who rewards and punishes?" Physician John Shackelford believed White's narrative completely undermined the Bible. He recognized that his "warfare" was ultimately "between the Old and the New Theology," and went on to caution White that if he rejected the Fall and the Atonement, "Christ's religion goes by the boards. A religion without the authority of God back of it is a religion without power." He concluded that if Christ was not "God manifest in the flesh," he either made a "tremendous mistake" or was an "imposter of the worst type."[53]

Other writers more explicitly noted the contradictions in White's religious opinions. Nevada Baptist minister William Phillips wondered what White really believed about Jesus Christ. "What think you of Christ? You write of him as 'the Lord Jesus Christ.' How? Intellectually? By virtue of His spiritual influence?" Phillips then reveals the core dilemma: "If He was not miraculously conceived; if He did not come in the power of the Holy Spirit at Pentecost; if He is not to come in judgment; if all the claims that the Gospels represent Him as making, for Himself, and all the claims that the epistles make for Him are accretions, additions to his teaching, superstitions, shall we then speak of Him as the Lord Jesus Christ?" Harry A. Miller, a "Truth Seeker" and tobacco leaf dealer, also questioned White's use of the phrase "Our Lord Jesus Christ" when it appeared he only saw Jesus as a man. He also questioned White's idiosyncratic views, asking if the story of the birth and life of Christ is myth and legend, then why does he refer to him as "our Lord Jesus Christ." "I want to know," Miller asked, "if by calling Him (Lord) you mean to express or refer to Him as being any thing besides a man or only as I believe a 'Great man.'"[54]

Perhaps White's most penetrating critic was an old family friend, Mary K. S. Eaton. After reading White's work, Eaton wrote that it left her feeling a "profound regret." Apparently not one to mince her words, she bluntly told White that his work was a wasted effort. "If I am not mistaken," she wrote, "the object is to prove the Christian religion a 'cunningly devised fable'; the Bible which Christians accept a tissue of falsehoods; its Divine author a myth; his son, that this same Bible pronounces 'God manifest in the flesh,' the greatest imposter the world has ever known; and then you inform us that *it is true* but in some high and mysterious sense!" She did not want to debate the finer details, but to simply ask—*cui bono?* "What if you succeed in creating doubts in the minds of men, in taking from them all trust in the Revelation they have accepted as coming from God. What then? What will you give them in its stead? Your poor starving theories? . . . A religion evolved from *human brains* stripped *of all* that is Divine? An image without a soul?"[55]

White attempted to answer these challenging questions. He told Eaton that he believed he was doing his best to "save the Bible," to strengthen Christianity

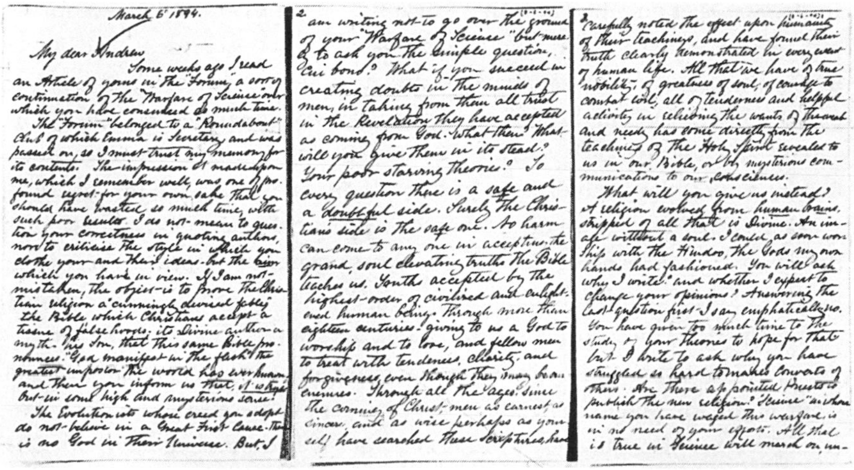

Figure 6.6. Mary K. S. Eaton's 1894 letter strongly reprimanding Andrew D. White's religious views. Courtesy of the Division of Rare and Manuscript Collections at Cornell University Library, Andrew Dickson White Papers, Reel 61.

by giving it a "new basis." Eaton, however, was astounded by such hubris. She responded mordantly, asking him "what do you suppose has saved it all these centuries without your helping hand? What means have you added in this great work?" Indeed, it seemed to Eaton that White's efforts had been to "prove it false, a *human* work subject to the accidents and full of imperfections that belong to all human works." From her perspective, "there was no God" in White's Bible. "You may not be an atheist in the sense that there is no God in the Universe, but when you declare that the Revelation he has made to man is fast crumbling away on account of what you call the 'human theological foundations'—that Rock which he says the gates of hell shall not prevail against—I take the liberty by repeating: there is no God in your Bible." Furthermore, Eaton believed Christianity required no new basis at all. "What in the name of poor, feeble, ignorant Christians, I ask, is this new basis?" According to Eaton, "the Rock on which Christianity is founded is in no danger from the ceaseless labors of its enemies." Since Christianity was not in danger, it did not need someone like White to protect it. Indeed, she noted that the attitude of the "truth destroyer is that of the 'Warfare of Science' man."[56]

They continued this awkward exchange for the next two years, with Eaton repeatedly charging him with naiveté, and in one letter even calling him "Sancta Simplicitas." White responded that she had misunderstood his "motives and statements," and called her critique "savage." She replied that one could respond in no other way to his claims of saving the Bible and strengthening Christianity with a new basis. For those who accept "Jesus as God manifest in the flesh," she wrote, "we need no other foundation than this secure one."[57] White eventually lost patience with Eaton. In 1896, he terminated the correspondence. He wrote that he had refused to read her most recent letters and would no longer read any others she sent. He had endured enough "torrents of misconception and objurgation," and was no longer interested in her "vitriolic criticism." He maintained that his work is the result of "long years of study and thought." In his "desire to speak the truth" he had "sacrificed much—very much." Importantly, he told her that he was not alone in his views. They are entertained by "those of a very large number of the foremost divines in the English Church, a considerable number in the American Episcopal Church, a very large number indeed in the Orthodox Congregational and other Churches, many even in the Roman Catholic Church, and their numbers are constantly increasing." Because these were "reverend, devoted, Christian men," he thus felt justified in holding his views. His sole motive, he added, "has been to prevent rash, hotheaded, men from throwing away Christianity altogether, and taking refuge in out and out atheism and materialism." Eaton's ridicule and mockery, he howled, "shows that you know nothing whatever of the problem involved." A mind so charged with ignorance and prejudice, he concluded, could and should never sit in judgment. He closed by saying that he "will not quarrel" with her any longer, for he was someone "not to be quarreled with."[58]

Unbelievers

The protests of Eaton and other religiously conservative readers were prescient, however, for soon the same secularists who appropriated Draper's narrative took up White's writings in their campaign to secularize society. The American *Free Thought Magazine*, for example, declared that White had "done immense service to the seeker after truth." His account revealed the "struggle which the liberal minded and honest churchmen have had in their efforts to give to Christianity the benefits of a reasonable, instead of an unbelievable, theology." White's narrative was not a mere history of "folly, cruelty, and crime," but a history of the "successive triumphs of science." It is a work which every freethinker "should have in his library."[59]

Across the Atlantic, the same secularists who included Draper in the pages of their histories of rationalism and freethought also included White. Other English secularists, Alfred W. Benn, for instance, reflected after reading White's book that "either there is greater freedom of thought in the United States, or the majority are more enlightened than with us." He argued that White's work was "continuing the labours of Buckle, Draper, and Mr. Lecky." But White, according to Benn, "disclaims any hostility to Christianity." Indeed, according to Benn, White seemed to indicate that such "revolutions in thought" have only served to advance

Christianity. Benn, however, thought the sentiment confusing. He called White a "Feuillant," after the constitutional monarchs of the Legislative Assembly of 1792. Whatever the ambiguity of White's religious views, Benn argued, "their general acceptance would involve the abandonment of what we now call Christianity."[60]

White also had to contend with atheists who were confused as to why he remained religious at all. In a revealing series of letters from husband and wife Edward Payson Evans (1831–1917) and Elizabeth Edson Gibson Evans (1832–1911), both expressed dismay over White's inconsistencies and contradictions. White first met the couple in Ann Arbor, where Edward was Professor of Modern Languages at the University of Michigan. Edward resigned in 1870 to become a freelance journalist, relocating to Munich, Germany. Elizabeth was also a frequent contributor to journals and magazines, in addition to authoring a number of radical books published by the Truth Seeker Company, including *A History of Religions* (1892) and *The Christ Myth* (1900), the latter denying the historical existence of Jesus altogether. Even before White had published his earlier *Warfare of Science* in 1876, Elizabeth had pressed him about his religious convictions, to which he replied in a long letter by telling her that he believed in God, the Psalms, and the Sermon on the Mount.[61]

After the appearance of *Warfare of Science,* White had asked the Evanses for their opinion. In an 1877 letter, Elizabeth criticized White's vagueness and loss of nerve, but hoped someday he would find the courage, joy, and peace of "disbelieving." She wrote, "considering your position, both public and private, it is a brave thing that you have done; but I cannot see, *I cannot see* how, having gone so far, you are able to stop where you do!" Indeed, she believed White had thoroughly "alienated the strictest sect of the Pharisees," and yet she could not fathom why he believed "there is no necessary antagonism between Science and Religion—that Science in its own victories has been fighting also for Religion—and that these two great Powers ought to defend and help each other." Religion and science, she added, have "no point of contact or agreement." See added that "conceptions of supernatural Beings and conditions date from the infancy of the human intellect, when the imagination had full play and before experience and experiment had taught men to reason correctly. Such conceptions have in the progress of civilization become refined; but they are no nearer the truth now than they ever were—*at least we do not know that they are.*" Every variety of science had destroyed religion, so "what is there left of the Bible as a record of inspired wisdom now that its geography and geology and astronomy and physiology and history have been proved to us false—its philosophy and theology traced to their primitive sources? And what is christianity without the Fall and the Atonement, the Trinity, and the incarnated Savior?"[62]

Edward was even more critical. He argued that White was "too generous to the discomfited foe in permitting him to retire from the field with the greater part of his weapons and war-material." He confessed that he belonged "to that class of persons who entertain what you characterize as 'that most mistaken of all mistaken ideas—the conviction that religion and science are enemies'; and the perusal of your book has strengthened me in this conviction." Indeed, reading White's book convinced Edward Evans that the "conflict between these two forces is fundamental and irrepressible and that the warfare will cease to be waged only when one of them shall be put *hors de combat* and cease to be a force in the world." He acknowledged that White was not "defending dogmatic Christianity, but standing up 'for the living kernel of religion.'" But Edward pressed him for this "kernel." He declared that whatever definition White gave, "there is no body or community of Christians on the face of the earth that would accept your definition of the Christian religion as full and adequate." Whatever the definition, in short, it will only lead to a "lamentable confusion of ideas." We must call "things by their right names," Edward forcefully demanded.[63]

White rather pitifully responded to their stringent critique several days later. He simply said "I think still that I could make a good defense of my use of the word 'religion.'" But in the numerous letters he wrote to the Evanses over the next fifty years, he seemed to have never again returned to defend his definition of religion. This did not prevent the Evanses, however, from insisting (perhaps even gloating) that White's work was having an effect counter to his intentions. As late as 1901, for example, Elizabeth reported to White that the Truth Seeker Company had listed his book in their *Catalogue of Free Thought Works*. "By the way," she gloated, "I hope you appreciate my self-sacrifice in omitting your book in my bibliography at the end of my '*The Christ Myth*.' I wanted to put it in, for though I did not consult it in writing my book, *still many* of the subjects coincide with your statements and for the sake of readers it would have been well to draw attention to your larger work."[64]

But while the Evanses criticized White for his lingering religious sentiments, other skeptics praised his work as a useful and powerful weapon against religion. His eldest son, Frederick White, had told him in 1889 that "several copies of the Freethinker's Magazine [i.e., *Freethought Magazine*] came last week for you with a note from Mr. H. L. Green the editor saying that he has read with great *pleasure* the later New Chapters in the Warfare of Science and hopes to see them all published complete some time."[65] Horace L. Green (1828–1903) himself wrote White later about Patterson's critique of his book in the *Chicago Tribune*, calling on him to "reply to what the doctor here says."[66]

Even the "great agnostic" Robert G. Ingersoll wrote White to tell him that he was reading with pleasure his "New Chapters." However, he opined that Christianity was not worth saving. "The Church pretended to have the word of God—

pretended to have all truth worth having—and as a consequence it was obliged to say 'Obey! Believe!' It could not say: 'Investigate—think!'" In his opinion, he could "not see how man can be free, or worthy of freedom, until he ceases to imagine that a Master exists." Turning White's argument on its head, Ingersoll claimed that "the only 'power in the universe strong enough to make truth-seeking safe' is man." Finding a perceptive contradiction in White's thinking, Ingersoll pointedly asked "if God makes truth-seeking safe now, why did he allow it for thousands of years to be dangerous? Was not truth as valuable then as now?" All orthodox churches then and now continue to deter investigation, according to Ingersoll, and the "power that makes for righteousness is exceedingly weak, where bigots are in the majority." Despite this criticism, Ingersoll maintained his work "will do good" to "increase intellectual hospitality," banish "provincialism of creeds," and to reveal the "egotism of ignorance." White has, according to Ingersoll, demonstrated that all supernatural religion is superstitious, and that the superstitious is untrue.[67]

Finally, secularist Charles Albert Watts even sent him a copy of the latest *Watts's Literary Guide*, which gave a lengthy review of White's book. He informed White that the publication of his book "has brought me dozens of letters urging me to endeavor to induce you to write for the Annual." White, however, declined, characteristically citing heavy pressure of work. This did not deter Watts from writing him again the following year, telling him he could write on whatever subject he chose, and mentioned that Thomas H. Huxley and others have contributed to the *Agnostic Annual*.[68]

Some of these letters must have deeply disturbed White, for he had always intended his work to silence religious "scoffers."[69] He told his personal secretary and librarian George Lincoln Burr (1857–1938) that he wanted to "give a fair and judicial yet hearty presentation of the *truth*—the *truth as it is in Jesus* one might very justly say. I think the world needs it, to take place of such gush as [John Henry] Newman on one side and such scoffing as [Robert G.] Ingersoll on the other."[70] But as we have seen, many of his readers, believers and unbelievers alike, interpreted White's work as nothing short of a direct assault on all religion.

Perhaps most disturbing for White, however, was his own son's growing irreligious attitude. Frail at birth and sickly throughout his adult life, Frederick White was always burdened by the heavy weight of his father's ambitions. In seeking his father's approval, Fred seemed to take White's nonsectarianism to extremes. While his letters were mostly concerned with business matters, there are occasional glimpses into his growing irreligious opinions. Early in 1886, for example, he told his father that he had heard Ingersoll argue a case "in a superb manner."[71] On another occasion, he told him that the religious revivals in New York were perfect examples of the "hypocrisy and salacity which are as inseparable from

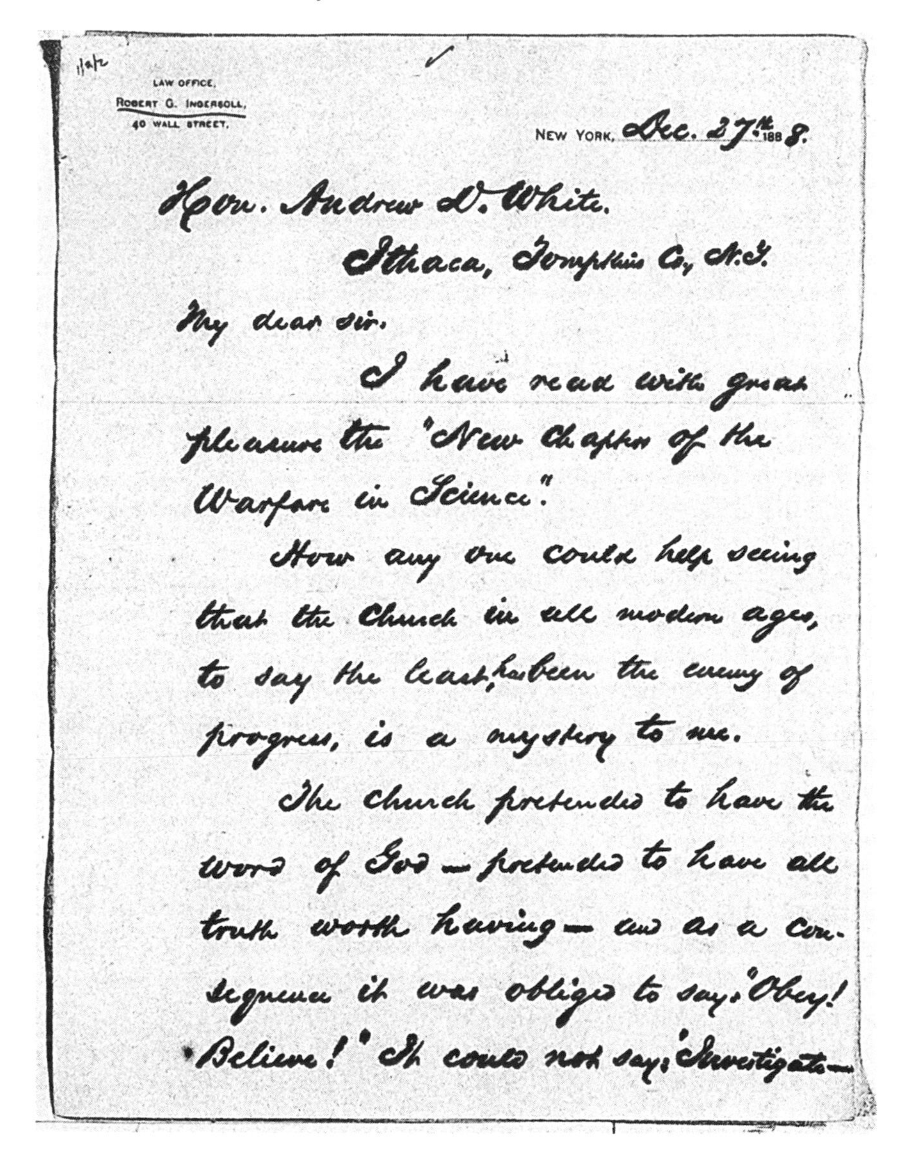

LAW OFFICE,
ROBERT G. INGERSOLL,
40 WALL STREET.

NEW YORK, Dec. 27th 1888.

Hon. Andrew D. White.
Ithaca, Tompkins Co, N.Y.

My dear sir.

I have read with great pleasure the "New Chapter of the Warfare in Science".

How any one could help seeing that the Church in all modern ages, to say the least, has been the enemy of progress, is a mystery to me.

The church pretended to have the word of God — pretended to have all truth worth having — and as a consequence it was obliged to say, "Obey! Believe!" It could not say, "Investigate—

Figure 6.7. In 1888 Robert G. Ingersoll praised Andrew D. White's "New Chapters" in the *Popular Science Monthly*. But to White's chagrin, he also urged him to completely abandon Christian belief. Courtesy of the Division of Rare and Manuscript Collections at Cornell University Library, Andrew Dickson White Papers, Reel 50.

orthodox Christianity as from Mohammedism." Ministers and the priesthood are "organized satyrs," Fred declared.[72] White replied that he did not want him talking and living like a "heathen," and instructed him to attend one of the more liberal services at home in Syracuse.[73] Fred responded that he occasionally attended a Presbyterian service, but at other times went to hear Unitarian Samuel R. Calthrop speak, who, although a "good and pure man," must "go to hell and burn eternally" according to the other ministers. He refused to attend the revival meetings at St. Paul's church in New York, where "absolution and confession" were held and the priest gave "nasty lectures" that scared ladies "half out of their wits." He told his father he "will go to church as often as you wish me to, but I cannot force myself to take any stock in clergy, dogma or creed." Even Calthrop's church, he said, was "a milder form of the same disease." Fred then unleashed a torrent of invectives toward the whole of Christianity:

> I am morally certain that the great majority of thinking practical men of the present time regard the present degeneration of religion as unworthy of any attention and the officers of the Church, Pope, Bishop, Priest and Minister, one and all as insincere hypocrites if not worse. What they want is money and power and—too lazy or stupid to gain either by work—they hoodwink females and children of both sexes with such phrases as immaculate lamb, incarnation, eucharist, aural confession, salvation, regeneration, try to repress all truth and progress—and so frighten the cowardly portion of mankind with their pictures of hell-fire, that they prefer to pay cash and believe impossible lies, to taking the risk of damnation. I hate to encourage these men in any way, they have a well organized conspiracy against truth and virtue now and I doubt if they can be exterminated in less than a century—and every man they can scare a little or get some cash out of adds just so much to the garbage-heap which future generations will have to clear away. I would not write this did I not know that among the intelligent men which I have talked with the sentiments which I have expressed on the subject are nearly universal.[74]

Fred attempted to reassure his father that "we all, I think, believe in an old-fashioned religion, a deity, and in the wonderful teachings of Christ." Fred had taken his father's critique of Christian theology to its logical conclusions. Unfortunately, Fred felt he failed to live up to his father's expectations. In 1901 Fred took his rifle, stepped into the bathroom of his home, and committed suicide.[75] Away at one of his diplomatic assignments, White received the telegram of Fred's death early in the morning. He later lamented in his diary, "my poor Fred committed suicide . . . my poor, suffering, unfortunate boy. Why could I not have been near him. Alas, alas that it should have come to this."[76] The response to his work must have been a constant source of frustration, chagrin, and perhaps even sadness for White—a persistent indication that his most integral purpose, the reconciliation of science and religion, was almost unanimously ignored.

CONCLUSION

The new challenges of the nineteenth century intensified the belief among many liberal Protestants that Christianity required another reformation. Responding to the advances of natural and historical sciences, they sought to combine progressivist conceptions of science with progressivist conceptions of religion. Some liberal theologians went as far as saying that rising "secular movements" were necessarily beneficial to theology, as "the pruning-knife are to the tree."[77] If religion would only rid itself of dogmatism, science and religion would be in harmony. Many liberal Protestants thus believed the separation of religion from theology was the best approach to bridging the perceived schism between modern thought and religious faith, and ultimately for bringing about reconciliation between science and religion.

These new conceptions came to influence late nineteenth-century discussions about the relationship between science and religion. The conflict occurred, liberals argued, not between scientific truth and religious truth, but between contesting theological ideas. As American theologian and historian of Christianity George P. Fisher (1821–1909) perceptively wrote in the *Princeton Review* near the end of the century, "It has not been a war of disbelievers and sceptics on the one side, who have been obliged to suffer at the hands of believers in Christianity for teaching scientific truth. It has commonly been a contest of Christian against Christian. Where there has been a combat of this sort it has been an intestine struggle."[78]

Fisher's insight is clearly reflected in the public response to Draper and White. While there was much overlap, inconsistency, and even contradiction in these responses, three general patterns emerge. The more liberal press praised their historical narratives as clearing the way for the reformation of religion. They embraced the rhetoric of "conflict" or "warfare" as effectively demonstrating the need to separate religion from theological traditions. Because liberal theologians had already independently fostered the distinction between the two, these reviewers interpreted Draper and White as providing the basis for a consensus that had eluded other attempts to reconcile modern thought with religious faith. As Jon Roberts has recently shown, at the turn of the twentieth century, Anglo-American liberal Protestants continued to implicitly and explicitly accept the narratives of Draper and White.[79] Many were thus sympathetic to their narratives, arguing that the progress of scientific knowledge revealed the need for Christians to make significant revisions to their faith. Failure to harmonize Christianity with modern thought, they warned, would convince the public to abandon Christianity altogether.

The more religiously orthodox, however, were unwilling to concede that religion could survive without theology. Rather than saving Christianity, orthodox believers warned that the approach of liberal Protestants like Draper and White and others would ultimately destroy it. The religiously conservative press thus criticized their narratives as unhistorical, misleading, and deceptive. We may draw a representative view from statements made by John Gresham Machen (1881–1937) in his *Christianity and Liberalism,* first published in 1923. Machen, who had left Princeton Seminary in protest of its nascent liberalism, argued that the liberal attempt at reconciling science and Christianity by separating religion from science is "open to objections of the most serious kind." Namely, that the result is both "un-Christian" and "un-scientific." According to Machen, liberalism creates "a religion which is so entirely different from Christianity as to belong in a distinct category" altogether. He described liberalism as a form of "naturalism" or "paganism" insofar as it denies that God's transforming, creative power is able to break into the natural course of events described by modern science. According to liberalism, "this world is really all in all." Moreover, while liberalism uses "traditional phraseology," it simply reduces divinity to nothing more than vague feelings. In doing so, liberalism abrogates the trustworthiness and integrity of language. Indeed, according to Machen, "the liberal attempt at reconciling Christianity with modern science has really relinquished everything distinctive of Christianity, so that what remains is in essentials only that same indefinite type of religious aspiration which was in the world before Christianity came upon the scene." Sounding remarkably like unbelieving critic Joseph McCabe, Machen declared that liberal Protestantism is the new Gnosticism. For Machen, then, "it is not the Christianity of the New Testament which is in conflict with science, but the supposed Christianity of the modern liberal Church, and that the real city of God, and that city alone, has defenses which are capable of warding off the assaults of modern unbelief."[80]

Indeed, and in fulfillment of these forewarnings, in the last decades of the nineteenth and early twentieth centuries, freethinkers, secularists, and atheists appropriated their narratives of "conflict" and "warfare" in their campaigns to secularize society. They interpreted these narratives as an assault against all religion. They employed the language of conflict and warfare to destroy religious belief rather than reform it. Indeed, many of these unbelievers credited Draper and White for their conversion to unbelief. At the same time, unbelievers "outed" men of science and liberal clergyman, arguing that they chose a half measure when they should have gone all the way. Unbelievers, however, were baffled that theists like Draper and White accepted such radical criticism of orthodoxy, and yet continued to believe that religious men and women remained somehow justified in their belief.

Thus, in one of those remarkable ironies of history, both orthodox and unbelievers found themselves in agreement. Draper and White, then, were attacked by the orthodox and unbelievers alike. Liberal Protestants had gambled on a risky strategy and lost. Unbelievers attacked the integrity of liberal theologians, censured their attempt to reconceive old doctrines, and castigated their refusal to make a clean break with the ancient religion. The unbeliever regarded such attempts at reconstructing religion as superfluous, believing that religion was indeed at war with science and, therefore, needed to be eliminated entirely from society. By appropriating their historical narratives, they left Draper and White with deeply troubled legacies.

Historians of science have much to gain by considering the early reception of the narratives of Draper and White. Such an approach sheds further light on how contemporaries reacted to those narratives that later scholars have so strongly condemned. But while Draper and White are regarded as the principal *casus belli* of the "conflict thesis," this early reception indicates that they were understood quite differently by many of their contemporaries. Liberal Protestants at the end of the century viewed themselves as protectors of "true Christianity" or "true religion" rather than its enemies. Indeed, they portrayed the orthodox theologian as a threat to science *and* religion. Many readers recognized that Draper and White were not secular critics of religion. Like other liberal Protestants, Draper and White offered their historical narratives as demonstrating the threat of theological dogmatism to both scientific and religious progress. In short, practitioners, promoters, and popularizers of such historical narratives played an important role in advancing the *cause* of liberal Protestantism. At the same time, these narratives had the unintended consequence of creating in the minds of contemporaries and later generations the belief that science and religion have been and are at war.

CONCLUSIONS

AT the beginning of the twentieth century, rationalists, freethinkers, secularists, and atheists seized upon the historical narratives of Draper, White, and other liberal Protestant historians and theologians, adopting them as weapons in their campaign to extinguish all religion. Skeptics declared war not only on traditional religious believers but on what Joseph McCabe caustically described as the "land of bunk," those obfuscating liberal religious thinkers who attempted to accommodate theology to fit the modern age. Writers for leading secularist publications imagined the progress of science as sweeping away all manifestations of religious belief. Religion was an illusion and therefore detrimental to the progress of society. The narratives of liberal Protestant thinkers were, in short, not merely appropriated but hijacked by skeptics, who reshaped them to justify and promote their own specific vision of a progressive and secular society.

This secular progressivism or humanism of skeptics like McCabe deeply informed the fledgling discipline of the history of science as it developed in the early twentieth century. Indeed, by mid-century, when history of science first emerged as an academic field of study, the belief that science and religion were in a constant state of conflict or warfare was held as a commonplace. During much of the twentieth century, this view dominated the historical interpretation of the relationship between religion and science. Interestingly, its leading advocates adhered to a positivist outlook, exemplified in the writings of such men as Paul Tannery (1843–1904), Jules Henri Poincaré (1854–1912), Charles Singer (1876–1960), and especially George Sarton (1884–1956), at one time the doyen of all historians of science. These early twentieth-century historians and philosophers of science viewed scientific advance as the progressive triumph of reason over superstition. They believed that positive scientific knowledge would replace earlier religious forms of understanding. History of science, in short, became the heart of secular faith in the progress and improvement of humanity.[1]

Perhaps more than anyone, the Belgian émigré George Sarton helped establish the history of science as a serious academic discipline. The German invasion of Belgium in 1914 forced Sarton to relocate to Britain and then to the United States, where he lived for the remainder of his life. Often called the "dean" or "father of the history of science," Sarton wrote what many still consider the most foundational works in the field, including his massive five-volume *Introduction to the History of Science* (1927–48). Other works include *The History of Science and the New Humanism* (1931), *The Study of the History of Science* (1936), *A History of Science* (1952–59), and *Horus: A Guide to the History of Science* (1952), among numerous articles and other writings. Sarton was also the founder and first editor of *Isis* (1912), perhaps the most important peer-reviewed academic journal on the history of science, medicine, and technology. In 1924 the History of Science Society was established for the specific purpose of furthering the study of the history of science as well as an international source of financial support for Sarton's work.[2] In 1936, he also founded *Osiris* to relieve *Isis* of the load of longer, more sustained studies. In 1940, Sarton became a professor of the history of science at Harvard University and was instrumental in making its program there one of the leading centers of the History of Science in the world.[3] But his influence extended beyond Cambridge, particularly in the History of Science Department at the University of Madison-Wisconsin, which was one of the largest and oldest academic programs of its kind in the United States.[4] Thus the professionalization of the history of science as a discipline was carried on in the United States largely by Sarton.[5]

But by the second half of the twentieth century, historians of science such as Alexandre Koyré, Charles Gillispie, John Greene, and numerous others were calling attention to the variety of connections between religion and scientific thought. As philosopher of science Thomas Kuhn wrote, while historians of science owe Sarton an immense debt for his role in establishing their profession, "the image of their specialty which he propagated continues to do much damage even though it has long since been rejected."[6] Cornell University historian of science Leslie Pearce Williams also complained that Sarton's view of history was a "painfully naïve" division of "good guys" and "bad guys," of "good practices" and "bad practices."[7] According to Alistair C. Crombie, "a hard critic might even say that Sarton's approach could easily have killed the study of the history of science."[8]

What needs to be emphasized here is that Sarton's vision of the history of science as the "new humanism" grew directly out of his commitment to Comtean positivism.[9] Arnold Thackray and Robert Merton, for example, have stated that what controlled Sarton's argument and guided his actions stemmed from a heritage of "positivism, progressivism, and Utopian socialism."[10] Tore Frängsmyer more aggressively argued that the kind of history of science Sarton promoted was "born out of the confident belief in progress of French Positivism."[11] More recently, John F. M. Clark suggests that Sarton "built on eighteenth- and nineteenth-century traditions of positivism and universal history."[12]

Sarton's dependence on positivist philosophy is unmistakable. As we have seen, Comte saw the tensions between science and religion as skirmishes in a long historical battle that science is destined to win. His positivist philosophy viewed all aspects of nature and society as conforming to immanent, natural laws. The positive stage will thus supersede both theological and metaphysical explanations, abandoning them as futile. But as we have seen, Comte's "stages" or periodization of history were not new. He no doubt was influenced by conceptions of progress found in the writings of Turgot, Saint-Simon, Fourier, and other "prophets of Paris." But as Christopher Dawson, Ernest L. Tuveson, Karl Löwith, and others have demonstrated, the idea of progress, the notion of an unfolding, cumulative advancement of mankind through stages or epochs, each reflecting some historic civilization or cultural development advancing by social reform and conflict, has a long Christian pedigree—particularly among Protestant historians and theologians.[13] This Protestant historiography of reformation, reason, and religious emancipation from Rome gradually transformed into a narrative of the progressive desacralization of the world.

Importantly, Comte's "Religion of Humanity" is exemplified in Sarton's call for a "New Humanism." Sarton was baptized in the Roman Catholic Church for his "mother's sake," who died shortly after his birth. Sarton was thus raised by a father who seems to have altogether neglected him.[14] Moreover, his father cared little for religion, and, as result, Sarton never "received the Holy Communion and never practiced any definite religion."[15] While studying at the University of Ghent, Sarton became familiar with the work of Austrian physicist, philosopher, and positivist Ernst Mach (1838–1916), who in his *The Science of Mechanics, a Critical and Historical Account of Its Development* (1883) claimed that the "conflict" between science and theology, or science and the church, was a "commonplace of history."[16]

Sarton's disdain for religion is evident in the majority of his writings. As a committed Comtean positivist, he viewed the gradual decline of religion as a mark of scientific progress. Indeed, his *Isis* was designed to spread the message of the positivists. In its opening article, for instance, Sarton readily admitted that his work adhered to the positivist school of Comte. He explicitly stated that Comte must be considered the "founder of the history of science." Unsurprisingly, then, he argued that "the interactions between science and religion have often had an aggressive character," and that "most of the time a real warfare" had existed between them. He found much heuristic value in this conception of the history of science. It reveals not only the "progress" of the human mind but also its "regressions," "sudden halts," "mishaps," and "superstitions," thus providing us with a "history of errors." The "progress of mankind," Sarton asserted, was an "intellectual unfolding." Sarton accorded science a preeminent place in the scale of human knowledge and mapped the "progress of civilization" through the history of science. In short, he wrote, "the purpose of the history of science, as I understand it, is to establish the genesis and the development of scientific facts and ideas, taking into account all intellectual exchanges and all influences brought into play by the very progress of civilization. It is indeed a history of human civilization, considered from its highest point of view. The center of interest is the evolution of science, but general history remains always in the background."[17]

The spirit of Comte could not be more alive than in statements such as these. While he blithely asserted that Comte was "crazy," he nevertheless praised his Positivist Calendar as a "remarkable document." On at least one occasion in 1919, Sarton even spoke at English positivist Richard Congreve's (1818–1899) Church of Humanity on Chapel Street in London. Some thirty years later, Sarton also visited Comte's "*domicile sacré*" in Paris and "communed" with his hero's spirit.[18] While he seemed to have dissociated himself from the Religion of Humanity, Sarton nevertheless seemed to think it appropriate that Comte the man was worthy of veneration.

Sarton's campaign, then, was launched under the aegis of positivism. For Sarton, the history of science unequivocally demonstrated the progress of humanity. The purpose of teaching the history of science was to establish a "New Humanism."[19] The "true humanist," he explained, "must know the life of science as he knows the life of art and the life of religion." He must not only "appreciate and admire what our ancestors did" but also must "take up their best traditions."[20] According to Sarton, "scientific activity is the only one which is obviously and undoubtedly cumulative and progressive." In addressing the matter of opposition or hostility to science from religious and "conservative people," he openly admitted that such groups were "undoubtedly right in their distrust and hatred of science, for the scientific spirit is the very spirit of innovation and adventure,—

the most reckless kind of adventure into the unknown. And such is its aggressive strength that its revolutionary activity can neither be restrained, nor restricted within its own field." He added that "sooner or later it will go out to conquer other fields and to throw floods of light into all the dark places where superstition and injustice are still rampant." The scientific spirit, he wrote, "is the greatest force for construction but also for destruction." Sarton summarized these views in a pithy statement: "The history of science is the story of an endless struggle against superstition and error; it is not a vivacious and spectacular struggle, but rather an obscure one—obscure, tenacious and slow. The resistance of science against every form of unreason or irrationality is so firm and yet so quiet, that it is almost as gentle as non-resistance would be, yet unshakable."[21]

Sarton believed that men of science are heroes fighting for truth against the forces of darkness. The enemies are religion and superstition in every shape and form, and the heroes of science have to fight with reason as their weapon. But the history of science, he contended, "cannot be an end in itself, but a means to a higher end: a deeper understanding of science, of nature, of life." All of this, according to Sarton, advances the New Humanism. Whereas the "old humanism" was the "revival of ancient knowledge," the "New Humanism is a revival of the knowledge patiently elaborated and accumulated for many centuries by men of science." This leads Sarton to the remarkable conclusion that "men without scientific knowledge are totally unfit to explain the progress of humanity."[22] Sarton was, then, a propagandist of the "new humanism"—and by "new" he undoubtedly meant "secular."

In line with his endorsement of positivism and his unequivocal commitment to a secularized ideology of progress, Sarton accepted the idea that a gradual and increasing separation from religion marked the progress of science. In his *Introduction*, for instance, he dismissed the "superstition and magic" that prevailed in his day. "The historian of science," he demanded, "cannot devote much attention to the study of superstition and magic, that is, of unreason, because this does not help him very much understand human progress." He added that "human folly being at once unprogressive, unchangeable, and unlimited, its study is a hopeless undertaking." The history of science must always be considered under two aspects, he said, "either positively as the gradual unfolding of truth, the increase of light, or negatively as the progressive triumph over error and superstition, the decrease of darkness." Ultimately, he argued that "the progress of science is absolutely dependent upon its emancipation from non-scientific issues, whatever they be, and in particular, upon its laicization."[23]

As the gatekeeper of the nascent discipline of history of science, Sarton also decided who and what was important to the field. Censorship was a common complaint by authors trying to publish in his journals. For instance, when Sarton offered to publish Robert Merton's work, he asked him to reduce his discussion of religion. Merton, who described Sarton as the "exigent and angry master and I the brooding and unruly apprentice," was astonished when Sarton asked him to condense his section on religion, a striking proposal indeed since Merton's presentation of Puritanism in relation to the rise of science became the most celebrated part of his monograph. But as Merton recollected, "this part was not condensed in the published version."[24]

Sarton relied on the work of countless others in constructing his narrative of the history of science. He listed many of them in his widely acclaimed *Guide to the History of Science,* a bibliographic reference guide for scholars interested in the history of science. Significantly, in a section devoted to early "treatises and handbooks on the history of science," Sarton cited Draper's *History of the Conflict* and White's *History of the Warfare* as important guides to the whole subject.[25] As White's younger contemporary, Sarton had actually developed something of a personal relationship with him. In the English reprint of his opening article for *Isis,* for example, he recommended to his readers White's two-volume masterpiece.[26] Moreover, Sarton had invited White to be a patron of *Isis* in 1912 and had even visited White at Cornell in 1916.[27] Further, in desperation Sarton wrote White several times in 1918, appealing to him for support in securing an appointment with the Carnegie Institution, where White was a trustee. Whether it was White's support or some other, Sarton was eventually appointed a "research associate" of the Carnegie Institution, which enabled him to devote full attention to his scholarship.[28]

It is worth noting that Sarton and other twentieth-century positivists informed the work of the Unity of Science movement, associated particularly with the logical positivists or empiricists of the Vienna Circle. Established in 1922 shortly after the Great War, the Unity of Science drew philosophers of science such as Hans Hahn (1879–1934), Otto Neurath (1882–1945), Moritz Schlick (1882–1936), Rudolf Carnap (1891–1970), Hans Reichenbach (1891–1953), Carl Gustav Hempel (1905–1997), and many others, who insisted that science and religion occupy separate, nonoverlapping domains. But according to their verificationist theory of knowledge, the logical positivists also argued that religious or metaphysical beliefs refer to unobservable entities, and as such lack meaning altogether. Hence, positivism not only distinguishes between science and religion, it does so on grounds that deny objective warrant to religious belief. While their approach was mainly philosophical, the logical positivists issued encyclopedic projects similar in scope to Sarton's, such as the *International Encyclopedia of Unified Science,* of which some twenty volumes were published between 1937 and 1969. Progressive, liberal, and sometimes socialist, these volumes presented the history of science as a disenchantment model of scientific progress.[29] Interestingly, Neurath, a founding figure of the Vienna Circle and editor-in-chief of *Unified Science*, believed that it was rational Protestantism that had reduced the power of priests and destroyed Christian theology.[30]

Further north, British philosophers Bertrand Russell (1872–1970) and Alfred Jules Ayer (1910–1989) were also influenced by the new positivism. Russell, for instance, argued that "between religion and science there has been a prolonged conflict, in which, . . . science has invariably proved victorious."[31] He also once referred to religion as "a disease born of fear and a source of untold misery to the human race." Science, on the other hand, will help humanity get over this "craven fear." Russell believed that science will teach humanity to "no longer look around for imaginary supports, no longer to invent allies in the sky, but rather to look to our own efforts here to make this world a fit place to live in, instead of the sort of place that the churches in all these centuries have made it."[32] Ayer, who provided the first English exposition of logical positivism in 1936, famously argued that "there is no opposition between the natural scientist and the theist who believes in a transcendent god," not because there is no inherent conflict, but rather because "religious utterances of the theist are not genuine propositions at all," and therefore "cannot stand in any logical relation to the propositions of science."[33] Religious propositions, in short, are meaningless. Not surprisingly, Ayer supported the secular humanist movements of his day, even becoming the first executive director of the British Humanist Association.[34]

We have dwelled on Sarton and early twentieth-century positivism because the so-called new atheists have taken up these arguments in their own writings. As we have seen, by the mid-twentieth century, intellectual historians had already reassessed the relationship between general history and the history of science, many ultimately dismissing the kind of "presentism" found in Sarton's body of work in favor of a more contextualist approach, with its focus on "particulars of local circumstances, people, epistemes, and politics," as historian Lynn K. Nyhart puts it.[35] We have now reached full circle in our study. The new atheists have adopted an outmoded approach to the history of science, advocating a "conflict thesis" that has been discredited by scholars working on the history of science and religion. Whether consciously or not, the new atheists draw from the positivists when they argue that not only has modern science made belief in God irrational, but that such religious faith has always obstructed the progress of science. Interestingly, a number of their critics have compared new atheism to Protestant fundamentalism.[36] Indeed, Christopher Hitchens once referred to himself as a "Protestant atheist."[37] While the comparison needs much qualification, there is nevertheless a great deal of truth in that description. That the new atheism is the product of a post-Protestant intellectual environment associated with questions of evidence, proof, and rationality, there is no doubt. But critics of the new atheism seem to fundamentally misunderstand the kind of Protestantism that emphasized such criteria of belief. This kind of Protestantism, for example, cannot be the kind of fundamental religion that John Gresham Machen and others supported and defended in the first half of the twentieth century.[38] Rather, as Edwin H. Wilson (1898–1993), American Unitarian who helped draft the "Humanist Manifesto" of 1933, wrote, "the modern humanist movement emerged from liberal religious change at the end of the previous century and the beginning of this one."[39]

While Sarton, the logical positivists, and even the new atheists may have, explicitly or implicitly, referred to the narratives of Draper, White, and other Protestant historiographers, their ultimate aim was radically different. As demonstrated in this book, a more careful reading of Draper and White reveals that they never intended their historical narratives to be used for such antireligious purposes. In each of their historical narratives, past conflict between science and religion offered an important guide for how, ultimately, these two contending powers might be reconciled. Draper presented a history of ecclesiastical corruption, arguing that the founding principles of the Protestant Reformation could not only correct this perversion of the simple and pure message of Christ, but also ameliorate what he believed was the coming religious and political crisis of the new century. For White, the history of science revealed the progressive character of both science and religion. He thus sought to provide Christianity with a "new basis," strengthening it by allowing it to grow more freely, uninhibited by theological dogmatism. Their publisher and popularizer Youmans, though he too rejected the idea that religion and science were in conflict, believed that Protestantism was now in decline. A new religion of the future, adapted to the prevailing intellectual fashions of the day, was required. Thus, rather than turning back to Protestant Christianity, Youmans looked forward to the next evolutionary stage in religious belief. Draper, White, and Youmans were committed theists who presupposed a worldview in which God had set up an independent and uniformly functioning system. Their narratives of conflict in the history of science and religion, therefore, were not meant to damage religion generally or theism specifically. Rather, they leveled their criticism against theological dogmatism or ecclesiasticism, not religious beliefs. In short, they intended to mitigate what they perceived as a coming religious crisis.

But while often praised by liberal theologians as providing the foundations for such a resolution, this attempted reconciliation ultimately failed. As we have seen, many freethinkers, skeptics, and nonbelievers articulated the logical conclusions of these narratives. Subsequently, in the first half of the twentieth century, many writers took it for granted that science and religion were natural enemies. Thus, when we discuss the origins of the "conflict thesis," the popular notion that science and religion are perennially at war or in conflict, we need to relocate its inception to a later century. That way of relating science and religion is a much more recent invention and one that lies ironically at the very foundations of the discipline of the history of science itself. Indeed, we find the forerunners of new atheism in this more recent body of work, not in the nineteenth-century writings of Draper and White, who strongly held the belief that science and religion could be ultimately reconciled.

One consequence of mislocating the origins of the "conflict thesis" is that despite the widening rejection of the conflict model for understanding science and religion relations, there still remains some dependence upon the very conflict meant to be undermined. As we have seen, the "conflict thesis" was never completely abandoned by historians of science—rather, it has been merely reconceptualized, redefined, or relocated to different areas. The irony here is this so-called "revisionist" work is remarkably similar to how Draper and White had conceived of the conflict in the first place. Some historians of science, for example, sought to eliminate the idea of "warfare" or "militant conflict" between science and Christianity by replacing it with a concept of a crisis of faith *within* religion. This is indeed the same position that Draper and White and countless others took at the end of the nineteenth century. By reading their narratives as histories to debunk rather than primary sources reflecting the nineteenth-century views of science and religion, modern historians of science have effectively created an interpretive cul-de-sac often perfunctorily repeating the same accusations again and again. It is one of those curious ironies of history that the revisionist historiographers of science who have so successfully debunked many myths about science and religion are themselves partly responsible for constructing another myth about the cofounders of the "conflict thesis."

But the "conflict," as Draper, White, and many of their contemporaries knew, can be traced back to Catholic-Protestant polemics of the sixteenth century. Scholars from various disciplines have recognized that Protestantism fostered a fundamental rejection of the sacramental worldview of medieval Catholicism and promoted removing from Christianity what was believed to be superstitious excrescences. By reconceptualizing religion as a set of internal beliefs, Protestant reformers were able to justify their succession from Rome. Perhaps more importantly, Protestants also constructed a new historiography that accused the Catholic Church of paganizing and thus corrupting Christianity. But this was not enough. Many Protestants subsequently adopted a rhetoric of rationality that overtly presented the movement as purging the superstition that had attached itself to the simple and pure teachings of Christ. History and reason had become a powerful and effective polemical tool used by Protestants against Catholicism.

Protestantism has proven to be a fissiparous movement. The vehemence with which Protestants attacked Catholicism had unintended consequences. A new generation of Protestants used similar strategies against other Protestants, opposing and accusing one another of irrationality and obstructing religious progress. This precedent was particularly popular among English liberal divines in the seventeenth and eighteenth centuries, especially among those inclined to examine the natural world. Protestant theologians and natural philosophers (who were frequently both) during those centuries increasingly denied an unpredictable, meddling God who worked through constant miraculous intervention. Rather, God worked through the regularities of an orderly universe, in the operation of natural laws. This religious conviction frequently presupposed, sanctioned, and even motivated scientific inquiry. These pious natural philosophers thus sought to protect, defend, and demonstrate the sovereignty of God in the uniformity of laws in nature rather than the revelation of Scripture.

Protestant historiography was further transformed during the late eighteenth and early nineteenth centuries into stadial histories of progressive liberation from religious superstition and theological dogmatism. By the mid-nineteenth century, histories and philosophies of science were used to legitimatize the emerging scientific disciplines. These early disciplines presupposed a Protestant heritage, and many of its advocates transferred values originally associated with clergymen to men of science. Many now claimed that nature pointed more clearly than revelation toward a Creator. As the "gentlemen of science" of the century contended, it was not science that created conflict, but rather those ignorant about the scientific process—namely, those religious conservatives or orthodox who continued to read the biblical text literally. Men of science thus denied that there was a conflict between science and religion. The conflict was against dogmatic assumptions about nature and Scripture. This was a story of scientific progress as the progress against institutionalized religion. When free inquiry into nature challenged the authoritarianism of established religious traditions, conflict was the result. Conflict could be overcome, however, with the abandonment of outdated theology. These histories and philosophies of science, therefore, were not meant to damage religion. Rather, it was leveled against theological dogmatism. The conflict between science and religion, then, was rather a conflict against long-standing components of Christian theology that many now deemed inessential to Christianity.

Draper and White were in this sense transitional figures. They stood between those early works of history and philosophy of science by nineteenth-century naturalist clergymen and the more pugilistic campaigns of the scientific naturalists. As this book hopes to have demonstrated, for Draper and White it was not science versus religion but religion versus the intellectual ideas of the modern age. They believed that if only Christianity could discard its accreted dogma and embrace a scientific worldview, its foundations would be strengthened rather than weakened. Science would modernize Christianity rather than replace it. While they rejected most traditional Christian doctrines and took up a more diffusive form of Christianity, both were self-acknowledged theists. This ambiguous hold to the Christian tradition was not uncommon among nineteenth-century liberal Protestants. Indeed, Draper and White cheered on the liberalization of Christian belief as the process necessary for the reconciliation to go forward. Science would liberalize belief, and help religion break away from its encrusted and unnecessary dogma.

At the turn of the century, the concerted effort to bring about a reconciliation between modern thought and religion continued. This effort was once again most conspicuously led by the more religiously liberal. But this attempted reconciliation was more of an accommodation by liberal Protestants to a naturalistic worldview. Early twentieth-century liberal Protestants actually further promulgated the narratives of Draper and White, as did their late nineteenth-century predecessors. Although more work is needed, such revelations suggest that the proposed reconciliation between science and religion by early twentieth-century scientists and liberal theologians was just as ambiguous and just as fragile as the ones proposed by Draper and White. While the modernists were anxious to forge a new theology purged of ancient dogma, the traditionalists felt that there was no point in preserving a church that no longer truly adhered to traditional Christian beliefs. Ironically, but perhaps unsurprisingly, nonbelievers agreed that religious modernism only further undermined historical Christianity. Thus, while they held opposing worldviews, conservatives and atheists rejected the proposed reconciliation for similar reasons. This failure testifies that what really was in conflict was not "science and religion" but enduring philosophical and theological questions about nature, humanity, knowledge, and God. It was, and remains, a conflict between two contending theological worldviews.

NOTES

INTRODUCTION

1. Sam Harris, "Science Must Destroy Religion," in *What is Your Dangerous Idea?* ed. John Brockman, 148–51. See also Harris, *The End of Faith*, 15, 109, 165; and Harris, *Letter to a Christian Nation*, 62–68.

2. Christopher Hitchens, *god is not Great*, 64–65, 260.

3. Victor J. Stenger, *God and the Folly of Faith*, 31–46. See also Stenger, *God: The Failed Hypothesis*; and Stenger, *The New Atheism*.

4. Standard critical assessments include, e.g., Alister McGrath and Joanna Collicutt McGrath, *The Dawkins Delusion?*; John F. Haught, *God and the New Atheism*; David Bentley Hart, *Atheist Delusions*; Karen Armstrong, *The Case for God*; Alister McGrath, *Why God Won't Go Away*; and more recently Borden Painter, *The New Atheist Denial of History*.

5. Alfred North Whitehead, *Science and the Modern World*, 180, 181, 182.

6. Whitehead, *Science and the Modern World*, 14.

7. See Pierre Duhem, *The Origins of Statics*, ix-xvii, xix-xxxv. For more on Duhem, see, e.g., R. N. D. Martin, "The Genesis of a Medieval Historian"; Roger Ariew and Peter Barker, trans. and eds., *Pierre Duhem: Essays in the History and Philosophy of Science*, esp. vii-xv. See also Stanley L. Jaki, *Uneasy Genius: The Life and Work of Pierre Duhem*.

8. See, e.g., Lynn Thorndike, *History of Magic and Experimental Science*; Charles H. Haskins, *Studies in the History of Medieval Science*, and *The Renaissance of the Twelfth Century*; Alexandre Koyré, *From the Closed World to the Infinite Universe*; and Marshall Clagett, *Greek Science in Antiquity*, and *The Science of Mechanics in the Middle Ages*. For more recent studies, see, e.g., Amos Funkenstein, *Theology and the Scientific Imagination*; David C. Lindberg, *The Beginnings of Western Science*; Edward Grant, *The Foundation of Modern Science in the Middle Ages*; and Marcia L. Colish, *Medieval Foundations of the Western Intellectual Tradition, 400–1400*. See also David C. Lindberg and Michael H. Shank, eds., *The Cambridge History of Science*, vol. 2: *Medieval Science*.

9. E. A. Burtt, *Metaphysical Foundations of Modern Physical Science*.

10. Burtt, *Metaphysical Foundations*, 23–60, 71, 103, 133, 144, 188, 258.

11. Burtt, *Metaphysical Foundations*, 226.

12. Dorothy Stimson, "Puritanism and the New Philosophy in 17th Century England"; Robert K. Merton, "Science, Technology and Society in Seventeenth Century England"; and Michael B. Foster, "The Christian Doctrine of Creation and the Rise of Modern Natural Science"; "Christian Theology and Modern Science of Nature," Parts 1 and 2.

13. Paul H. Kocher, *Science and Religion in Elizabethan England*; Richard S. Westfall, *Science and Religion in Seventeenth-Century England*; John Dillenberger, *Protestant Thought and Natural Science*; R. Hooykaas, *Religion and the Rise of Modern Science*. See also Charles Webster, *The Great Instauration:*; Margaret C. Jacob, *The Newtonians and the English Revolution.*

14. R. Hooykaas, *Religion and the Rise of Modern Science.*

15. Hebert Butterfield, *The Whig Interpretation of History.*

16. C. T. McIntire, *Herbert Butterfield: Historian as Dissenter*, 205.

17. Herbert Butterfield, *The Origins of Modern Science: 1300–1800.*

18. Butterfield, *Origins of Modern Science*, 13.

19. Charles E. Raven, *Science, Religion, and the Future*, 33–50.

20. Charles Coulston Gillispie, *Genesis and Geology*, xi.

21. Owen Chadwick, *The Victorian Church*, 1.2; 2.3.

22. Susan Faye Cannon, *Science in Culture.*

23. Robert M. Young, *Darwin's Metaphor.*

24. Frank M. Turner, *Contesting Cultural Authority*, esp. 171–200. See also Turner, *Between Science and Religion*, esp. 8–37.

25. James R. Moore, *The Post-Darwinian Controversies.*

26. David N. Livingstone, *Darwin's Forgotten Defenders*; Jon H. Roberts, *Darwinism and the Divine in America*; Ronald L. Numbers, *Darwinism Comes to America.*

27. David C. Lindberg and Ronald L. Numbers, eds., *God and Nature*, 1–18.

28. John Hedley Brooke, *Science and Religion*, 5, 321.

29. See John Hedley Brooke, "Presidential Address: Does the History of Science have a Future?" *British Journal for the History of Science*, vol. 32, no. 1 (1999): 1–20.

30. The literature is extensive. A small sampling of more general studies includes: Gary B. Ferngren, ed., *Science and Religion*; John Hedley Brooke, Margaret J. Osler, and Jitse van der Meer, eds., *Science in Theistic Contexts*; David C. Lindberg and Ronald L. Numbers, eds., *When Science and Christianity Meet*; Edward Grant, *Science & Religion, 400 B.C.–A.D. 1550*; Richard G. Olson, *Science and Religion, 1450—1900*; Thomas Dixon, *Science and Religion: A Very Short Introduction*; Ronald L. Numbers, ed., *Galileo Goes to Jail and Other Myths about Science and Religion*; Donald A. Yerxa, *Recent Themes in The History of Science and Religion*; Peter Harrison, ed., *The Cambridge Companion to Science and Religion*; Thomas Dixon et al., eds., *Science and Religion: New Historical Perspectives*; John Hedley Brooke and Ronald L. Numbers, eds., *Science and Religion Around the World*; Peter Harrison et al., eds., *Wrestling with Nature*; Ronald L. Numbers and Kostas Kampourakis, eds., *Newton's Apple and Other Myths about Science*; Donald A. Yerxa, ed., *Religion and Innovation.*

31. Peter Harrison, *The Bible, Protestantism, and the Rise of Natural Science*, 107.

32. Peter Harrison, *The Fall of Man and the Foundations of Science*, 3, 87–88.

33. Peter Harrison, *The Territories of Science and Religion*, x, 3, 14–15, 16, 197. Harrison has summarized all these views in his "Protestantism and the Making of Modern Science," in Thomas Albert Howard and Mark A. Noll, eds., *Protestantism after 500 Years* (Oxford: Oxford University Press, 2016), 98–120.

34. Robert K. Merton, *Science, Technology and Society in Seventeenth-Century England*, xvi.

35. Colin A. Russell, R. Hooykaas, and David C. Goodman, *The 'Conflict Thesis' and Cosmology*.

36. Moore, *Post-Darwinian Controversies*, 19–49, 50.

37. Ronald L. Numbers, "Science and Religion," 59–80.

38. David C. Lindberg and Ronald L. Numbers, "Beyond War and Peace"; see also *God and Nature*, 1–3; and *When Science & Christianity Meet*, 1.

39. Colin A. Russell, *Cross-Currents*, esp. 190–96; and "The Conflict Metaphor and Its Social Origins," 3–26.

40. Brooke, *Science and Religion*, 34–36.

41. See, e.g., Peter Harrison, *The Cambridge Companion to Science and Religion*, 4; *The Territories of Science and Religion*, 172; "'Science' and 'Religion': Constructing the Boundaries," in Dixon et al., *Science and Religion*, 27; "Myth 24: That Religion has Typically Impeded the Progress of Science," in Numbers et al., *Newton's Apple*, 199; "Religion, Scientific Naturalism and Historical Progress," in Yerxa, *Religion and Innovation*, 87–99; and most recently, "Science and Secularization," *Intellectual History Review*, vol. 27, no. 1 (2017): 47–70, 51–52.

42. Numbers, *Galileo Goes to Jail*, 1–2.

43. See, e.g., Numbers, "Science and Religion," 61; Lindberg and Numbers, *God and Nature*, 1–3; Brooke, *Science and Religion*, 34–35; Lindberg and Numbers, *When Science & Christianity Meet*, 1; Numbers, *Galileo Goes to Jail*, 1–2.

44. See Moore, *Post-Darwinian Controversies*, 16, 68–76, 102–10, 346–51.

45. Whitehead, *Science and the Modern World*, 188.

46. Whitehead, *Science and the Modern World*, 191.

47. Dixon et al., *Science and Religion*, 263–82, 283–98.

48. Numbers, *Galileo Goes to Jail*, 6–7.

49. See, e.g., James A. Secord, *Victorian Sensation*, 9–40; Bernard Lightman, *Victorian Popularizers of Science*, 1–38; and most recently Aileen Fyfe, *Steam-Powered Knowledge*, 1–11.

50. A recent collection of essays, edited by Jeff Hardin, Ronald L. Numbers, and Ronald A. Binzley, *The Warfare between Science & Religion*, has begun correcting this conventional interpretation by taking this religious context more seriously. But much work remains to be done.

51. Cf. David N. Livingstone, "Science, Religion, and the Cartographies of Complexity," in Yerxa, *Recent Themes in the History of Science and Religion*, 51–55.

52. See, e.g., James Turner, *Without God, Without Creed*; Michael J. Buckley, *At the Origins of Modern Atheism*; Charles Taylor, *A Secular Age*; Michael Allen Gillespie, *The Theological Origins of Modernity*; Christopher Lane, *The Age of Doubt*; Brad S. Gregory, *The Unintended Reformation*; and Dominic Erdozain, *The Soul of Doubt*.

53. Frank E. Manuel, *The Prophets of Paris*.

54. See, e.g., David A. Pailin, *Attitudes to Other Religions*; Tomoko Masuzawa, *The Invention of World Religions*; Guy G. Stroumsa, *A New Science*; Lynn Hunt, Margaret C. Jacob, and Wijnad Mijnhardt, *The Book That Changed Europe*; and Brent Nongbri, *Before Religion*.

55. See Donald Fleming, *John William Draper and the Religion of Science*; and Glenn C. Altschuler, *Andrew D. White—Educator, Historian, Diplomat*.

ONE: Draper and the New Protestant Historiography

1. John William Draper, *History of the Conflict between Religion and Science*, vi–xvi.

2. Draper, *History of the Conflict*, 52–56; 58–67; 62–64, 152–60; 160–62; 167–69; 169–73; 177–81; 182–200; 218–19.

3. Draper, *History of the Conflict*, 327–67.

4. Draper, *Scientific Memoirs*, xii.

5. The most complete biography of John W. Draper remains Donald Fleming, *John William Draper and the Religion of Science*. But there were also numerous contemporary accounts, including, e.g., George F. Barker, *Memoir of John William Draper*, and the tribute by colleagues and students in "Draper Centenary Issue: New York Pays Tribute to John William Draper—One Hundredth Anniversary of the Birth of Famous Scientists and Historian Celebrated at University Heights," *The New Yorker: The New York University Weekly*, vol. 4, no. 26 (9 May, 1911): 151–58.

6. J. S. Featherstone, A *Tribute of Grateful Remembrance to the Memory of the Rev. John Christopher Draper*.

7. Birth Certificate of John William Draper, June 23, 1811, container 18, John William Draper Family Papers, Library of Congress, Manuscript Division, Washington D.C. (hereafter, "Draper Family Papers").

8. J. T. Slugg, *Woodhouse Grove School*, 253–55.

9. Barker, *Memoir of John William Draper*, 351. On science and Dissent, see Paul Wood, ed., *Science and Dissent in England*, esp. 1–18 and 19–37.

10. Benjamin N. Martin, "A Sketch of John W. Draper," *Magazine of American History with Notes and Queries*, vol. 8 (New York and Chicago: A.S. Barnes & Company, 1882), 240. See also Edward L. Youmans, "Sketch of Dr. J. W. Draper, *Popular Science Monthly*, 4 (Jan., 1874): 361–67.

11. See, e.g., Walter E. Houghton, *The Victorian Frame of Mind*; Gertrude Himmelfarb, *Victorian Minds*; J. Wesley Bready, *England before and after Wesley*; W. R. Ward, *Religion and Society in England 1790–1850*; George Kitson Clark, *Churchmen and the Condition of England*, (London: Methuen, 1973); and G. M. Young, *Victorian England: Portrait of an Age*. See also more recently David Bebbington, *Victorian Religious Revivals*, and D. Bruce Hindmarsh, *The Spirit of Early Evangelicalism*.

12. See, e.g., William Connor Sydney, *England and the English in the Eighteenth Century*.

13. Élie Halévy, *A History of the English People in the Nineteenth Century*, 1.410.

14. Bebbington, *Victorian Religious Revivals*, 9.

15. See, e.g., Thomas Coke and Henry Moore, *The Life of the Rev. John Wesley* (London: G. Paramore, 1792), 51–53; Robert Southey, *The Life of Wesley; and the Rise and Progress of Methodism*, 2 vols. (London: Longman, 1820), 1.47–48. See also Luke Tyerman, *The Life and Times of the Rev. John Wesley*, 3 vols. (New York: Harper & Brothers, 1872).

16. Coke and Moore, *The Life of the Rev. John Wesley*, 53.

17. Southey, *The Life of Wesley*, 1.168.

18. Tyerman, *The Life and Times of the Rev. John Wesley*, 1.289.

19. Tyerman, *The Life and Times of the Rev. John Wesley*, 1.289–90.

20. See James Turner, *Without God, Without Creed*, 35–72, 73–113. See also Elisabeth Jay, *The Religion of the Heart*, 16–50, 51–105, and 106–205.

21. See Hindmarsh, *The Spirit of Early Evangelicalism*, 102–42.

22. On evangelicals and science, see David N. Livingstone, D.G. Hart, and Mark A. Noll, eds., *Evangelicals and Science in Historical Perspective*, esp. 120–41. On science and the Wesleyan tradition in particular, see Robert E. Schofield, "John Wesley and Science in 18th Century England," *Isis*, vol. 44, no. 4 (1953): 331–40. See also the series of articles by John W. Haas, Jr.: "John Wesley's Views on Science and Christianity: An Examination of the Charge of Antiscience," *Church History*, vol. 63, no. 3 (1994): 378–92; "Eighteenth Century Evangelical Responses to Science: John Wesley's Enduring Legacy," *Science and Christian Belief*, vol. 6, no. 2 (1994): 83–102; and, "John Wesley's Vision of Science in the Service of Christ," *Perspectives on Science and Christian Faith*, vol. 47, no. 4 (1995): 234–43. More recently, see Randy L. Maddox, "John Wesley's Precedent for Theological Engagement with the Natural Sciences," *Wesleyan Theological Journal*, 44.1 (2009): 23–54.

23. Barker, *Memoir of John William Draper*, 352.

24. Henry Terry, "Edward Turner, M.D., F.R.S. (1798–1837)," *Annals of Science*, vol. 2, no. 2 (1937): 137–52.

25. "Edward Turner, M.D., F.R.S.," *Gentlemen's Magazine*, vol. 7 (Apr. 1837): 434–35.

26. English historian H. Hale Bellot listed Draper as one of Austin's students in *University College London 1826–1926*, 188.

27. "Mrs. Austin," *Times*, no. 25887 (Aug. 12, 1867): 10.

28. See Bellot, *University College London 1826–1926*, 187–88.

29. Robert Campbell, ed., *Lectures on Jurisprudence or the Philosophy of Positive Law by the late John Austin*, 2 vols. (London: John Murray, 1885), 1.420.

30. Barker, *Memoir of John William Draper*, 380.

31. Fleming, *John William Draper*, 8.

32. See Edgar Fahs Smith, *The Life of Robert Hare: An American Chemist (1781–1858)* (Philadelphia: J. B. Lippincott Co., 1917), 481.

33. "Prayers against the Cholera," *Pall Mall Gazette* (Oct. 9, 1865): 649.

34. Matthew Stanley, *Huxley's Church and Maxwell's Demon: From Theistic Science to Naturalistic Science* (Chicago: University of Chicago Press, 2015), esp. 34–79.

35. See William Mullinger Higgins and John William Draper, *Magazine of Natural History and Journal of Zoology, Botany, Mineralogy, Geology, and Meteorology*, vol. 5 (1832): 164–72, 264–72, 532–34, 632–67; and vol. 6 (1833): 344–50.

36. Higgins and Draper, *Magazine of Natural History*, vol. 5, 165.

37. On Higgins, see, e.g., comments in Jonathan Topham, "Science and Popular Education in the 1830s"; and Ralph O'Connor, "Young-earth Creationists in Early Nineteenth-century Britain? Towards a Reassessment of 'Scriptural Geology.'"

38. W. M. Higgins, *Mosaical and Mineral Geologies*, 128, 137.

39. "Draper's Lecture," *Southern Literary Messenger*, vol. 2, no. 9 (Aug. 1836): 596; "Lecture: The Last Course of Lectures Delivered during the Years 1836–1837," *Southern Literary Magazine*, vol. 3, no. 11 (Nov. 1837): 693–98.

40. Draper, *Introductory Lecture in the Course of Chemistry*, 9, 11, 13–14.

41. Draper, *Introductory Lecture to the Course of Chemistry: Relations of Atmospheric Air to Animals and Plants*, 5, 6, 7–8, 11, 13.

42. Draper, *Introductory Lecture to the Course of Chemistry: Relations and Nature of Water*, 5, 9, 13.

43. Draper, *Introductory Lecture on Oxygen Gas*, 4, 6–7, 13, 14 (my emphasis).

44. Draper, *The Influences of Physical Agents on Life*, 6, 7–8, 20, 11.

45. See, e.g., Draper, *Treatise on the Forces which Produce the Organization of Plants; A Text-book on Chemistry*; and *A Text-Book on Natural Philosophy*.

46. Draper, *Human Physiology, Statical and Dynamical*.

47. Draper, *Human Physiology*, iv-vii.

48. Draper, *Human Physiology*, 600–602.

49. Draper, *Human Physiology*, 624–34.

50. See discussion in Charles Taylor, *A Secular Age*, 221ff.

51. Nicholas Jardine, "*Naturphilosophie* and the kingdoms of nature," in N. Jardine et al., eds., *Cultures of Natural History*, 230–45.

52. Adrian Desmond, *The Politics of Evolution*, esp. 25–100. For more detailed studies on the "godless institution in Gower Street," see, e.g., Negley Harte, *The University of London, 1836–1986*, and, esp., Rosemary Ashton, *Victorian Bloomsbury*.

53. Draper, *Human Physiology*, 472, 580–81, 583, 324, 514.

54. Draper, *Human Physiology*, 539. On Buffon's natural history, see Peter J. Bowler, *Evolution: The History of an Idea*, 75–80. On Buffon's religious beliefs, see Jacques Roger, *Buffon: A Life in Natural History*, trans. Sarah Lucille Bennefoim ed. L. Pearce Williams (Ithaca, NY: Cornell University Press, 1997), 431.

55. John William Draper, *Human Physiology, Statical and Dynamical*, viii.

56. "Address by Professor Draper," *New York Herald*, no. 8459 (Nov. 3, 1859): 5.

57. Henry Brougham, *Inaugural Discourse on being installed Lord Rector of the University of Glasgow, Wednesday, April 6, 1825* (London: Andrew and John M. Ducan, 1825), 47–50 (my emphasis).

58. Henry Brougham, *Practical Observations upon the Education of the People, Addressed to the Working Classes and their Employers* (London: Printed by Richard Taylor, 1825).

59. James A. Secord, *Visions of Science*, 14.

60. Henry Brougham, *A Discourse on the Objects, Advantages, and Pleasures of Science*.

61. For a recent study that questions the conventional view, see James C. Ungureanu, "A Yankee at Oxford: John William Draper at the British Association for the Advancement of Science at Oxford, 30 June 1860," 135–50.

62. John William Draper, "The Intellectual Development of Europe (considered with reference to the views of Mr. Darwin and others) that the Progression of Organisms is Determined by Law," container 8, Draper Family Papers.

63. Darwin did not explicitly identify apes as the ancestors of humans until he published his *The Descent of Man, and Selection in Relation to Sex*, 2 vols. (London: John Murray, 1871), where he announced that humans had "descended from a hairy quadruped, furnished with a tail and pointed ears" (2.389).

64. John William Draper, "Dr. Draper's Lecture on Evolution: Its Origin, Progress, and Consequences," *Popular Science Monthly*, vol. 12 (Dec 1877): 175–92.

65. Draper, "Dr. Draper's Lecture," 175–76.

66. Draper, "Dr. Draper's Lecture," 186–91.

67. John Stuart Mill, "The Spirit of the Age," *The Examiner*, 1197 (Jan. 9, 1831): 20–21.

68. See discussion in Aileen Fyfe, *Steam-Powered Knowledge: William Chambers and the Business of Publishing, 1820–1860*, 1–12.

69. Bernard Lightman, *Victorian Popularizers of Science*, see esp. 1–38.

70. Sir Humphry Davy, *Consolations in Travel, Or the Last Days of a Philosopher*, 37, 229–80.

71. For a recent biography, see Jan Golinski, *The Experimental Self: Humphry Davy and the Making of a Man of Science.*

72. See Peter J. Bowler, *The Invention of Progress.*

73. See, e.g., Walter F. Cannon, "Scientists and Broad Churchmen"; Cannon, "The Problem of Miracles in the 1830s"; and Michael Ruse, "The Relationship between Science and Religion in Britain, 1830–1870." See also the comprehensive studies by Jack Morrell and Arnold Thackray, *Gentlemen of Science;* Richard Yeo, *Defining Science;* Pietro Corsi, *Science and Religion;* and Laura J. Snyder, *The Philosophical Breakfast Club.*

74. Draper, *Scientific Memoirs,* 76, 246, 271, 284, and 322.

75. Draper, *A Text-Book on Natural Philosophy,* iv.

76. Thomas Dick, *The Christian Philosopher,* 133, 132, 277.

77. Dick, *Christian Philosopher,* 245–47, 123.

78. Dick, *Christian Philosopher,* 12, 160–200. For a detailed study on Dick, see William J. Astore, *Observing God: Thomas Dick, Evangelicalism, and Popular Science in Victorian Britain and America.*

79. See Charles Babbage, *Reflections on the Decline of Science in England* (London: B. Fellowes, 1830); *The Ninth Bridgewater Treatise: A Fragment* (London: John Murray, 1837); and *The Exposition of 1851; or, Views of the Industry, the Science, and the Government of England* (London: John Murray, 1851).

80. John F. W. Herschel to John W. Draper, Oct. 6, 1840, Draper Family Papers, Box 4. For details on Draper's photochemistry, see Sarah Kate Gillespie "John William Draper and the Reception of Early Scientific Photography," *History of Photography,* vol. 36, no. 3 (2012): 241–54; and Howard R. McManus, "The Most Famous Daguerreian Portrait: Exploring the History of the Dorothy Catherine Draper Daguerreotype," *The Daguerreian Annual* (1995): 148–71.

81. See Ungureanu, "A Yankee at Oxford," 144.

82. John F. W. Herschel, *A Preliminary Discourse on the Study of Natural Philosophy* (London: Longman, 1830), 1–8; new ed. (London: Longman, Brown, Green & Longmans, 1851), 6, 27–28, 59, 108.

83. Herschel, *A Preliminary Discourse,* new ed., 9, 16, 17.

84. "Science and Scripture," *Athenaeum,* no. 1925 (Sept. 17, 1864): 375. For a detailed discussion of this document, see, W. H. Brock and R. M. Macleod, "The Scientists' Declaration: Reflexions on Science and Belief in the Wake of *Essays and Reviews,* 1864–5," *British Journal for the History of Science,* vol. 9, no. 1 (1976): 39–66; and, more recently, Hannah Gay, "'The Declaration of Students of the Natural and Physical Sciences,' revisited: Youth, Science, and Religion in mid-Victorian Britain," in William Sweet and Richard Feist, eds., *Religion and the Challenges of Science,* 19–38.

85. See S. S. Schweber, "John Herschel and Charles Darwin: A Study of Parallel Lives," 1–71.

86. Herschel, *A Preliminary Discourse,* new ed., 72.

87. Charles Lyell, *Principles of Geology,* 1.44.

88. William Whewell, *History of the Inductive Sciences,* 3.638, 3.653, 1.268.

89. Robert Chambers to John W. Draper, June 23, 1864; John W. Draper to Robert Chambers, July 19, 1864, container 2, Draper Family Papers.

90. See James A. Secord, *Victorian Sensation,* esp. 17–24.

91. Robert Chambers, *Vestiges of the Natural History of Creation,* 8.

92. Chambers, *Vestiges of the Natural History of Creation*, 8, 152, 153, 155, 157, 159, 198, 233, 235, 312, 324, 326, 332, 341, 348, 355, 357, 361, 362, 365, 372, 376, 377, 381, 382, 383, 385.

93. Robert Chambers, *Explanations*, 3.

94. Secord, *Victorian Sensation*, 85–87.

95. See his reflections on "true religion" in Charles Gibbon, ed., *The Life of George Combe*, 1.223–25.

96. George Combe, *The Constitution of Man*, 19, 23–32, 395–96 (my emphasis).

97. George Combe, *On the Relation between Science and Religion*, v-xxxi, 11–14. For a study of Combe, see John van Wyhe, *Phrenology and the Origins of Victorian Scientific Naturalism*.

98. Draper, *A History of the Intellectual Development of Europe*, iii-iv; 15, 17, 258.

99. Draper, *A History of the Intellectual Development of Europe*, 193–96, 198, 201.

100. Draper, *A History of the Intellectual Development of Europe*, 206–27, 230.

101. Draper, *A History of the Intellectual Development of Europe*, 382–402.

102. Draper, *A History of the Intellectual Development of Europe*, 429, 465–93.

103. Draper, *A History of the Intellectual Development of Europe*, 511–41, 556, 616.

104. One recent exception being Geoffrey Cantor, "What shall we do with the 'Conflict Thesis'?" in Thomas Dixon et al., eds., *Science and Religion*, 283–98.

105. Draper, *Thoughts on the Future Civil Policy of America*, iii-iv, 61, 205, 236, 280–81 (my emphasis).

106. Draper, *History of the American Civil War*, 1.iii; 1.21; 1.31, 37.

107. Draper, *History of the American Civil War*, 1.39–109; 1.419; 1.152; 3.676.

108. Draper, *History of the Conflict*, v.

109. Draper, *History of the Conflict*, 1–38, 47–73.

110. Draper, *History of the Conflict*, 78–101, 102–18, 138–51.

111. Whewell, *History of the Inductive Sciences*, 3.298–99.

112. Draper, *History of the Conflict*, 139–40.

113. Ernest Renan, *The Life of Jesus*, 291, 301–11.

114. Ernest Renan, *Averroes et L'Averroïsme*, 18–19.

115. Draper, *History of the Conflict*, 97, 114.

116. See Edward Gibbon, *The History of the Decline and Fall of the Roman Empire*, 10.43–48.

117. Gibbon, *The History of the Decline*, 1.482–513.

118. See B. W. Young, "'Scepticism in Excess': Gibbon and Eighteenth-Century Christianity," *The Historical Journal*, vol. 41, no. 1 (1998): 179–99.

119. Draper, *History of the Conflict*, 152–200.

120. Draper, *History of the Conflict*, 201–85.

121. Draper, *History of the Conflict*, 327–64. See Philip Schaff and S. Irenaeus Prime, eds., *History, Essays, Orations and Other Documents of the Sixth General Conference of the Evangelical Alliance* (New York: Harper & Brothers, 1874). On the significance of the meeting, see George M. Marsden, *Fundamentalism and American Culture: The Shaping of Twentieth-Century Evangelicalism 1870–1925* (New York: Oxford University Press, 1980), 11–21.

122. John W. Draper, "Political Effect of the Decline of Faith in Continental Europe," *Princeton Review*, vol. 1 (Jan.–June 1879): 78–96.

123. See most recently, e.g., Lawrence M. Principe, "The Warfare Thesis," in Jeff Har-

din, Ronald L. Numbers, and Ronald A. Binzley, eds., *The Warfare between Science & Religion*, 6–26.

124. See L. S. Jacyna, "Immanence or Transcendence: Theories of Life and Organization in Britain, 1790–1835," *Isis*, vol. 74, no. 3 (1983): 311–29.

125. William Clark, "Report on Animal Physiology; comprising a Review of the Progress and Present State of Theory, and of our Information respecting the Blood, and the Powers which circulate it," in *Report of the Fourth Meeting of the British Association for the Advancement of Science; Held at Edinburgh in 1834* (London: John Murray, 1835), 95.

126. Thomas Hun, *Medical Systems, Medical Science and Empiricism: An Introductory Lecture, before the Albany Medical College, Delivered October 3, 1848* (Albany: Joel Munsell, 1849), 14.

127. Patrick Edward Dove, *The Theory of Human Progress, and Natural Probability of a Reign of Justice* (Boston: Benjamin B. Mussey & Co., 1851); *Romanism, Rationalism, and Protestantism, Viewed Historically: In Relation to National Freedom and National Welfare* (Edinburgh: Shepard & Elliot, 1855).

128. See, e.g., W. M. Simon, *European Positivism in the Nineteenth Century*; T. R. Wright, *The Religion of Humanity: the Impact of Comtean Positivism on Victorian Britain* (Cambridge: Cambridge University Press, 1986); Charles D. Cashdollar, *The Transformation of Theology, 1830–1890*; and Gillis J. Harp, *Positivist Republic.*

129. Cashdollar, *The Transformation of Theology*, 21–205.

130. John Stuart Mill, *Autobiography*, 163.

131. For a sampling of Saint-Simon's ideas, see *Social Organization, The Science of Man and Other Writings*, ed. and trans. Felix Markham (New York: Harper, 1964). See also discussion in Frank E. Manuel, *The Prophets of Paris*, 103–94; and Richard G. Olson, *Science and Scientism in Nineteenth-Century Europe*, 41–61.

132. Gerturd Lenzer, ed., *Auguste Comte and Positivism: The Essential Writings* (New York: Harper, 1975), lxvii. For detailed discussion, see Manuel, *The Prophets of Paris*; 249–96; Olson, *Science and Scientism in Nineteenth-Century Europe*, 62–84. See also a recent and very accessible account of Comte's philosophy in Michel Bourdeau, Mary Pickering, and Warren Schmaus, eds., *Love, Order, and Progress: The Science, Philosophy, and Politics of Auguste Comte.*

133. Mill, *Autobiography*, 277.

134. See Timothy Larsen, *John Stuart Mill: A Secular Life*, esp. 60–67. See also Alan P. F. Sell, *Mill on God: The Pervasiveness and Elusiveness of Mill's Religious Thought*; Linda C. Raeder, *John Stuart Mill and the Religion of Humanity*; and Harry Settanni, *The Probabilist Theism of John Stuart Mill.*

135. Larsen, *John Stuart Mill*, 15.

136. Larsen, *John Stuart Mill*, 95–104, 131–45, 149–66, 200–15.

137. John Stuart Mill, *Three Essays on Religion.*

138. Mill, *Three Essays on Religion*, 136, 242–57.

139. Larsen, *John Stuart Mill*, 210–14.

140. Draper, *Intellectual Development*, 382–83. On Joachim, see, e.g., Karl Löwith, *Meaning in History*, 145–59; and Jacob Taubes, *Occidental Eschatology*, trans. David Ratmoko (Stanford: Stanford University Press, 2009; 1947), 85–98. See also Ernest Lee Tuveson, *Millennium and Utopia*; and Norman Cohn, *The Pursuit of the Millennium.*

141. Draper, *Intellectual Development*, 5; *History of the American Civil War*, 1.93. On

Bodin, see Ann Blair, *The Theater of Nature: Jean Bodin and Renaissance Science* (Princeton, NJ: Princeton University Press, 1997).

142. M. H. Abrams, *Natural Supernaturalism*, 183.

143. On Hegel's philosophy of history and religion, see the collected essays in Stephen Houlgate and Michael Baur, eds., *A Companion to Hegel* (Malden, MA: Blackwell, 2011), esp. Robert Bernasconi, "'The Ruling Categories of the World': The Trinity in Hegel's Philosophy of History and the Rise and Fall of Peoples," 315–31.

144. John W. Draper to Nathan Appleton, May 12, 1875, container 1, Draper Family Papers.

145. Draper, *Intellectual Development*, 466, 469, 471, 474, 479, 490.

146. Draper, *History of the Conflict*, 219–24.

147. Draper, *History of the Conflict*, 224.

148. See J. G. A. Pocock, "Johann Lorenz von Mosheim: modern ecclesiastical historian," in *Barbarism and Religion*, vol. 5, 163–212.

149. See Thomas Buchan, "John Wesley and the Constantinian Fall of the Church," in Christian T. Collins Winn et al., eds., *The Pietist Impulse in Christianity* (Cambridge: James Clarke, 2012), 146–60.

150. Quoted in Pocock, "Johann Lorenz von Mosheim," 205–6.

151. Draper, *History of the Conflict*, 204.

152. See B. W. Young, "John Jortin, Ecclesiastical History, and the Christian Republic of Letters," 961–81.

153. Draper, *Intellectual Development*, 200.

154. Draper, *History of the Conflict*, 50–51.

155. Draper, *Intellectual Development*, 523; *History of the Conflict*, 211.

156. Draper, *Intellectual Development*, 292, 294; *Thoughts on the Future Civil Policy of America*, 282, 290–91.

157. Draper, *History of the American Civil War*, 3.642.

158. Draper, *Intellectual Development*, 494; *Thoughts on the Future Civil Policy of America*, 282.

159. On Froude, see, e.g., Basil Willey, *More Nineteenth Century Studies*, 106–36; Jeffrey Paul von Arx, *Progress and Pessimism*, 173–200; and Rosemary Jann, *The Art and Science of Victorian History*, 105–40. See also the definitive biography by Ciaran Brady, *James Anthony Froude: An Intellectual Biography of a Victorian Prophet*.

160. Froude's self-critique of Protestantism can be found in the numerous essays he published in popular periodicals, some of which were later collected in his *Short Studies on Great Subjects*.

161. Michael Madden, "Curious Paradoxes," 199–216.

162. Froude, "Conditions and Prospects of Protestantism," in *Short Studies on Great Subjects*, 2.165.

163. Froude, "The Oxford Counter-Reformation," in *Short Studies on Great Subjects*, 4.155.

164. See, e.g., Draper, *Intellectual Development*, 37, 156, 228, 278, 309, 326–81, 382–402, 429, 465–93; *History of the Conflict*, 68–101, 201–54, 298, 363.

165. Draper, *History of the Conflict*, xv.

166. Draper, *History of the Conflict*, 363.

167. Taylor, *A Secular Age*, 221.

TWO: White and the Search for a "Religion Pure and Undefiled"

1. Draper, *History of the Conflict*, ix.

2. Samuel D. Tillman to A. D. White, Oct. 12, 1869, Andrew Dickson White Papers, Division of Rare and Manuscript Collections at Cornell University, reel 9 (hereafter cited as: White Collection, and reel number).

3. "First of the Course of Scientific Lectures—Prof. White on 'The Battle-Fields of Science,'" *New York Daily Tribune* (Dec. 18, 1869): 4. White's lecture was also printed in the *Thirtieth Annual Report of the American Institute of the City of New York, for the Year 1869–70* (Albany: The Argus Company, 1870), 199–218.

4. Andrew D. White, "The Warfare of Science," *Popular Science Monthly*, vol. 8, no. 25 (Feb. 1876): 385–409; vol. 8, no. 33 (March 1876): 553–70.

5. Andrew D. White, "New Chapters in the Warfare of Science," *Popular Science Monthly*, running throughout from vol. 27 (Oct. 1885) to vol. 47 (Oct. 1895).

6. Andrew Dickson White, *A History of the Warfare of Science with Theology in Christendom*.

7. White, *Warfare of Science*, 20, 23, 24, 65, 74, 75, 92, and 101.

8. White, *History of the Warfare*, 1.ix.

9. White, *History of the Warfare*, 1.xii, 325, 410; 2.393, 394.

10. Glenn C. Altschuler, *Andrew D. White—Educator, Historian, Diplomat*, 153.

11. The most important sources to White's life and work are his own *Autobiography* and *Diaries of Andrew D. White*, the latter edited by Robert Morris Ogden. An equally important source is the Andrew Dickson White Papers, Division of Rare and Manuscript Collections at Cornell University library, which contains White's voluminous correspondence, lectures, speeches, articles, and other unpublished manuscripts.

12. White, *Autobiography*, 1.287.

13. White, *Autobiography*, 1.288.

14. Andrew D. White, *Outlines of a Course of Lectures on History* (Detroit: H. Barns & Co., 1861), White Collection, reel 139.

15. On the life and career of Cornell, see Alonzo Cornell, *True and Firm: Biography of Ezra Cornell*.

16. White, *Autobiography*, 1.291.

17. See, e.g., the classic studies by Henry A. Pochmann, *German Culture in America*; Jurgen Herbst, *The German Historical School in American Scholarship*; and Laurence R. Veysey, *The Emergence of the American University*. More recently, see Thomas Albert Howard, *Protestant Theology and the Making of the Modern German University*; Elizabeth A. Clark, *Founding the Fathers*; and Annette G. Aubert, *The German Roots of Nineteenth-Century American Theology*. See also the recent collection of essays in Louis Menand, Paul Reitter, and Chad Wellmon, eds., *The Rise of the Research University: A Sourcebook*.

18. Howard, *Protestant Theology and the Making of the Modern German University*, 12, 29–38, 131.

19. *Report of the Committee on Organization, Presented to the Trustees of the Cornell University. October 21st, 1866*, 20, 47–48.

20. *The Cornell University. First General Announcement* (Albany: Weed, Parsons and Co., 1868), 21. White had repeated the claim that Cornell had "Christian trustees and pro-

fessors earnestly devoted to building up Christian civilization" in his Cooper Union "The Battle-Fields of Science" speech.

21. *Account of the Proceedings at the Inauguration October 7th 1868*, 8, 13, 15.

22. Andrew D. White to Ezra Cornell, Aug. 3, 1869, White Collection, reel 9.

23. "Annual Report to Board of Trustees, June 27, 1873," White Collection, reel 136.

24. See "Reply to attack on Cornell University for Failure to Provide Religious Instruction," "Reply to Editor of Northern Christian Advocate on Cornell's Education Policy on Religion," and "Cornell University: What It Has Done, What It Is Doing, and What It Hopes to Do"—all in White Collection, reel 146.

25. *Cornell Era*, vol. 16, no. 13 (Jan. 18, 1884): 124–27.

26. White, *Autobiography*, 1.397–411.

27. *Cornell Era*, vol. 16, no. 13 (Jan. 18, 1884): 126.

28. White, *Autobiography*, 1.426.

29. White, *Autobiography*, 1.24.

30. See John B. Roney, *The Inside of History*, esp. 13–21.

31. *Catalogue of the Historical Library of Andrew Dickson White:* The Protestant Reformation and its Forerunners.

32. Ogden, *Diaries of Andrew D. White*, 79.

33. White, *Autobiography*, 1.39.

34. On Lessing, see the still useful introductory essay by Henry Chadwick in *Lessing's Theological Writings*, 9–49. See also Henry E. Allison, *Lessing and the Enlightenment*; Toshimasa Yasukata, *Lessing's Philosophy of Religion and the German Enlightenment*; and most recently the definitive biography by H. B. Nisbet, *Gotthold Ephraim Lessing: His Life, Works, and Thought.*

35. See Chadwick, *Lessing's Theological Writings*, 51–56; Yasukata, *Lessing's Philosophy of Religion*, 57–60; Nisbet, *Gotthold Ephraim Lessing*, 537–70.

36. Yasukata, *Lessing's Philosophy of Religion*, 72–88; Nisbet, *Gotthold Ephraim Lessing*, 601–23.

37. Yasukata, *Lessing's Philosophy of Religion*, 89–116; Nisbet, *Gotthold Ephraim Lessing*, 571–600.

38. For a fuller treatment of Schleiermacher's thought, see, e.g., Karl Barth, *Protestant Theology in the Nineteenth Century*, 411–59; Claude Welch, *Protestant Thought in the Nineteenth Century*, 1.57–85; and Alister McGrath, *The Making of Modern German Christology, 1750–1990*, 36–49.

39. Barth, *Protestant Theology*, 440.

40. See Peter Vogt, "Nicholas Ludwig von Zinzendorf (1700–1760)," in Carter Lindberg, ed., *The Pietist Theologians: An Introduction to Theology in the Seventeenth and Eighteenth Centuries* (Oxford: Blackwell, 2005), 207–23.

41. Schleiermacher, *On Religion: Speeches to its Cultured Despisers*, trans. Richard Crouter, 19, 21, 50.

42. Schleiermacher, *Sendschreiben an Dr. Lücke*, ed. Hermann Mulert (Gissen, 1908), 40; quoted in Welch, *Protestant Thought in the Nineteenth Century*, 1.63.

43. White, *Autobiography*, 1.39.

44. See Leopold von Ranke, "Über die Idee der Universalhistorie," in Fritz Stern, ed., *The Varieties of History: From Voltaire to the Present*, 59.

45. See discussion in Georg G. Iggers, *The German Conception of History*, 63–89; and Michael Bentley, *Modern Historiography* (London: Routledge, 1999), 36–42.

46. See Albrecht Ritschl, *The Christian Doctrine of Justification and Reconciliation*, trans. and eds. H. R. Mackingtosh and A. B. Macaulay, 199, 203–13. On Ritschl and the shape and influence of German theology in the nineteenth century, see Johannes Zachhuber, *Theology as Science in Nineteenth-Century Germany*.

47. Andrew D. White, "Glimpses of Universal History," *The New Englander*, vol. 15, no. 59 (Aug. 1857): 398–427.

48. Andrew D. White, *The Most Bitter Foe of Nations, and the Way to its Permanent Overthrow*.

49. Andrew D. White, *Outlines of Lectures on History*.

50. Andrew D. White, *The Message of the Nineteenth Century to the Twentieth*.

51. Andrew D. White, *Some Practical Influences of German Thought upon the United States*.

52. Andrew D. White, *On Studies in General History and the History of Civilization*.

53. Andrew D. White, *European Schools of History and Politics*.

54. Andrew D. White, *Evolution and Revolution*.

55. White, "The Battle-Fields of Science," 4.

56. White, *Warfare of Science*, 9–10; cf. 42, 63, 70–72, 110–11, 120, 122, 144–45, 148–51.

57. A. D. White to Daniel C. Gilman, July 24, 1878, White Collection, reel 23.

58. A. D. White to Charles Kendall Adams, Apr. 25, 1879, White Collection, reel 24.

59. Excerpts of this interview were printed in the Cornell student newspaper, *Cornell Daily Sun*, vol. 5, no. 74 (Feb. 5, 1885): 1, 4.

60. White, *Autobiography*, 2.495.

61. White, *History of the Warfare*, 1.v-vi.

62. White, *History of the Warfare*, 1.xii, 113, 167, 410.

63. White, *History of the Warfare*, 2.168.

64. White, *History of the Warfare*, 2.207, 208.

65. White, *History of the Warfare*, 2.219, 250, 263.

66. White, *History of the Warfare*, 2.288.

67. White, *History of the Warfare*, 2.311–48.

68. White, *Autobiography*, 2.566–67.

69. White, *History of the Warfare*, 2.385.

70. White, *History of the Warfare*, 2.390.

71. On Arnold's religious beliefs, see, e.g., Basil Willey, *Nineteenth Century Studies*, 251–83; Ruth apRoberts, *Arnold and God*; and James C. Livingston, *Matthew Arnold and Christianity*.

72. Matthew Arnold, *St. Paul and Protestantism*, 30, 32, 35, 49, 51, 71, 159.

73. Matthew Arnold, *Literature and Dogma*, xii, 22, 61–78, 81, 185–86, 277, 280.

74. Matthew Arnold, *God and the Bible*, xiv, 19, 90, 107, 130, 141–42, 389.

75. White, *History of the Warfare*, 2.393.

76. White, *History of the Warfare*, 2.394.

77. White, *History of the Warfare*, 2.395.

78. White, *Autobiography*, 2.513–28.

79. White, *Autobiography*, 2.495.

80. White, *Autobiography*, 2.533.

81. A. D. White to Willard Fiske, Oct. 19, 1880, White Collection, reel 27.

82. More comprehensive studies on nineteenth-century American liberal Protestantism can be found in, e.g., Stow Persons, *Free Religion*; Earl Morse Wilbur, *Three Prophets*

of American Liberalism: Channing, Emerson, Parker; Lloyd J. Averill, *American Theology in the Liberal Tradition*; Paul A. Carter, *The Spiritual Crisis of the Gilded Age*; William R. Hutchison, *The Modernist Impulse in American Protestantism*; Gary Dorrien, *The Making of American Liberal Theology*; and W. Creighton Peden, *Empirical Tradition in American Liberal Religious Thought, 1860–1960*.

83. T. W. Higginson to A. D. White, Apr. 27, 1876, White Collection, reel 21.

84. Quoted in *Proceedings at the Ninth Annual Meeting of the Free Religious Association*, 105.

85. T. W. Higginson to A. D. White, Dec. 6, 1876, White Collection, reel 21.

86. See Jenkin Lloyd Jones to A. D. White, Apr. 17, 1894, White Collection, reel 61; William Howell Reed to A. D. White, Jan. 31, 1896; D. W. Morehouse to A. D. White, Apr. 8, 1896, White Collection, reel 66; L. H. Stone to A. D. White, Apr. 18, 1896; May 4, 1896, White Collection, reel 66.

87. See, e.g., Hutchison, *The Modernist Impulse in American Protestantism*, 95–105; Dorrien, *The Making of American Liberal Theology*, 1.294–304.

88. See, e.g., Clark, *Founding the Fathers*, 43–50.

89. Philip Schaff, *America: A Sketch of the Political, Social, and Religious Character of the United States of North America*, 97.

90. Philip Schaff, *Theological Propaedeutic: A General Introduction to the Study of Theology*, 403.

91. Aubert, *The German Roots of Nineteenth-Century American Theology*, 15–35.

92. White, *A History of Warfare*, 2. 258–59, 263.

93. White, *Autobiography*, 1.11.

94. “Unitarian Christianity,” in *The Works of William E. Channing*, 3.59–103.

95. “Christian Worship,” *Works of William E. Channing*, 4.343–44.

96. “Unitarian Christianity,” *Works of William E. Channing*, 3.102–3.

97. White, *Autobiography*, 2.535. See also *History of the Warfare*, 2.366–68.

98. “A Discourse of the Transient and Permanent in Christianity,” in *The Critical and Miscellaneous Writings of Theodore Parker*, 152–89.

99. “Primitive Christianity,” in *Critical and Miscellaneous Writings of Theodore Parker*, 247–75.

100. “Discourses on Matters Pertaining to Religion,” in *The Collected Works of Theodore Parker*, vol. 1.

101. “Truth and the Intellect,” in *Collected Works of Theodore Parker*, vol. 2.

102. Horace Bushnell, *Discourses on Christian Nurture*.

103. Horace Bushnell, *God in Christ*, esp. 72, 80, 81, and 93.

104. Horace Bushnell, *Nature and the Supernatural*, 1–35, 422–25.

105. White, *Autobiography*, 2.535–37.

106. On the significance of Beecher, see, e.g., William G. McLoughlin, *The Meaning of Henry Ward Beecher*, and, more recently, Debbie Applegate, *The Most Famous Man in America: The Biography of Henry Ward Beecher*. See also remarks in Dorrien, *The Making of American Liberal Theology*, 1.179–260.

107. Dorrien, *The Making of American Liberal Theology*, 1.179, 1.181.

108. “Beecher on Evolution,” *New York Times* (Jan. 7, 1883): 2. This lecture was also reprinted by Lyman Abbot (1835–1922), Beecher’s successor at Plymouth Church, in his *Henry Ward Beecher: A Sketch of His Career*, 566–73.

109. Henry Ward Beecher, *Evolution and Religion*, Parts I and II.

110. Beecher, *Evolution and Religion*, Part I, 6–10.
111. Beecher, *Evolution and Religion*, Part I, 44–55.
112. Beecher, *Evolution and Religion*, Part I, 113–15.
113. Beecher, *Evolution and Religion*, Part I, 140.
114. White, *Autobiography*, 2.559–67, 562.
115. White, *Autobiography*, 2.533.
116. White, *Autobiography*, 2.494–95.
117. White, *Autobiography*, 2.510.
118. White, *Autobiography*, 2.568–73.

THREE: English Protestantism and the History of Conflict

1. Herbert Butterfield, *The Whig Interpretation of History*, 3.

2. See the classic studies by E. R. Norman, *Anti-Catholicism in Victorian England*; John Wolffe, *The Protestant Crusade in Great Britain, 1829–1860*; and D. G. Paz, *Popular Anti-Catholicism in Mid-Victorian England*. More recently, see, e.g., Michael Wheeler, *The Old Enemies: Catholic and Protestant in Nineteenth-Century English Culture*, and Jon Gjerde, *Catholicism and the Shaping of Nineteenth-Century America*.

3. Paz, *Popular Anti-Catholicism in Mid-Victorian England*, 299.

4. See, e.g., Rev. T. Mozley, *Reminiscences Chiefly of Oriel College and the Oxford Movement* (Boston: Houghton Mifflin, 1882). More recently, see also C. Brad Faught, *The Oxford Movement: A Thematic History of the Tractarians and Their Times* (University Park: Pennsylvania State University Press, 2003).

5. See Wheeler, *The Old Enemies*, 1–48.

6. See the Appendix in Wolffe, *The Protestant Crusade in Great Britain*, 318–19.

7. Paz, *Popular Anti-Catholicism in Mid-Victorian England*, 60–61.

8. Robert H. Ellison and Carol Marie Engelhardt, "Prophecy and Anti-Popery in Victorian London: John Cumming Reconsidered," *Victorian Literature and Culture*, vol. 31, no. 1 (2003): 337–89. See also Miriam Elizabeth Burstein, "Anti-Catholic Sermons in Victorian Britain," in Robert H. Ellison, ed., *A New History of the Sermon*, 233–67.

9. See, e.g., Susan M. Griffin, *Anti-Catholicism and Nineteenth-Century Fiction*; Mark Knight and Emma Mason, *Nineteenth-Century Religion and Literature*, esp. 189–216.

10. Gregory D. Dodds "An Accidental Historian: Erasmus and the English History of the Reformation."

11. On Protestant identity and the role of history, see Bruce Gordon, ed., *Protestant History and Identity in Sixteenth-Century Europe*. See also the discussion in A. G. Dickens and John M. Tonkin, *The Reformation in Historical Thought*, 7–57; Donald R. Kelly, *Faces of History: Historical Inquiry from Herodotus to Herder*, 130–35, 162–87; and Ernst Breisach, *Historiography: Ancient, Medieval, and Modern*, 3rd ed., 153–70, 171–98.

12. See the classic study by John M. Headley, *Luther's View of Church History*.

13. Glanmor Williams, *Reformation Views of Church History*, esp. 7–21.

14. On the idea of progress among Protestants, see, e.g., Robert Nisbet, *History of the Idea of Progress*; Karl Löwith, *Meaning in History*; and Ernest Lee Tuveson, *Millennium and Utopia: A Study in the Background of the Idea of Progress*.

15. Quoted in Korey D. Maas, *The Reformation and Robert Barnes: History, Theology and Polemic in Early Modern England*, 175.

16. Pearson, Rev. George, ed., *Writings and Translations of Myles Coverdale*, 82.

17. Christmas, Rev. Henry, ed., *Select Works of John Bale*, 251, 514.

18. Colin Haydon, *Anti-Catholicism in Eighteenth-Century England*, 28; see also Norman, *Anti-Catholicism in Victorian England*, 13–14; Wolffe, *The Protestant Crusade in Great Britain*, 112; Wheeler, *The Old Enemies*, 78–80.

19. See the works compiled by Peter Milward, *Religious Controversies of the Elizabethan Age*, and *Religious Controversies of the Jacobean Age*.

20. Alexandra Walsham, "History, Memory, and the English Reformation," 907.

21. Alexandra Walsham, "The Reformation of the Generations: Youth, Age and Religious Change in England, c. 1500–1700," 99.

22. While no comprehensive study on "pagano-papism" as yet exists, several authors have offered helpful comments, including David A. Pailin, *Attitudes to Other Religions*, 121–36; Jonathan Z. Smith, *Drudgery Divine*, 1–35; Peter Harrison, *'Religion' and the Religions in the English Enlightenment*, 139–46; and perhaps the most substantial treatment, S. J. Barnett, *Idol Temples and Crafty Priests*, 22–45, 105–28.

23. On the Puritan movement, see Patrick Collinson, *The Elizabethan Puritan Movement*. On the historiography of the Puritans, see also Paul C. H. Lim, "Puritans and the Church of England: historiography and ecclesiology," in John Coffey and Paul C. H. Lim, eds., *The Cambridge Companion to Puritanism*, 223–37.

24. David Hartley, *Observations on Man, His Frame, His Duty, and His Expectations*, 1.490.

25. See W. J. Torrance Kirby, "Reason and Law," and Egil Grislis, "Scriptural Hermeneutics," in W. J. Torrance Kirby, ed., *A Companion to Richard Hooker*, 251–71, 273–304.

26. Francis Bacon, "An Advertisement touching the Controversies of the Church of England," in Brian Vickers, ed., *Francis Bacon: The Major Works*, 1–19.

27. See, e.g., John Henry, *Knowledge is Power*; Stephen A. McKnight, *The Religious Foundations of Francis Bacon's Thought*; and Steven Matthews, *Theology and Science in the Thought of Francis Bacon*.

28. Bacon, "The Advancement of Learning," in Vickers, ed., *The Major Works*, 138, 141–42, 152–53.

29. Bacon, "The Advancement of Learning," in Vickers, ed., *The Major Works*, 126.

30. Bacon, "Valerius Terminus," in James Spedding, Robert Leslie Ellis, and Douglas Denon Heath, eds., *The Works of Francis Bacon*, 6.33. For a detailed study of this text, see Benjamin Milner, "Francis Bacon: The Theological Foundations of *Valerius Terminus*," *Journal of the History of Ideas*, vol. 58, no. 2 (1997): 245–64.

31. Bacon, "New Atlantis," in Vickers, ed., *The Major Works*, 457–90.

32. James Spedding, ed., *The Letters and the Life of Francis Bacon*, 3.253.

33. Paul Nelles, "The Uses of Orthodoxy and Jacobean Erudition: Thomas James and the Bodleian Library," *History of Universities*, vol. 22, no. 1 (2007): 21–70.

34. Anthony Grafton, *Worlds Made by Words*, 4.

35. Paul H. Kocher, *Science and Religion in Elizabethan England*; Richard S. Westfall, *Science and Religion in Seventeenth-Century England*.

36. Thomas Sprat, *The History of the Royal-Society of London, For the Improving of Natural Knowledge* (London: Printed by T. R. for J. Martyn at the Bell, 1667).

37. Sprat, *History of the Royal Society*, 35.

38. Sprat, *History of the Royal Society*, 349.

39. Sprat, *History of the Royal Society*, 113, 257–59, 349.

40. Sprat, *History of the Royal Society*, 362–63.

41. This point is also implicitly made in Paul B. Wood, "Methodology and Apologetics: Thomas Sprat's 'History of the Royal Society,'" *British Journal for the History of Science*, vol. 13, no. 1 (1980): 1–26.

42. It should be noted that this work was republished in the mid-nineteenth century by controversial preacher John Cumming (1807–1881), *Supplement to Gibson's Preservative from Popery: Being Important Treatises on the Romish Controversy*, 8 vols. (London: British Association for Promoting the Principles of the Reformation, 1850).

43. William C. Placher, *The Domestication of Transcendence.*

44. See John Gascoigne, *Cambridge in the Age of the Enlightenment*, esp. 40–51.

45. C. A. Patrides, ed., *The Cambridge Platonists*, 4.

46. Nathaniel Culverwell, *An Elegant and Learned Discourse of the Light of Nature*, 17–18, 213, 216, 218.

47. John Smith, *Select Discourses Treating Theological Topics*, 3, 42, 50, 377.

48. Benjamin Whichcote, *Moral and Religious Aphorisms*, 42, 47, 52.

49. Ralph Cudworth, *Mr. Cudworth's Sermon Preached before the Noble House of Commons, at Westminster, March 31st, 1647* (Cambridge: J. Talboys Wheeler, 1852), 2, 3, 10, 11.

50. Ralph Cudworth, *The True Intellectual System of the Universe*, 2.328, 2.447.

51. It should be no surprise, then, that Andrew Dickson White glowingly praised Cudworth's work as "one of the greatest glories of the English Church." He added that Cudworth "purposed to build a fortress which should protect Christianity against all dangerous theories of the universe, ancient and modern. The foundations of the structure were laid with old thoughts thrown often into new and striking forms; but, as the superstructure arose more and more into view, while genius marked every part of it, features appeared which gave the rigidly orthodox serious misgivings." See *A History of the Warfare*, 1.16–17.

52. Henry More, "The Epistle Dedicatory," *An Antidote against Atheism.*

53. Henry More, "*Enthusiasmus Triumphatus*; or A Brief Discourse of the Nature, Causes, Kinds, and Cure of Enthusiasm," in *A Collection of Several Philosophical Writings of Dr. Henry More.*

54. More, *An Antidote against Atheism*, 49.

55. See, e.g., overviews by B. J. Shapiro, "Latitudinarianism and Science in Seventeenth-Century England"; John Marshall, "The Ecclesiology of the Latitude-men 1660–1689"; John Spurr, "'Rational Religion' in Restoration England"; and Raymond D. Tumbleson, "'Reason and Religion': The Science of Anglicanism." For full-length studies, see Martin I. J. Griffin, *Latitudinarianism in the Seventeenth-Century Church of England*; W. M. Spellman, *The Latitudinarians and the Church of England, 1660–1700*; and Arthur F. Marotti, *Religious Ideology and Cultural Fantasy.*

56. Gilbert Burnet, *History of My Own Time*, 1.334.

57. Gilbert Burnet, "Autobiography," in H. C. Foxcroft, *A supplement to Burnet's history of my own time*, 463.

58. Gilbert Burnet, *History of the Reformation of the Church of England.*

59. William Chillingworth, *The Religion of Protestants*, 464–65.

60. Chillingworth, *The Religion of Protestants*, 463.

61. P. Des Maizeaux, *An Historical and Critical Account of the Life and Writings of William Chillingworth*, 56. See also discussion in Robert Orr, *Reason and Authority: The Thought of William Chillingworth.*

62. Chillingworth, *The Religion of Protestants*, 412–13.

63. Edward Stillingfleet, "Irenicum," in *The Works of that Eminent and most Learned Prelate, Dr. Edw. Stillingfleet*, 2.148–49, 2.151, 2.152.

64. Stillingfleet, "Origines Sacræ," in *Works*, 2.382.

65. Stillingfleet, "A Rational Account of the Grounds of Protestant Religion," in *Works*, 4.135, 4.196, 4.197.

66. Thomas Birch, *The Life of the Most Reverend Dr. John Tillotson*, 29.

67. Thomas Birch, *The Works of Dr. John Tillotson*, 6.425, 6.435–36.

68. Tillotson, "The Advantages of Religion to Societies," in *Works*, 1.420–21.

69. Tillotson, "The Hazard of Being Saved in the Church of Rome," in *Works*, 2.39, 2.54, 2.58–59.

70. Tillotson, "The Protestant Religion Vindicated from the Charge of Singularity and Novelty," in *Works*, 2.463, 2.467–68.

71. Tillotson, "Preached at Whitehall, April 4, 1679," in *Works*, 2.255–81.

72. Tillotson, "Of the Great Duties of Natural Religion, with the Ways and Means of Knowing Them," in *Works*, 5.273–97.

73. Tillotson, "Instituted Religion Not Intended to Undermine Natural," in *Works*, 5.298–322.

74. Tillotson, "A Thanksgiving Sermon for the late Victory at Sea," in *Works*, 3.236.

75. Gilbert Burnet, *A sermon preached at the funeral of the most reverend Father in God, John Tillotson, Lord Archbishop of Canterbury*, 30.

76. John Wilkins, *Of the Principles and Duties of Natural Religion*, 34.

77. Wilkins, *Principles and Duties*, 57.

78. Wilkins, *Principles and Duties*, 49.

79. Wilkins, *Principles and Duties*, 356.

80. Wilkins, *Principles and Duties*, vi.

81. On these figures, see Margaret C. Jacob, *The Newtonians and the English Revolution*, and, more recently, Katherine Calloway, *Natural Theology in the Scientific Revolution*.

82. Robert Boyle, "The Christian Virtuoso," in Thomas Birch, *The Works of the Honourable Robert Boyle*, 5.715.

83. See Maurice Wiles, *Archetypal Heresy: Arianism through the Centuries*, esp. 62–164.

84. Peter Gay, *The Enlightenment: An Interpretation*, 1.327.

85. Harrison, *'Religion' and the Religions in the English Enlightenment*, 62.

86. Barnett, *Idol Temples and Crafty Priests*, 1–21.

87. Jeffrey R. Wigelsworth, *Deism in Enlightenment England*; Wayne Hudson, *The English Deists*, and *Enlightenment and Modernity*; and Wayne Hudson, Deigo Lucci, and Jeffrey R. Wigelsworth, eds., *Atheism and Deism Revalued*. Most of this recent work draws inspiration from Justin Champion's original *Pillars of Priestcraft Shaken: The Church of England and Its Enemies, 1660–1730*, which is still worth perusing.

88. Robert E. Sullivan, *John Toland and the Deist Controversy*, 255.

89. Joseph Priestley, *Memoirs of the Rev. Dr. Joseph Priestley* (London, 1809), 17. The most detailed examination of Priestley's life and work are found in Robert E. Schofield two volumes, *The Enlightenment of Joseph Priestley* and *The Enlightened Joseph Priestley*.

90. Basil Willey, *The Eighteenth-Century Background*, 182–83.

91. Dennis G. Wigmore-Beddoes, *Yesterday's Radicals*.

92. Priestley, *Memoirs*, 67–68.

93. Joseph Priestley, *A Catechism, for Children, and Young Persons*, 43.

94. Joseph Priestley, "A Free Address to Protestant Dissenters on the Subject of the Lord's Supper," in *The Theological and Miscellaneous Works of Joseph Priestley*, 21.249–92 (hereafter, *Works*).

95. Priestley, "Appendix," *Works*, 25.375.

96. Priestley, "An Appeal to the Serious and Candid Professors of Christianity," *Works*, 2.383–416.

97. Priestley, "Institutes of Natural and Revealed Religion," *Works*, 2.280.

98. Priestley, "The History of the Corruptions of Christianity," *Works*, 5.13.

99. Priestley, "The History of the Corruptions of Christianity," *Works*, 5.8, 5.4, 5.19.

100. Priestley, "The History of the Corruptions of Christianity," *Works*, 5.503.

101. Joseph Priestley, *A General History of the Christian Church*, 4.400.

102. For a succinct summary of views, see Peter J. Bowler, *The Invention of Progress*.

103. Thomas A. Howard, *Religion and the Rise of Historicism*, 17–18, 23–50.

104. Friedrich Schleiermacher, "Am zweiten Tage des Reformations-Jubelfestes, 1817," *Sämmtliche Werke*, ii/4 (Berlin, 1835), 67–8; quoted in Thomas A. Howard, *Remembering the Reformation*, 48.

105. Quoted in Howard, *Remembering the Reformation*, 58.

106. See, e.g., the classic studies by Karl Hillebrand, *Six Lectures on the History of German Thought*; John Tulloch, *Movements of Religious Thought in Britain during the Nineteenth Century*; Frédéric A. Litchtenberger, *History of German Theology in the Nineteenth Century*; and Otto Pfleiderer, *The Development of Theology in Germany since Kant and its Progress in Great Britain since 1825*. See also Walter F. Schirmer, *Der Einfluss der deutschen Literatur auf die englische im 19. Jahrhundert*, and, more recently, John R. Davis, *The Victorians and Germany*.

107. "Prospectus of the Leader," *The Leader*, vol. 1, no. 1 (Mar. 30, 1850): 22.

108. "The New Reformation," *The Leader*, vol. 1, no. 5 (Apr. 27, 1850): 105–6; see also "The Progress of the New Reformation," *The Leader*, vol. 1, no. 7 (May 11, 1850): 153.

109. G. H. Lewes, "Social Reform," *The Leader*, vol. 1, no. 20 (Aug. 10, 1850): 469–70.

110. "The Creed of Christendom," *The Leader*, vol. 2, no. 78 (Sept. 20, 1851): 897–99.

111. William Rathbone Greg, *The Creed of Christendom*, viii-xvii. A more detailed account of Greg and his work can be found in Richard J. Helmstadter, "W. R. Greg: A Manchester Creed," in Richard J. Helmstadter and Bernard Lightman, eds., *Victorian Faith in Crisis*, 187–222.

112. Greg, *The Creed of Christendom*, viii.

113. "Literature," *The Leader*, vol. 1, no. 8 (May 18, 1850): 181; see also subsequent reviews, "Newman's Phases of Faith," *The Leader*, vol. 1, no. 9 (May 25, 1850): 206–07; vol. 1, no. 10 (June 1, 1850): 232–33; vol. 1, no. 11 (June 8, 1850): 256–58; vol. 1, no. 12 (June 15, 1850): 281–82; and vol. 1, no. 14 (June 29, 1850): 329–30.

114. Francis William Newman, *Phases of Faith*, 44, 76, 188.

115. Francis William Newman, *The Religious Weakness of Protestantism*, and *Thoughts on a Free and Comprehensive Christianity*. On Newman, see David Hempton, *Evangelical Disenchantment*, 41–69.

116. "Mackay's Progress of the Intellect," *Westminster Review*, vol. 54, no. 2 (Jan. 1851): 353–68. A more detailed account of these themes in Eliot's thought can be found in Hempton, *Evangelical Disenchantment*, 19–40.

117. Robert William Mackay, *The Progress of the Intellect*, 1.20, 1.22, 1.40.

118. See Bernard Lightman, "'Robert Elsmere' and the Agnostic Crisis of Faith," in Helmstadter and Lightman, *Victorian Faith in Crisis*, 283–311, on 300.

119. Mary A. Ward, "The New Reformation," *Nineteenth Century*, vol. 25, no. 145 (Mar. 1889): 454–80.

120. W. E. H. Lecky, *Religious Tendencies of the Age*, 1.

121. Jeffrey Paul von Arx, *Progress and Pessimism*, 71, 78.

122. Lecky, *Religious Tendencies of the Age*, 137–38.

123. Lecky, *Religious Tendencies of the Age*, 27, 148, 192–93, 196–97.

124. W. E. H. Lecky, *History of the Rise and Influence of the Spirit of Rationalism in Europe*.

125. Lecky, *History of the Rise and Influence of the Spirit of Rationalism*, 1.200–01.

126. Lecky, *History of the Rise and Influence of the Spirit of Rationalism*, 1.191.

127. Charles Voysey, "The New Reformation: Part I," *Fortnightly Review*, 41.241 (Jan. 1887): 124–138; W. H. Fremantle, "The New Reformation: Part II—Theology Under Its Changed Conditions," *Fortnightly Review*, 41.243 (Mar. 1887): 442–58; John W. Burgon, "The New Reformation: 'Theology Under Its Changed Conditions'—A Reply to Canon Fremantle," *Fortnightly Review*, 41.244 (Apr. 1887): 587–612; and W. Benham, "The New Reformation: Dean Burgon and Mr. Fremantle," *Fortnightly Review*, 41.245 (May 1887): 743–52.

128. On Voysey, see, e.g., M. A. Crowther, *Church Embattled*, 127–37.

129. Voysey, "The New Reformation," 132.

130. For a history of the Broad Church, see, e.g., Tod E. Jones, *The Broad Church: A Biography of a Movement*.

131. Duncan Forbes, *The Liberal Anglican Idea of History*.

132. Forbes, *The Liberal Anglican Idea of History*, 20.

133. Draper, *Intellectual Development*, 1, 11, 15.

134. See the classic studies by John Tulloch, *Movements of Religious Thought in Britain During the Nineteenth Century*, and William Tuckwell, *Pre-Tractarian Oxford: A Reminiscence of the Oriel 'Noetics'*. But see also more recently the excellent collection of essays in Jeremy Catto, ed., *Oriel College: A History*.

135. Thomas Arnold, "The Oxford Malignants and Dr. Hampden," *Edinburgh Review*, vol. 63, no. 127 (1836): 225–39.

136. Frederick Denison Maurice, *Theological Essays* (London: Macmillan, 1853).

137. Frederick Denison Maurice, *The Claims of the Bible and of Science* (London: Macmillan, 1863), 117.

138. *Charles Kingsley, His Letters and Memories of his Life*, 2 vols., ed. Mrs. Charles Kingsley, 2.155.

139. Francis Darwin, *The Life and Letters of Charles Darwin*, 2.81–82. On Kingsley's support of Darwin, see Piers J. Hale, "Darwin's Other Bulldog."

140. Charles Kingsley, "The Natural Theology of the Future," *Macmillan's Magazine*, vol. 23, no. 137 (1871): 369–78.

141. See discussion in Wheeler, *The Old Enemies*, 68–70, 105–10, and 219–20.

142. *Essays and Reviews* (London: John W. Parker and Son, 1860). For detailed discussion, see the two major monographs by Ieuan Ellis, *Seven against Christ*, and Josef L. Altholz, *Anatomy of a Controversy: The Debate over "Essays and Reviews," 1860–1864*. See

also the superbly annotated edition by Victor Shea and William Whitla, eds., *Essays and Reviews: The 1860 Text and Its Reading.*

143. James C. Ungureanu, "A Yankee at Oxford: John William Draper at the British Association for the Advancement of Science at Oxford, 30 June 1860," 135–50.

144. Frederick Temple, *The Present Relations of Science to Religion.*

145. Frederick Maurice, *The Life of Frederick Denison Maurice, Chiefly Told in His Own Letters,* 2.383.

146. A. P. Stanley, "Theology of the Nineteenth Century," *Fraser's Magazine,* vol. 71, no. 422 (1865): 252–68.

147. A. P. Stanley, "The Hopes of Theology," *Macmillan's Magazine,* vol. 36, no. 211 (1877): 1–14.

148. Paul White, *Thomas Huxley,* 104.

149. Quoted in full in Shea and Whitla, *Essays and Reviews,* 657–59.

150. [Anon.] "The Edinburgh and the Quarterly on the New Theology," *Reader,* vol. 4, no. 83 (July 30, 1864): 123–24. It should be noted that the *Reader* was the forerunner of *Nature.* See Melinda Baldwin, *Making "Nature": The History of a Scientific Journal,* esp. 21–47.

151. Ieuan Ellis, *Seven against Christ,* 99–101.

152. Rowland E. Prothero and G. Bradley, *Life and Correspondence of Arthur Penrhyn Stanley,* 2 vols. (London: John Murray, 1893), 2.238–39.

153. Alexandra Walsham, "The Reformation and 'The Disenchantment of the World' Reassessed," *Historical Review,* vol. 51, no. 2 (2008): 497–528.

FOUR: American New Theology and the Evolution of Religion

1. Annette G. Aubert, *The German Roots of Nineteenth-Century American Theology.*

2. See James Hastings Nichols, *Romanticism in American Theology.*

3. Philip Schaff, *The Principles of Protestantism,* 183.

4. Samuel Taylor Coleridge, *Aids to Reflection,* 26.

5. For Priestley's influence on American religion, see J. D. Bowers, *Joseph Priestley and English Unitarianism in America.* See also the general studies by Robert Fuller, *Religious Revolutionaries;* Amanda Porterfield, *Conceived in Doubt;* Eric R. Schlereth, *An Age of Infidels;* and John Fea, *Was America Founded as a Christian Nation?*

6. Benjamin Franklin, *The Private Life of the Late Benjamin Franklin,* 105.

7. Thomas S. Kidd, *Benjamin Franklin.*

8. Ethan Allen, *Reason the Only Oracle of Man,* 51.

9. Thomas Paine, "The Age of Reason," in Philip S. Foner, ed., *The Complete Writings of Thomas Paine,* 1.464, 1.487, 1.490, 1.495.

10. Herbert Hovenkamp, *Science and Religion in America,* 23.

11. Quoted in Perry Miller, *The New England Mind,* 97.

12. Cotton Mather, *The Christian Philosopher,* 1, 221, 222, 283, 299–300.

13. Richard F. Lovelace, *The American Pietism of Cotton Mather,* 241.

14. Andrew J. Lewis, *A Democracy of Facts,* esp. 107–28.

15. Thaddeus Mason Harris, *The Natural History of the Bible,* i.

16. Thaddeus Mason Harris, *The Beauties of Nature Delineated.*

17. Jacob Friedrich Feddersen, *Christoph Christian Sturms, gewesenen Hauptpastors zu St. Petri und Scholarchen in Hamburg, Leben und Charakter* (Hamburg: Johann Henrich Herold, 1786), 51.

18. See Charles Christopher Reiche, *Fifteen discourses on the marvelous works in nature, delivered by a father to his children*; George Riley, *Beauties of the Creation.*

19. John Toogood, *The Book of Nature*; James Fisher, *A Spring Day; or, Contemplations on Several Occurrences which Naturally Strike the Eye in that Delightful Season.*

20. Harris, *The Beauties of Nature Delineated*, 13.

21. For general studies, see, e.g., Stanley M. Guralnick, *Science and the Ante-Bellum American College*; Theodore Dwight Bozeman, *Protestants in an Age of Science*; Herbert Hovenkamp, *Science and Religion in America, 1800–1860*; John C. Greene, *American Science in the Age of Jefferson*; A. Hunter Dupree, *Science in the Federal Government*; Walter H. Conser, Jr.; and George H. Daniels, *American Science in the Age of Jackson.*

22. Benjamin Silliman, *Outline of the Course of Geological Lectures Given in Yale College*, 7.

23. See John C. Greene, "Protestantism, Science and American Enterprise: Benjamin Silliman's Moral Universe," in Leonard G. Wilson, ed., *Benjamin Silliman and His Circle*, 11–27.

24. George P. Fisher, *The Life of Benjamin Silliman*, 1.108, 1.372.

25. B. Silliman to E. Hitchcock, March 11, 1830, Fisher, *The Life of Benjamin Silliman*, 2.136, 2.139. On Hitchcock's own religious views, see Stanley M. Guralnick, "Geology and Religion before Darwin: The Case of Edward Hitchcock, Theologian and Geologist (1793–1864)," *Isis*, vol. 63, no. 4 (1972): 529–43.

26. Benjamin Silliman, "Address before the Association of American Geologists and Naturalists, assembled at Boston, April 24, 1842," *American Journal of Science*, vol. 43, no. 2 (1842): 217–50.

27. See Jon H. Roberts, *Darwinism and the Divine in America.*

28. Asa Gray, *Darwiniana*, 87, 129–77, 253–66, 266–82.

29. Asa Gray, *Natural Science and Religion*, 77.

30. A. Hunter Dupree, *Asa Gray.*

31. Joseph Le Conte, *Religion and Science*; *Evolution and its Relation to Religious Thought*; *Evolution: Its Nature, Its Evidences, and Its Relation to Religious Thought*; and *The Autobiography of Joseph Le Conte*, ed. William Dallam Armes. For a more detailed study, see Lester D. Stephens, *Joseph Le Conte: Gentile Prophet of Evolution.*

32. See, e.g., Earl Morse Wilbur, *Three Prophets of American Liberalism: Channing, Emerson, Parker*; William R. Hutchison, *The Transcendentalist Ministers*; Walter Donald Kring, *Liberals among the Orthodox*; John Allen Macaulay, *Unitarianism in Antebellum South*; and E. Brooks Holified, *Theology in America.*

33. For Emerson's thought, see Edward W. Emerson, ed., *The Works of Ralph Waldo Emerson.*

34. Emerson, "Nature," in *Works*, 1.16.

35. Emerson, "An Address delivered before the Senior Class in Divinity College," in *Works*, 1.142.

36. Emerson, "An Address," 1.127, 1.130, 1.148.

37. Clarence L. F. Gohdes, *The Periodicals of American Transcendentalism*, 10.

38. Mark Noll, *America's God*, 3–18.

39. Jerry Wayne Brown, *The Rise of Biblical Criticism in America*, 179.

40. See, e.g., Francis Cogliano, *No King, No Popery*; Ray Allen Billington, *The Protestant Crusade*; John Higham, *Strangers in the Land*; and Jon Gjerde, *Catholicism and the Shaping of Nineteenth-Century America.*

41. See Elizabeth A. Clark, *Founding the Fathers: Early Church History and Protestant Professors in Nineteenth-Century America.*

42. See Samuel Miller's "Introductory Essay" in *A History of Popery*, 3–18; and Miller, *The Dangers of Education in Roman Catholic Seminaries*, 11. See also Clark, *Founding the Fathers*, 239–41.

43. Samuel F. B. Morse, *Foreign Conspiracy Against the Liberties of the United States*, 89, 137–39.

44. Lyman Beecher, *A Plea for the West*, 136, 137.

45. Rev. S. R. Calthrop, "Religion and Evolution," *The Religious Magazine and Monthly Review*, vol. 1, no. 3 (1873): 193–227.

46. S. R. Calthrop, "Religion and Science," *Unitarian Review and Religious Magazine*, vol. 2, no. 4 (1874): 309–35.

47. Newman Smyth, *The Religious Feeling*, esp. 29–52, 166.

48. Newman Smyth, *Old Faiths in New Light*, 15, 16, 62–127.

49. Newman Smyth, *The Orthodox Theology of To-day*, 22, 39, 40.

50. Theodore T. Munger, *Freedom of Faith*, 3–44.

51. Lyman Abbott, *The Evolution of Christianity*, 66, 206, 227, 249, 250–51, 258.

52. Lyman Abbott, *The Theology of an Evolutionist*, 9–10.

53. Abbott, *The Theology of an Evolutionist*, 73, 188–89.

54. Abbott, *The Theology of an Evolutionist*, 21.

55. M. J. Savage, *The Religion of Evolution*, 5, 6, 13–15, 20, 27.

56. Savage, *The Religion of Evolution*, 42, 44, 46, 60, 149, 216, 231.

57. M. J. Savage, *The Evolution of Christianity*, 147, 165–78.

58. M. J. Savage, *My Creed*, 10–11, 27.

59. M. J. Savage, *Religious Reconstruction.*

60. M. J. Savage, *Signs of the Times*, 64, 68–69, 173, 180, 187.

61. C. A. Briggs, "The Theological Crisis," *North American Review*, vol. 153, no. 416 (1891): 99–114.

62. Charles Augustus Briggs, *The Authority of Holy Scripture*, 27, 33–34, 66–67.

63. Quoted in Walter P. Rogers, *Andrew D. White and the Modern University*, 82.

64. See Frederick Rudolph, *The American College and University*, 68–85.

65. See, e.g., Henry K. Rowe, *History of Andover Theological Seminary*; and Daniel Day Williams, *The Andover Liberals.*

66. See Guralnick, *Science and the Ante-Bellum American College.*

67. Thomas Paine, *Collected writings*, ed. Eric Foner, 825.

68. Benjamin Franklin, "Something of my religion . . ." in *The Portable Enlightenment Reader*, ed. Isaac Kramnick, 166–67.

69. Thomas Jefferson, "Religion . . . my views of it . . ." in Kramnick, ed., *The Portable Enlightenment Reader*, 160–66.

70. See George Marsden, *The Soul of the American University*, esp. 408–28.

71. Julie A. Reuben, *The Making of the Modern University*, 4.

72. Charles William Eliot, *Educational Reform*, 8.

73. Eliot, *Educational Reform*, 61–86.

74. Charles W. Eliot, *The Religion of the Future.*

75. See John W. Boyer, *"Broad and Christian in the Fullest Sense": William Rainey Harper and the University of Chicago.*

76. See Thomas Wakefield Goodspeed, *A History of the University of Chicago*, 98–129.

77. Conrad Cherry, *Hurrying Toward Zion*, 13.

78. James P. Wind, *The Bible and the University: The Messianic Vision of William Rainey Harper*, 7.

79. William Rainey Harper, *Religion and the Higher Life*, 1–20.

80. Harper, *Religion and the Higher Life*, 17.

81. Harper, *The Trend in Higher Education*.

82. Edward McNall Burns, *David Starr Jordan*.

83. David Starr Jordan, *The Days of a Man*, 1.46–47.

84. David Starr Jordan, *Standeth God Within the Shadow*, 8.

85. David Starr Jordan, *The Foundation Ideals of Stanford University*, 20.

86. David Starr Jordan, *The Voice of the Scholar*, 272.

87. David Starr Jordan, *College and the Man*, 14.

88. David Starr Jordan, *The Religion of a Sensible American*.

89. See Jon H. Roberts and James Turner, *The Sacred and the Secular University*, esp. 19–41.

90. William C. Placher, *The Domestication of Transcendence*.

91. Reuben, *The Making of the Modern University*, 113.

92. See now classic studies by Max Weber, *The Protestant Ethic and the Spirit of Capitalism*, trans. Talcott Parsons; James Turner, *Without God, Without Creed*; Michael J. Buckley, *At the Origins of Modern Atheism*; Charles Taylor, *A Secular Age*; Brad S. Gregory, *The Unintended Reformation*. See also most recently the study by James Simpson, *Permanent Revolution*.

FIVE: Youmans and the "Peacemakers"

1. Bernard Lightman, "Victorian Sciences and Religions: Discordant Harmonies," 346.

2. John Wolffe, *God and Greater Britain*, 6.

3. Lance St. John Butler, *Victorian Doubt*, 86. See also Timothy Larsen, *Crisis of Doubt*; Christopher Lane, *The Age of Doubt*; and Dominic Erdozain, *The Soul of Doubt*.

4. See, e.g., Tess Cosslett, *Science and Religion in the Nineteenth Century*; Adrian Desmond, *Huxley: From Devil's Disciple to Evolution's High Priest*; Paul White, *Thomas Huxley: Making the "Man of Science"*; Stephen S. Kim, *John Tyndall's Transcendental Materialism and the Conflict between Religion and Science in Victorian England*; Mark Francis, *Herbert Spencer and the Invention of Modern Life*; and Michael Taylor, *The Philosophy of Herbert Spencer*. See also the recent publication of Ruth Barton, *The X Club: Power and Authority in Victorian Science*, esp. 362–444.

5. Robert M. Young, *Darwin's Metaphor*, 191.

6. James R. Moore, "Theodicy and Society: The Crisis of the Intelligentsia," in Richard J. Helmstadter and Bernard Lightman, eds., *Victorian Faith in Crisis*, 153–86.

7. See, e.g., Bernard Lightman, *The Origins of Agnosticism*; Lightman, *Evolutionary Naturalism in Victorian Britain*. More recently, see the collection of essays in Bernard Lightman and Michael S. Reidy, eds., *The Age of Scientific Naturalism: Tyndall and his Contemporaries*, and Gowan Dawson and Bernard Lightman, eds., *Victorian Scientific Naturalism*. For a succinct statement, see also "The Theology of Victorian Scientific Naturalists," in *Science Without God? Rethinking the History of Scientific Naturalism*, eds. Peter Harrison and Jon H. Roberts, 235–53.

8. *Evangelical Christendom, Christian Work, and the news of the Churches. Also a Monthly*

Record of the Transactions of the Evangelical Alliance, vol. 16 (London: William John Johnson, 1875), 116.

9. Thomas H. Huxley, "Agnosticism," in *Collected Essays: Science and Christian Tradition*, vol. 5 (London: Macmillan, 1894), 255.

10. Thomas H. Huxley, "On the Physical Basis of Life," in *Collected Essays: Method and Results*, vol. 1 (London: Macmillan, 1894), 156.

11. The main source for the life of Youmans remains the one by his close friend and fellow popularizer of science, American philosopher John Fiske, *Edward Livingston Youmans, Interpreter of Science for the People*. Good biographical material on Youmans can also be found in Charles M. Haar, "E. L. Youmans: A Chapter in the Diffusion of Science in America"; and William Leverette, Jr., "E. L. Youmans' Crusade for Scientific Autonomy and Respectability."

12. Fiske, *Edward Livingston Youmans*, 9.

13. Fiske, *Edward Livingston Youmans*, 14.

14. Fiske, *Edward Livingston Youmans*, 58.

15. On the Appletons, see Gerard R. Wolfe, *The House of Appleton*.

16. Edward L. Youmans, *A Class-Book of Chemistry*, 6–19.

17. See Wolfe, *The House of Appleton*, 184–206.

18. Edward L. Youmans, *The Correlation and Conservation of Forces*, xi-xii (my emphasis).

19. Edward L. Youmans, ed., *The Culture Demanded by Modern Life*, v-viii, 1–56.

20. Fiske, *Edward Livingston Youmans*, 116–40.

21. See Wolfe, *The House of Appleton*, 41–52.

22. Fiske, *Edward Livingston Youmans*, 255.

23. Fiske, *Edward Livingston Youmans*, 255–65.

24. "What We Mean by Science," *Appleton's Journal of Literature, Science, and Art*, vol. 1, no. 1 (April 3, 1869): 22–23.

25. "Editor's Table: Purpose and Plan of Our Enterprise," *Popular Science Monthly*, vol. 1, no. 1 (May 1872): 113–15 (hereafter as *PSM*).

26. Leverette, "E. L. Youmans' Crusade for Scientific Autonomy and Respectability," 13.

27. "Editor's Table: Charles Robert Darwin," *PSM*, vol. 21 (June 1882): 266–68.

28. "Editor's Table: The Discoverer of Oxygen," *PSM*, vol. 1 (July 1872): 368–70; John W. Draper, "Priestley's Discovery of Oxygen Gas," *PSM*, vol. 5 (Aug. 1874): 385–98.

29. "Sketch of Dr. J. W. Draper," *PSM*, vol. 4 (Jan. 1874): 361–67.

30. "Literary Notices: A Work of Great Importance," *PSM*, vol. 6 (Nov. 1874): 121.

31. John W. Draper, "The Great Conflict," *PSM*, vol. 6 (Dec. 1874): 227–32.

32. Bernard Lightman, "The International Scientific Series and the Communication of Darwinism," 27–38. The series has also been studied closely by Roy M. MacLeod, "Evolutionism, Internationalism and Commercial Enterprise in Science: The International Scientific Series 1871–1910," in A. J. Meadows, ed., *Development of Science Publishing in Europe*, 63–93; and Leslie Howsam, "An Experiment with Science for the Nineteenth-Century Book Trade: The International Scientific Series," 187–207. See also Fiske's account, *Edward Livingston Youmans*, 266–294.

33. E. L. Youmans to Herbert Spencer, April 21, 1871; quoted in Fiske, *Edward Livingston Youmans*, 266.

34. Youmans to Spencer, quoted in Fiske, 268.

35. E. L. Youmans to Catherine Schofield Youmans, June 3, 1871; quoted in Fiske, 270–71.

36. "Prospectus of the International Scientific Series," *PSM*, vol. 1 May–October, 1872, 775.

37. See Fiske, *Edward Livingston Youmans*, 278; MacLeod, "Evolutionism, Internationalism and Commercial Enterprise in Science," 70; Lightman, "The International Scientific Series," 31–35.

38. MacLeod, "Evolutionism, Internationalism and Commercial Enterprise in Science," 74.

39. "Editor's Table: The Conflict of Religion and Science," *PSM*, vol. 6 (Jan. 1875): 361–64; "Literary Notices: History of the Conflict between Religion and Science," *PSM*, vol. 6 (Jan. 1875): 371–72.

40. "Editor's Table: Draper and his Critics," *PSM*, vol. 7 (June 1875): 230–33.

41. "Editor's Table: 'The Conflict of Ages,'" *PSM*, vol. 8 (Feb. 1876): 493–94; "Literary Notices," *PSM*, vol. 9 (May 1876): 112–14; "Correspondence: 'What Constitutes Religion?'" *PSM*, vol. 9 (June 1876): 239–41.

42. Andrew D. White, "Science and Public Affairs," *PSM*, vol. 2 (April 1873): 736–39; "Sanitary Science and Public Instruction," *PSM*, vol. 4 (Feb. 1874): 421–29; "Scientific and Industrial Education in the United States," *PSM*, vol. 5 (June 1874): 170–91.

43. Andrew D. White, "The Warfare of Science I," *PSM*, vol. 8 (Feb. 1876): 385–409; and "The Warfare of Science II," *PSM*, vol. 8 (March 1876): 553–70.

44. White, "The Warfare of Science I," 391; "The Warfare of Science II," 567.

45. "'The Conflict of Ages,'" 493; "Literary Notice: The Warfare of Science," *PSM*, vol. 9 (July 1876): 370–71.

46. "Editor's Table: Conclusion of Dr. White's New Chapters," *PSM*, vol. 47 (June 1895): 267–68.

47. Fiske, *Edward Livingston Youmans*, 129. See also Leonard Huxley, *Life and Letters of Thomas Henry Huxley*, 2.279.

48. See Barton, *The X Club*, 409.

49. Professor Huxley, "Scientific Education," *Appleton's Journal*, vol. 1, no. 20 (Aug. 14, 1869): 627–30.

50. Fiske, *Edward Livingston Youmans*, 333–35.

51. "Editor's Table: Prof. Huxley's Lectures," *PSM*, vol. 9 (Aug. 1876): 500; "Editor's Table: Prof. Huxley," *PSM*, vol. 9 (Sept. 1876): 621–22.

52. For Huxley's lectures, see: "Professor Huxley's Lectures I," *PSM*, vol. 10 (Nov. 1876): 43–56; "Professor Huxley's Lectures II," *PSM*, vol. 10 (Dec. 1876): 207–23; "Professor Huxley's Lectures III," *PSM*, vol. 10 (Jan. 1877): 285–98. For Youmans's defense, see: "Editor's Table: Professors Huxley's Lectures," *PSM*, vol. 10 (Nov. 1876): 103–6; "Miscellany: Evolution of the Horse," *PSM*, vol. 10 (Nov. 1876): 118; "Editor's Table: Evolution and Copernican Theory," *PSM*, vol. 10 (Dec. 1876): 236–40; "Editor's Table: Professor Huxley on the Horse," *PSM*, vol. 10 (Jan. 1877): 370–71; "Editor's Table: Some Questions Answered," *PSM*, vol. 10 (March 1877): 616–19; and "Editor's Table: The Order of Nature," *PSM*, vol. 10 (April 1877): 748–51.

53. T. H. Huxley, "Joseph Priestley," *PSM*, vol. 6 (Nov. 1874): 90–107.

54. For relevant articles, see Gladstone, "Dawn of Creation and of Worship," *PSM*, vol. 28 (April 1886): 865–79; Huxley, "The Interpreters of Genesis and the Interpreters of Nature," *PSM*, vol. 28 (Feb. 1886): 449–60; Gladstone, "Proem to Genesis: A Reply

to Professor Huxley," *PSM*, vol. 28 (March 1886): 614–34; Huxley, "Mr. Gladstone and Genesis," *PSM*, vol. 28 (April 1886): 788–803. Youmans also published an editorial on the debate, "Editor's Table: The Gladstone-Huxley Controversy," vol. 28 (April 1886): 840, in which he sided, unsurprisingly, with Huxley.

55. T. H. Huxley, "Science and the Bishops," *PSM*, vol. 32 (Jan. 1888): 352–68.

56. T. H. Huxley, "The Lights of the Church and the Light of Science," *PSM*, vol. 37 (Sept. 1890): 631–50.

57. Huxley, "Lights of the Church," 647.

58. T. H. Huxley, "The Decline of Bibliolatry," *PSM*, vol. 41 (Sept. 1892): 594–603. This article is an excerpt from the Prologue of Huxley's *Essays Upon Some Controverted Questions*.

59. See Bernard Lightman, "Interpreting Agnosticism as a Nonconformist Sect: T. H. Huxley's 'New Reformation,'" in Paul Wood, ed., *Science and Dissent in England, 1688–1945*, 197–214.

60. Huxley, *Collected Essays, Science and Christian Tradition*, 5.228.

61. Quoted in Cyril Bibby, *T. H. Huxley: Scientist, Humanist and Educator* (New York: Horizon Press, 1960), xxii.

62. Huxley, *Collected Essays, Science and Christian Tradition*, 5.239, 5.267. On the origins of the term "agnosticism," see Bernard Lightman, "Huxley and Scientific Agnosticism: The Strange History of a Failed Rhetorical Strategy," *British Journal for the History of Science*, vol. 35, no. 3 (2002): 271–89.

63. Huxley, *Collected Essays, Darwiniana*, 2.147.

64. Leonard Huxley, *Life and Letters of Thomas Henry Huxley*, 2.98.

65. L. Huxley, *Life and Letters*, 2.243.

66. Thomas H. Huxley, "Science and Religion," *Builder*, 17 (Jan. 1859): 35–36.

67. T. H. Huxley to F. D. Dyster, Jan. 30, 1859; quoted in Desmond, *Huxley*, 253.

68. L. Huxley, *Life and Letters*, 1.233–39.

69. Frank M. Turner, "Victorian Scientific Naturalism and Thomas Carlyle," *Victorian Studies*, vol. 18, no. 3 (1975): 325–43, on 330–31.

70. James Anthony Froude, *Thomas Carlyle*, 2.2.

71. Thomas H. Huxley, "Science and 'Church Policy,'" *Reader*, vol. 4, no. 105 (1864): 821.

72. L. Huxley, *Life and Letters*, 1.427–28.

73. Huxley, *Collected Essays, Science and Education*, 3.191–92.

74. L. Huxley, *Life and Letters*, 2.361.

75. John Tyndall, "Science and Religion," *PSM*, vol. 2 (Nov. 1872): 79–82. This essay first appeared as "On Prayer," *Contemporary Review*, vol. 20 (June 1872): 763–66. See also Robert Bruce Mullin, "Science, Miracles, and the Prayer-Gauge Debate," in David C. Lindberg and Ronald L. Numbers, eds., *When Science and Christianity Meet*, 203–24.

76. "Literary Notices: Prof. Tyndall's New Book," *PSM*, vol. 1 (June 1872): 245; "Editor's Table: Professor Tyndall," *PSM*, vol. 1 (Oct. 1872): 751–52; "Sketch of Professor Tyndall," *PSM*, vol. 2 (Nov. 1872): 103–9: "Editor's Table: Tyndall and Froude," vol. 2 (Jan. 1873): 373–76. See also "Editor's Table: Tyndall's Lectures in New York," *PSM*, vol. 2 (Feb. 1873): 499–500.

77. Fiske, *Edward Livingston Youmans*, 317; Overton, *Portrait of a Publisher*, 57–58. See also "Editor's Table: The Farewell Banquet to Professor Tyndall," *PSM*, vol. 2 (March 1873): 626. A number of these speeches appeared in the pages of *PSM*, which were later

collected and published as *Proceedings at the Farewell Banquet to Professor Tyndall*, 1873. For a detailed account of this banquet, see Katherine R. Sopka, "John Tyndall: International Popularizer of Science," in W. H. Brock, N. D. McMillan, and R. C. Mollan, eds., *John Tyndall: Essays on a Natural Philosopher*, 193–203.

78. Tyndall, "Inaugural Address before the British Association," *PSM*, vol. 5 (Oct. 1874): 652–86.

79. See Bernard Lightman, "Scientists as Materialists in the Periodical Press: Tyndall's Belfast Address," in Geoffrey Cantor and Sally Shuttleworth, eds., *Science Serialized: Representation of the Sciences in Nineteenth-Century Periodicals*, 199–237.

80. Hermann von Helmholtz, "Tyndall's Relation to Popular Science," *PSM*, vol. 5 (Oct. 1874): 734–40.

81. John Trowbridge, "Science from the Pulpit," *PSM*, vol. 6 (April 1875): 734–39.

82. "Editor's Table: Professor Tyndall's Address," *PSM*, vol. 5 (Oct. 1874): 746–48; "Editor's Table: Matter and Life," *PSM*, vol. 6 (Nov. 1874): 110–12; "Editor's Table: The Conflict of Religion and Science," *PSM*, vol. 6 (Jan. 1875): 361–64. Youmans also reprinted Tyndall's own "Apology for the Belfast Address," which he described as "masterly" and "conclusive"—John Tyndall, "Reply to the Critics of the Belfast Address," *PSM*, vol. 6 (Feb. 1875): 422–40; "Editor's Table: Tyndall and His Reviewers," *PSM*, vol. 6 (Feb. 1875): 500–504.

83. John Tyndall, "Martineau and Materialism," *PSM*, vol. 8 (Dec. 1875): 129–48; "Virchow and Evolution," *PSM*, vol. 14 (1879): 266–90.

84. Tyndall, "The Sabbath I," *PSM*, vol. 18 (Dec. 1880): 246–56; "The Sabbath II," *PSM*, vol. 18 (Jan. 1881): 310–23.

85. Leonard Huxley, "John Tyndall: A Centenary Sketch," *Cornhill Magazine*, vol. 122, no. 293 (1920): 627–40.

86. William T. Jeans, *Lives of the Electricians: Professors Tyndall, Wheatstone, and Morse*, 5.

87. Many of these letters are collected in G. Cantor and G. Dawson, eds., *The Correspondence of John Tyndall*, vol. 1 (Pittsburgh, PA: University of Pittsburgh Press, 2016), and M. Baldwin and J. Browne, eds., *The Correspondence of John Tyndall*, vol. 2 (Pittsburgh, PA: University of Pittsburgh Press, 2016). Here I will cite them as "Letter," followed by letter number. On Catholicism, see, e.g., J. Tyndall to J. Tyndall Sr., April 10, 1841: Letter 0054; J. Tyndall to J. Tyndall Sr., April 25, 1841: Letter 0055; J. Tyndall Sr. to J. Tyndall, April 30, 1841: Letter 0056; and J. Tyndall Sr. to J. Tyndall, Sept. 10, 1841: Letter 0092.

88. See John Tyndall, *New Fragments* (New York: D. Appleton and Co., 1896), 347–91.

89. J. Tyndall to T. Hirst, Feb. 1, 1848: Letter 0342.

90. J. Tyndall to T. Hirst, Nov. 21, 1848: Letter 0366.

91. Royal Institution, Journals of John Tyndall, June 26, 1847, quoted in Geoffrey Cantor, "John Tyndall's Religion: A Fragment," *Notes and Records of the Royal Society*, vol. 69, no. 4 (2015): 419–36, on 430.

92. Journals, June 26, 1847, quoted in Cantor, 429.

93. Journals, May 12, 1844, quoted in Cantor.

94. Journals, April 7, 1850. I wish to thank Bernie Lightman for sharing with me scans of Tyndall's journals he obtained from the Royal Institution.

95. John Tyndall, *Fragments of Science*, 2.191, 2.194, 2.196, 2.202–23.

96. See, e.g., Richard Hofstadter, *Social Darwinism in American Thought*, esp. 31–50.

More recently, see Bernard Lightman, "Spencer's American Disciples," in *Global Spencerism*, by Lightman, 137–47.

97. Haar, "E. L. Youmans: A Chapter in the Diffusion of Science in America," 200.

98. Fiske, *Edward Livingston Youmans*, 78.

99. Hebert Spencer, *Autobiography*, 1.171–72.

100. J. D. Y. Peel, *Herbert Spencer: The Evolution of a Sociologist* (New York: Basic Books, 1971), viii.

101. Spencer, *Autobiography*, 2.477.

102. See, e.g., Bernard Lightman, "Spencer's British Disciples," in Mark Francis and Michael W. Taylor, eds., *Herbert Spencer Legacies*, 222–44; and "Spencer's American Disciples: Fiske, Youmans, and the Appropriation of the System," in Bernard Lightman, ed., *Global Spencerism*, 123–48.

103. Herbert Spencer, "What Knowledge is of Most Worth?" *Westminster Review*, vol. 16, no. 1 (1859): 1–41.

104. See Spencer, *Autobiography*, 1.645–55.

105. Herbert Spencer, *First Principles of a New System of Philosophy*, 11, 21, 22, 46, 99, 107.

106. Herbert Spencer, "The Study of Sociology I," *PSM*, vol. 1 (May 1872): 1–17.

107. Herbert Spencer, *An Autobiography*, 2.61. See also remarks at 2.95, and 2.111.

108. "Editor's Table: The Charges Against 'The Popular Science Monthly,'" *PSM*, vol. 20 (Jan. 1882): 404–9.

109. See, e.g., Herbert Spencer, "Mr. Martineau on Evolution," *PSM*, vol. 1 (July 1872): 313–23; "Replies to Criticism I," *PSM*, vol. 4 (Jan. 1874): 295–309; "Replies to Criticism II," *PSM*, vol. 4 (Feb. 1874): 402–15; "Replies to the Quarterly Reviewers," *PSM*, vol. 4 (March 1874): 541–52; "Criticism Corrected I," *PSM*, vol. 17 (Oct. 1880): 795–801; "Criticism Corrected II," *PSM*, vol. 18 (Nov. 1880): 101–7; "Criticism Corrected III," *PSM*, vol. 18 (Jan. 1881): 387–96.

110. See "A Quadrangular Duel," *Saturday Review*, vol. 57, no. 1493 (June 1884): 747–48. The only scholarly study devoted to the debate is Sydney Eisen, "Frederic Harrison and Herbert Spencer: Embattled Unbelievers," *Victorian Studies*, vol. 12, no. 1 (1968): 33–56.

111. H. Spencer to E. L. Youmans, Nov. 11, 1883, in David Duncan, ed., *The Life and Letters of Herbert Spencer*, 252–53.

112. Herbert Spencer, "Religious Retrospect and Prospect," *PSM*, vol. 24 (Jan. 1884): 340–51. This article was also published as the closing chapter of Part VI, "Ecclesiastical Institutions," of Spencer's *The Principles of Sociology*, 3 vols. (New York: D. Appleton and Co., 1898).

113. Spencer, *First Principles of a New System of Philosophy*, 3–123. See also discussion in Lightman, "Herbert Spencer and the Worship of the Unknowable," in *The Origins of Agnosticism*, 68–90.

114. H. Spencer to E. L. Youmans, Jan. 19, 1884, in Duncan, *Life and Letters of Herbert Spencer*, 253–54.

115. See Frederic Harrison, "Neo-Christianity," *Westminster Review*, vol. 18, no. 2 (Oct. 1860): 293–332.

116. Frederic Harrison, "The Ghost of Religion," *PSM*, vol. 25 (Aug. 1884): 440–51.

117. Hebert Spencer, "Retrogressive Religion," *PSM*, vol. 25 (Aug. 1884): 451–74.

118. Frederic Harrison, "Agnostic Metaphysics," *PSM*, vol. 26 (Jan. 1885): 299–310.

119. H. Spencer to E. L. Youmans, Oct. 6, 1884, in Duncan, *Life and Letters of Herbert Spencer*, 257.

120. Herbert Spencer, "Last Words about Agnosticism," *PSM*, vol. 26 (Jan. 1885): 310–23.

121. "Editor's Table: Harrison and Spencer on Religion," *PSM*, vol. 26 (Jan. 1885): 407–9.

122. James Fitzjames Stephen, "The Unknowable and the Unknown," *Nineteenth Century*, vol. 15, no. 88 (June 1884): 905–19.

123. Stephen, "The Unknowable and the Unknown."

124. Wilfrid Ward, "The Clothes of Religion," *National Review*, vol. 3, no. 16 (June 1884): 554–73; see also "A Pickwickian Positivist," *National Review*, vol. 4, no. 20 (Oct. 1884): 222–37.

125. "A Quadrangular Duel," 748; see also "More Ghosts," *Saturday Review*, vol. 57, no. 1471 (Jan. 5, 1884): 15–16; and "The Religion of Humanity," *Saturday Review*, vol. 58, no. 1506 (Sept. 6, 1884): 308–9.

126. S. Rowe Bennett, "Spencer—Harrison—Arnold: An Eclectic Essay," *Contemporary Review*, 48 (Aug. 1885): 200–210.

127. Rev. Canon Curteis, "Christian Agnosticism," *PSM*, vol. 25 (May 1884): 78–86.

128. Henry Drummond, *Natural Law in the Spiritual Universe*, xiii-xiv.

129. H. Spencer to E. L. Youmans, April 12, 1883, in Fiske, *Edward Livingston Youmans*, 378.

130. "Literary Notices," *PSM*, vol. 26 (Nov. 1884): 131. For a more detailed discussion of Drummond, see James R. Moore, "Evangelicals and Evolution: Henry Drummond, Herbert Spencer, and the Naturalisation of the Spiritual World," *Scottish Journal of Theology*, vol. 38, no. 3 (1985): 383–418.

131. Hofstadter, *Social Darwinism in American Thought*, 31–50.

132. H. W. Beecher to H. Spencer, June 18, 1866, in Duncan, *The Life and Letters of Herbert Spencer*, 128–29.

133. "Pope Huxley," *Spectator*, vol. 43, no. 2170 (Jan. 29, 1870): 135–36; "Science in its Condescending Mood," *Spectator*, vol. 43, no. 2205 (Oct. 1, 1870): 1169–71.

134. "Agnostic Dreamers," *Spectator*, vol. 57, no. 2897 (Jan. 5, 1884): 9–11.

135. "The Positivist Dream," *Spectator*, vol. 45, no. 2315 (Nov. 9, 1872): 1422–23.

136. "Agnostic Dreamers"; "News of the Week," *Spectator*, vol. 57, no. 2897 (Jan. 5, 1884): 1–2.

137. "Mr. Frederic Harrison on Religion," *Spectator*, vol. 61, no. 3150 (Nov. 10, 1888): 1549–50.

138. "'Unreal Words' in Religious Belief," *Spectator*, vol. 59, no. 3030 (July 24, 1886): 982–84.

139. "Humanist Theology," *Spectator*, vol. 60, no. 3063 (March 12, 1887): 350–51.

140. "The New Reformation," *Spectator*, vol. 62, no. 3166 (March 2, 1889): 294–95; "Dr. Liddon on the 'New Reformation,'" *Spectator*, vol. 63, no. 3195 (Sept. 21, 1889): 371–72.

141. This address is printed in full in Fiske, *Edward Livingston Youmans*, 491–501.

142. An important source to the Free Religious Association can be found in the recollections of one of its founding presidents, Octavius Brooks Frothingham, *Recollections and Impressions, 1822–1890*, 115–32. For more scholarly treatments, see, e.g., Sidney Warren, *American Freethought, 1860–1914*; Stow Persons, *Free Religion: An American Faith*; and Leigh Eric Schmidt, *Restless Souls: The Making of American Spirituality*.

143. *Proceedings at the First Annual Meeting of the Free Religious Association, Held in Boston, May 28th and 29th, 1868*, 9.

144. *Proceedings at the First Annual Meeting of the Free Religious Association*, 20.

145. Frothingham, *Recollections and Impressions*, 120.

146. Clarence L. F. Gohdes, *The Periodicals of American Transcendentalism*, 210–28, 229–54. See also the collection of essays in *Freedom and Fellowship in Religion.*

147. Schmidt, *Restless Souls*, 84.

148. "Fifty Affirmations," *The Index*, vol. 1, no. 1 (Jan. 1, 1870): 1.

149. "The Impeachment of Christianity," *The Index*, vol. 3, no. 106 (Jan. 6, 1872): 4–5.

150. See, e.g., "Literary Notices: Proceedings of the Eleventh Annual Meeting of the Free Religious Association," *PSM*, vol. 14 (Nov. 1878): 116; and "Literary Notices: Proceedings at the Twelfth Annual Meeting of the Free Religious Association," *PSM*, vol. 15 (Oct. 1879): 851–52.

151. "Literary Notices," *PSM*, vol. 15 (Oct. 1879): 851.

152. "Editor's Table: The Progress of Theology," *PSM*, vol. 5 (May 1874): 114.

153. Charles Kingsley, "The Study of Physical Science: A Lecture to Young Men," *PSM*, vol. 1 (Aug. 1872): 451–57; "Popular Geology," *PSM*, vol. 1 (Sept. 1872): 613–17; "Literary Notices," *PSM*, vol. 5 (Sept. 1874): 632–33.

154. George Henslow, "Genesis, Geology, and Evolution," *PSM*, vol. 4 (Jan. 1874): 324–28.

155. William Stanley Jevons, "Evolution and the Doctrine of Design," *PSM*, vol. 5 (May 1874): 98–103.

156. Dean Stanley, "Discourse on the Death of Lyell," *PSM*, vol. 7 (May 1875): 90–93.

157. "Editor's Table: Dean Stanley's Sermon," *PSM*, vol. 7 (May 1875): 113–14.

158. "Editor's Table: Who are the Propagators of Atheism?" *PSM*, vol. 5 (July 1874): 365–67.

159. "Literary Notices: The Doctrine of Evolution," *PSM*, vol. 5 (July 1874): 375–76. See Alexander Winchell, *The Doctrine of Evolution*, 105.

160. "Literary Notices: Darwiniana," *PSM*, vol. 9 (Sept. 1876): 624–27.

161. "Literary Notices: Reconciliation of Science and Religion," *PSM*, vol. 11 (Aug. 1877): 498–500.

162. James Thompson Bixby, "Science and Religion as Allies," *PSM*, vol. 9 (Oct. 1876): 690–702.

163. "Literary Notices," *PSM*, vol. 9 (Oct. 1876): 759.

164. John William Dawson, "The So-Called 'Conflict of Science and Religion,'" *PSM*, vol. 10 (Nov. 1876): 72–74. This article was an extract from his *The Origin of the World, According to Revelation and Science.* Youmans also provided readers of his magazine a glowing "Sketch of Dr. John W. Dawson," *PSM*, vol. 8 (Dec. 1875): 231–34. Dawson, it should be noted, was also an ardent anti-Darwinian. For Dawson's complicated understanding of science–religion relations, see Susan Sheets-Pyenson, *John William Dawson: Faith, Hope, and Science.*

165. Seeley published his "Natural Religion" in serial form in *Macmillan's Magazine* between February 1875 and January 1878. Youmans, however, published only the first four installments. Later Seeley collected and revised these essays, publishing them in *Natural Religion*, 1882. Those installments that appeared in the *Popular Science Monthly* were: "The Deeper Harmonies of Science and Religion," *PSM*, vol. 7 (May 1875): 66–80; vol. 7 (July 1875): 301–15; vol. 7 (Sept. 1875): 573–88; and vol. 8 (Dec. 1875): 225–31. For a recent study

of Seeley, see Ian Hesketh, *Victorian Jesus: J. R. Seeley, Religion, and the Cultural Significance of Anonymity*.

166. J. R. Seeley, "The church as a teacher of morality," in W. L. Clay, ed., *Essays on Church Policy*, 247–91.

167. "The Deeper Harmonies of Science and Religion I," 66, 68, 76, 78.

168. "Literary Notices: Natural Religion," *PSM*, vol. 21 (Oct. 1882): 847–50.

169. "Editor's Table: The Religious Recognition of Nature," *PSM*, vol. 14 (Jan. 1879): 392–95.

170. Fiske, *Edward Livingston Youmans*, 201–2, footnote.

171. Fiske, *Edward Livingston Youmans*, 321.

172. H. W. Beecher, "The Study of Human Nature," *PSM*, vol. 1 (July 1872): 327–35.

173. Henry Ward Beecher, "The Progress of Thought in the Church," *North American Review*, vol. 135, no. 309 (Aug. 1882): 99–117.

174. "Beecher on Theology and Evolution," *PSM*, vol. 21 (Sept. 1882): 697–99.

175. See "Beecher on Evolution," *PSM*, vol. 27 (July 1885): 412–13; "Beecher's Position on Evolution," *PSM*, vol. 28 (Feb. 1886): 554–56.

SIX: Reading Draper and White

1. Linda Woodhead, ed., *Reinventing Christianity*, see esp. 1–21.

2. Woodhead, *Reinventing Christianity*, 7.

3. "Theology and Religion," *Spectator*, no. 2055 (Nov. 16, 1867): 1281–83.

4. See, e.g., "The Review: Human Physiology," *National Era*, vol. 10 (Oct. 23, 1856): 170; "Draper's Physiology," *North American Review*, vol. 83, no. 173 (Oct. 1856): 561; "Draper, *Human Physiology*," *American Journal of Medical Sciences*, vol. 65, no. 1 (Jan. 1857): 76–92; "Human Physiology, Statical and Dynamical," *Athenaeum*, 1524 (Jan. 10, 1857): 51; "Draper's Physiology," *Methodist Quarterly Review*, vol. 9, no. 23 (July 1857): 419–28; "Science," *Westminster Review*, vol. 63, no. 134 (Oct. 1857): 558–68; "Reviews: Todd, Bowman, Dunglison, and Draper on *Physiology*," *British and Foreign Medico-Chirurgical Review*, vol. 21, no. 41 (Jan. 1858): 19–21; and "Human Physiology," *Spectator*, vol. 32, no. 1626 (Aug. 27, 1859): 886–88; "New Publications," *New York Times* (Sept. 14, 1878): 3; "Science: Draper's Memoirs," *The Academy*, 338 (Oct. 26, 1878): 407–9; "Draper's Scientific Memoirs," *Nature*, vol. 19, no. 472 (Nov. 14, 1878): 26–28; and "Dr. Draper's Scientific Memoirs," *Spectator*, vol. 52, no. 2652 (April 26, 1879): 536–37.

5. All in the Draper Family Papers, Library of Congress, Washington D.C. (hereafter "Draper Family Papers"): R. Chambers to J. W. Draper, June 23, 1864, container 2; J. Tyndall to J. W. Draper, July 15, 1863, container 7; and G. Bancroft to J. W. Draper, October 1, 1865, container 1.

6. Ethel F. Fisk, ed., *The Letters of John Fiske*, 117, 118.

7. "Literary Notices," *Harper's New Monthly Magazine*, vol. 27, no. 157 (June 1863): 128–29; "Critical Notices: Draper's Intellectual Development of Europe," *North American Review*, vol. 97, no. 1 (July 1, 1863): 291; "Review of Current Literature," *Christian Examiner*, vol. 75, no. 238 (1863): 137–39.

8. "Literature: History of the Intellectual Development of Europe," *Athenaeum*, 1914 (July 2, 1864): 9–10; "The Literature of the United States in 1870," *Athenaeum*, 2254 (Jan. 7, 1871): 12–15; [H. B. Wilson], "Theology and Philosophy," *Westminster Review*, vol. 26, no. 2 (Oct. 1864): 470–95; [Sheldon Amos], "The Intellectual Development of Europe," *Westminster Review*, vol. 27, no. 1 (Jan. 1865): 94–142.

9. G. Bancroft to J. W. Draper, Dec. 4, 1874, container 1, Draper Family Papers.

10. J. W. Draper to J. Tyndall, July 24, 1865; J. W. Draper to J. Tyndall, Nov. 6, 1865; and J. Tyndall to J. W. Draper, Jan. 11, 1866; container 7, Draper Family Papers.

11. John Tyndall, "Apology for the Belfast Address," *Fragments of Science*, 2.235. For more on this important correspondence, see the short note in James C. Ungureanu, "Tyndall and Draper," *Notes and Queries*, vol. 64, no. 1 (2017): 125–28.

12. "Literary," *Appleton's Journal*, vol. 12, no. 301 (Dec. 26, 1874): 827–28; "Our Book Table," *New Orleans Bulletin*, vol. 1, no. 256 (Jan. 17, 1875): 5; A. S. Gardiner, "Religion and Science," *New York Evangelist*, vol. 48 (1877): 2.

13. [William H. Williams], "The Church and Science," *New Orleans Monthly Review*, vol. 2, no. 2 (March 1875): 132–45.; "Contemporary Literature," *Universalist Quarterly and General Review*, vol. 12 (April 1875): 251–53 ; Thomas Hill, "The Struggles of Science," *Unitarian Review and Religious Magazine*, vol. 3, no. 4 (April 1875): 339–56; [Sheldon Amos, Emilia Frances Dilke], "Science," *Westminster Review*, vol. 47, no. 2 (April 1875): 522–41.

14. "History of Scientific Progress," *Christian Union*, vol. 11, no. 7 (Feb. 17, 1875): 135–36; Tayler Lewie, "An Impersonal God," *Christian Union*, vol. 11, no. 7 (Feb. 17, 1875): 146; Charles L. Brace, "The Great Conflict," *Christian Union*, vol. 11, no. 12 (March 24, 1875): 238–39; and Adam Stwin, "Dr. Draper's 'Conflict.'" *Christian Union*, vol. 11, no. 15 (April 14, 1875): 301–2.

15. Henry B. Smith, "Draper's Intellectual Development of Europe," *American Presbyterian and Theological Review*, New Series, vol. 1, no. 4 (Oct. 1863): 615–31; and "Short Notices," *Biblical Repertory and Princeton Review*, vol. 38, no. 1 (Jan. 1866): 144–50. See also similar reviews in "Quarterly Book-Table," *Methodist Quarterly Review*, vol. 16, no. 1 (Jan. 1864): 164–67; and "Positivism," *American Quarterly Church Review, and Ecclesiastical Register*, vol. 16, no. 1 (April 1864): 35–56.

16. Undated newspaper and journal clippings, container 14, Draper Family Papers; "Quarterly Book-Table," *Methodist Quarterly Review*, vol. 27, no. 1 (Jan. 1875): 159–63; "The Conflict between Religion and Science," *Southern Review*, vol. 18, no. 35 (July 1875): 122–53; "Culture and Progress: The Conflict between Religion and Science," *Scribner's Monthly*, vol. 9, no. 5 (March 1875): 635–37; "Draper's Religion and Science," *Presbyterian Quarterly and Princeton Review*, New Series, vol. 14, no. 13 (1875): 158–65.

17. "Professor Draper's Books," *Catholic World*, vol. 7, no. 38 (May 1868): 155–74.

18. "Professor Draper's Books," 169.

19. "The Conflict of Science and Religion," *Brownson's Quarterly Review*, Last Series, vol. 3, no. 2 (April 1875): 153–73.

20. "Draper's Conflict between Religion and Science," *Catholic World*, vol. 21, no. 122 (May 1875): 178–200.

21. Joseph Treat to John W. Draper, Jan. 5, 1875; Joseph Treat to John W. Draper, Jan. 30, 1875, in container 7, Draper Family Papers.

22. See Edward Royle, *Radicals, Secularists, and Republicans: Popular Freethought in Britain, 1866–1915*, 78–80.

23. T. D. Hall, "Can Christianity be made to Harmonize with Science," *An Essay Read before the Liberal League, at Minneapolis, Sunday, March 7th, 1875*; container 14, Draper Family Papers.

24. See, e.g., Edward Royle, "Freethought: The Religion of Irreligion," in D. G. Paz, ed., *Nineteenth-Century English Religious Traditions: Retrospect and Prospect*, 171–96.

25. Leigh Eric Schmidt, *Village Atheists*; Christopher Grasso, *Skepticism and American Faith.*

26. For a brief study, see Bernard Lightmam, "Ideology, evolution and late-Victorian agnostic popularizers," in James R. Moore, ed., *History, Humanity and Evolution*, 285–309. See also F. J. Gould, *The Pioneers of Johnson's Court: A History of the Rationalist Press Association from 1899 Onwards*; A. G. Whyte, *The Story of the R.P.A., 1899–1949*; and more recently Bill Cooke, *The Gathering of Infidels: A Hundred Years of the Rationalist Press Association.*

27. See Gould, *The Pioneers of Johnson's Court*, 8–12; Whyte, *The Story of the R.P.A.*, 24.

28. J. M. Robertson, *The Dynamics of Religion: An Essay in English Culture History*, 249.

29. J. M. Robertson, *A Short History of Freethought*, 1.13, 2.380, 2.407.

30. J. M. Robertson, *A History of Freethought in the Nineteenth Century*, 1.261–62.

31. J. M. Wheeler, *Frauds and Follies of the Fathers*; *A Biographical Dictionary of Freethinkers of All Ages and Nations*, 112, 332; Samuel P. Putnam, *Religion a Curse, Religion a Disease, Religion a Lie*; *400 Years of Freethought*; George E. MacDonald, *Fifty Years of Freethought, Being the Story of the Truth Seeker*, 168.

32. Joseph McCabe, *A Biographical Dictionary of Modern Rationalists*, 221–22.

33. Emanuel Haldeman-Julius, *The First Hundred Million*, esp. 45–55. See also the recent study by R. Alton Lee, *Publisher for the Masses, Emanuel Haldeman-Julius.*

34. Joseph McCabe, *The Conflict Between Science and Religion.*

35. McCabe, *Conflict*, 15–25, 26–38.

36. Edward B. Davis, "Science and Religious Fundamentalism in the 1920s: Religious pamphlets by leading scientists of the Scopes era provide insight into public debates about science and religion," *American Scientist*, vol. 93, no. 3 (2005): 253–60.

37. "The Warfare of Science," *Popular Science Review*, vol. 15, no. 58 (Jan. 1876): 414–15; "Literary Notices," *Dublin University Magazine*, vol. 88, no. 525 (Sept. 1876): 382–84; and [J. H. B. Browne], "The Warfare of Science," *Westminster Review*, vol. 51 (Jan. 1877): 19–36. Slightly more critical, however, were the *Academy*, vol. 10, no. 228 (Sept. 16, 1876): 287; *Spectator*, vol. 49, no. 2524 (Nov. 11, 1876): 1415; and *Examiner*, no. 3592 (Dec. 2, 1876): 1360.

38. "The Book of the Wars of the Lord," *Outlook*, vol. 53, no. 25 (June 20, 1896): 1153; and "Literature: History of the Warfare of Science with Theology," *Independent*, vol. 48, no. 2480 (June 11, 1896): 16–17. See also Walton Battershall, "The Warfare of Science with Theology," *North American Review*, vol. 165, no. 1 (July 1, 1897): 87–98; and William W. McLane, "The Case of Theology Versus Science," *Homiletic Review*, vol. 34, no. 1 (July 1897): 8–13.

39. David Starr Jordan, "The Stability of Truth I," *Popular Science Monthly*, vol. 50 (March 1897): 642–53; "The Stability of Truth II," *Popular Science Monthly*, vol. 50 (April 1897): 749–57.

40. David Starr Jordan, "The Conflict of Science," in *Addresses at the Thirtieth Meeting of the Unitarian Club of California*, 19–30.

41. David Starr Jordan, "The Warfare of Science," *Dial*, vol. 21, no. 246 (Sept. 16, 1896): 146–48.

42. T. T. Munger to Andrew D. White, June 27, 1895; Oct. 13, 1895; Oct. 21, 1895; Jan. 3, 1896; Jan. 15, 1896; Jan. 23, 1896; Feb. 27, 1896; March 13, 1896; March 25, 1896; March 30, 1896; Nov. 10, 1896; Dec. 8, 1896— all in Division of Rare and Manuscript Collections at

Cornell University, Ithaca, New York, reels 63, 64, 65, and 68 (hereafter, "White Collection").

43. Andrew Carnegie to Andrew D. White, May 6, 1896, White Collection, reel 66.

44. E. D. Jackson to Andrew D. White, March 9, 1894, White Collection, reel 61.

45. James Gorton to Andrew D. White, March 30, 1894; James Gorton to Andrew D. White, May 4, 1894, White Collection, reel 61.

46. Charles Kendall Adams, "Mr. White's 'Warfare of Science with Theology,'" *Forum*, vol. 22 (Sept. 1896): 65–78; Jacob Gould Schurman, "Scientific Literature," *Science*, vol. 4, no. 102 (Dec. 11, 1896): 879–81.

47. Augustine F. Hewitt, "The Warfare of Science," *Catholic World*, vol. 53, no. 315 (June 1891): 393–99; vol. 53, no. 316 (July 1891): 567–76; vol. 53, no. 317 (Aug. 1891): 678–87; and vol. 54, no. 320 (Nov. 1891): 194–203.

48. "Books of the Week," *Manchester Guardian* (June 23, 1896): 4; "Science and Theology," *Speaker*, vol. 14 (Aug. 8, 1896): 149–50; "The Rights and Limits of Theology," *Quarterly Review*, vol. 203, no. 405 (Oct. 1905): 461–91.

49. "Old Testament Miracles which are being attacked in the New Theology," *Chicago Tribune*, vol. 56, no. 80 (March 21, 1897): 45–46; Alexander Patterson, "Holds by the Bible Lore," *Chicago Tribune*, vol. 56, no. 87 (March 28, 1897): 16; Elizabeth A. Reed, "Flaws in White's Book: Facts upon which he based his deductions disputed," *Chicago Tribune*, vol. 56, no. 187 (July 6, 1897): 6.

50. "White: The Warfare of Science with Theology," *American Historical Review*, vol. 2, no. 1 (Oct. 1896): 107–13; "A History of the Warfare of Science with Theology in Christendom," *Nation*, vol. 62 (May 28, 1896): 421–22; Elizabeth Cady Stanton, "Reading the Bible in the Public Schools," *The Arena*, vol. 17 (1897): 1033–37.

51. James McCosh, *The Development Hypothesis: Is it Sufficient?* 75.

52. [B. B. Warfield], "Recent Theological Literature: *A History of the Warfare of Science with Theology in Christendom*," *The Presbyterian and Reformed Review* vol. 9, no. 35 (1898): 510-12.

53. Albert Britt to A. D. White, Jan. 9, 1905, White Collection, reel 93; John Shackelford to A. D. White, May 23, 1909, White Collection, reel 105.

54. William Philips to A. D. White, Nov. 6, 1895, White Collection, reel 64; Harry A. Miller to A. D. White, Jan. 16, 1917, White Collection, reel 123.

55. Mary K. S. Eaton to A. D. White, March 6, 1894, White Collection, reel 61.

56. Mary K. S. Eaton to A. D. White, April 16, 1894, White Collection, reel 61.

57. Mary K. S. Eaton to A. D. White, Jan. 21, 1896, White Collection, reel 65.

58. A. D. White to Mary K. S. Eaton, Jan. 30, 1896, White Collection, reel 65.

59. Henry M. Taber, "Liberalized Christianity," *Free Thought Magazine*, vol. 14 (Jan. 1896): 9–20; see also "Book Review," *Free Thought Magazine*, vol. 14 (Aug. 1896): 534–35.

60. Alfred W. Benn, "A History of the Warfare of Science with Theology in Christendom," *Academy*, no. 1275 (Oct. 10, 1896): 255–56.

61. A. D. White to E. E. Evans, Feb. 22, 1874, White Collection, reel 15.

62. E. E. Evans to A. D. White, Feb. 22, 1877, White Collection, reel 22.

63. E. P. Evans to A. D. White, Feb. 22, 1877, White Collection, reel 22.

64. E. E. Evans to A. D. White, Jan. 4, 1901, White Collection, reel 82.

65. Frederick White to A. D. White, Feb. 2, 1889, White Collection, reel 51.

66. H. L. Green to A. D. White, March 28, 1897, White Collection, reel 70.

67. R. G. Ingersoll to A. D. White, Dec. 27, 1888, White Collection, reel 50.

68. Charles A. Watts to A. D. White, July 1, 1896, White Collection, reel 67; Charles A. Watts to A. D. White, March 29, 1897, White Collection, reel 70.

69. White provided extensive reflections on his religious beliefs in his *Autobiography*, quotation from 2.495.

70. A. D. White to George Lincoln Burr, Aug. 26, 1885, White Collection, reel 44.

71. Frederick White to A. D. White, Jan. 5, 1886, White Collection, reel 45.

72. Frederick White to A. D. White, April 13, 1886, White Collection, reel 45.

73. A. D. White to Frederick White, April 29, 1886, White Collection, reel 45.

74. Frederick White to A. D. White, May 13, 1886, White Collection, reel 45.

75. "F. D. White A Suicide: Son of Ambassador White Shoots Himself Through the Head," *New-York Tribune*, vol. 61, no. 19959 (July 9, 1901), 1.

76. Diary entry, July 11, 1901, White Collection, reel 132.

77. See, e.g., J. Llewelyn Davies, "The Debts of Theology to Secular Movements," *Contemporary Review*, vol. 16 (Dec. 1870): 189–206.

78. George P. Fisher, "The Alleged Conflict of Natural Science and Religion," *Princeton Review* (1883).

79. Jon H. Roberts, "Liberal Protestants," in Jeff Hardin, Ronald L. Numbers, and Ronald A. Binzley, eds., *The Warfare Between Science & Religion*, 143–62.

80. J. Gresham Machen, *Christianity and Liberalism*.

CONCLUSIONS

1. See the recent review essays by John F. M. Clark, "Intellectual History and the History of Science," in Richard Whatmore and Brian Young, eds., *A Companion to Intellectual History*, 155–69; and Lynn K. Nyhart, "Historiography of the History of Science," in Bernard Lightman, ed., *A Companion to the History of Science*, 7–22. See also the classic work by Helge Kragh, *An Introduction to the Historiography of Science*.

2. I. Bernard Cohen, "The Isis Crises and the Coming of Age of the History of Science Society."

3. James B. Conant, "George Sarton and Harvard University."

4. Victor L. Hilts, "History of Science at the University of Wisconsin."

5. There are a number of helpful collected biographical studies on Sarton and the History of Science Society—see, e.g., "The George Sarton Memorial Issue"; "Sarton, Science, and History"; and most recently "Focus: 100 Volumes of *Isis:* The Vision of George Sarton."

6. Thomas S. Kuhn, *The Essential Tension*, 146.

7. L. Pearce Williams, "A History of Science."

8. A. C. Crombie, "The Appreciation of Ancient and Medieval Science" and "Six Wings."

9. For Sarton's positivistic outlook, see, e.g., Arnold Thackray and Robert K. Merton, "On Discipline Building: The Paradoxes of George Sarton"; and esp. Tore Frängsmyr, "Science of History: George Sarton and the Positivist Tradition in the History of Science."

10. Thackray and Merton, "On Discipline Building," 479.

11. Frängsmyer, "Science or History," 104.

12. Clark, "Intellectual History and the History of Science," 159.

13. Christopher Dawson, *Progress and Religion*; Karl Löwith, *Meaning in History*;

Ernest L. Tuveson, *Millennium and Utopia*. See also Robert Nisbet, *History of the Idea of Progress*.

14. According to his daughter, May Sarton, when Sarton was allowed to be present at the dinner table, "if he so much as babbled a single word, his father, without raising his head from his newspaper, reached forward to touch the bell . . . and when the maid appeared, said simply, '*Enlevez-le*' (Remove it)." See her other childhood memories in *I Knew a Phoenix*, 15. See also the somewhat idiosyncratic biography by Lewis Pyenson, *The Passion of George Sarton*.

15. Quoted in I. Bernard Cohen, "George Sarton," 286.

16. Ernst Mach, *The Science of Mechanics, A Critical and Historical Account of its Development*, trans. Thomas J. McCormack (Chicago: Open Court Publishing, 1919), 446. For Mach's influence on Sarton, see Hayo Siemsen, "Ernst Mach, George Sarton and the Empiry of Teaching Science Part I," *Science & Education* 21 (2012): 447–84; and "Ernst Mach and George Sarton's Successors: The Implicit Role Model of Teaching Science in USA and Elsewhere, Part II," *Science & Education* 22 (2013): 951–1000.

17. George Sarton, "L'histoire de la science,"; reprinted in English as "The History of Science."

18. George Sarton, "Auguste Comte, Historian of Science."

19. See, e.g., George Sarton, "The Teaching of the History of Science," *Scientific Monthly*, vol. 7, no. 13 (1918): 193–211; "The Faith of the Humanist," *Isis*, vol. 3, no. 1 (1920): 3–6; "The Teaching of the History of Science," *Isis*, vol. 4, no. 2 (1921): 225–49; "The New Humanism," *Isis*, vol. 6, no. 1 (1924): 9–42; and "The Teaching of the History of Science," *Isis*, vol. 13, no. 2 (1930): 272–97.

20. Sarton, "The Faith of the Humanist," 3–6.

21. George Sarton, *The History of Science and the New Humanism*, esp. 10, 43–48, 179.

22. Sarton, "The New Humanism," 33.

23. George Sarton, *Introduction to the History of Science, Volume 1: From Homer to Omar Kahayyam*, 19, 25, 28, 32.

24. Robert K. Merton, "George Sarton: Episodic Recollection by an Unruly Apprentice". See also Merton's prefatory "Recollections & Reflections" in George Sarton, *The History of Science and the New Humanism*.

25. George Sarton, *Horus: A Guide to the History of Science*, 118, 121.

26. Sarton, "The History of Science," 339.

27. Thackray and Merton, "On Discipline Building," 485, fn. 32.

28. See George Sarton to A. D. White, March 31, 1918; April 5, 1918; April 10, 1918; and May 15, 1918, all in White Collection, reel 124.

29. See, e.g., George A. Reisch, "Planning Science: Otto Neurath and the *International Encyclopedia of Unified Science*," *British Journal of the History of Science*, vol. 27, no. 2 (1994): 153–75.

30. Otto Neurath, "Ways of the Scientific World-Conception," in *Philosophical Papers, 1913–1946*, trans. Robert S. Cohen and Marie Neurath (Dordrecht: Reidel, 1983), 39.

31. Bertrand Russell, *Religion and Science*, with a new introduction by Michael Ruse (Oxford: Oxford University Press, 1997), 7.

32. Bertrand Russell, *Why I Am Not a Christian and Other Essays on Religion and Related Subjects* (London: George Allen and Unwin, 1957), 22, 24.

33. Alfred Jules Ayer, *Language, Truth and Logic*, 117.

34. See Ben Rogers, *A. J. Ayer: A Life.*

35. Nyhart, "Historiography of the History of Science," 8.

36. See most recently Liam Jerrold Fraser, *Atheism, Fundamentalism and the Protestant Reformation.*

37. Mike Brown, "Godless in Tumourville: Christopher Hitchens interview," https://www.telegraph.co.uk/culture/books/8388695/Godless-in-Tumourville-Christopher-Hitchens-interview.html.

38. See, e.g., D. G. Hart, *Defending the Faith.* For the broader context of American religious fundamentalism, see George M. Marsden, *Fundamentalism and American Culture*; Mark A. Noll, *Between Faith and Criticism*; and the collection of essays in Martin E. Marty and R. Scott Appleby, eds., *Fundamentalism and Society.*

39. Edwin H. Wilson, *The Genesis of a Humanist Manifesto.*

BIBLIOGRAPHY

PRIMARY SOURCES

Manuscript Material

John William Draper Family Papers, Manuscript Division, Library of Congress, Washington, DC.

Andrew Dickson White Papers at Cornell University, Library Division of Rare and Manuscript Collections, 2B Carl A. Kroch Library, Cornell University, Ithaca, New York.

The Correspondence of John Tyndall. 6 vols. General Editors, James Elwick, Bernard Lightman, and Michael S. Reidy. Pittsburgh, PA: University of Pittsburgh Press.

Newspapers and Periodicals

Academy

Agnostic Annual

American Historical Review

American Journal of Medical Sciences

American Journal of Science

American Presbyterian and Theological Review

American Quarterly Church Review, and Ecclesiastical Register

Appleton's Journal of Literature, Science, and Art

Arena

Athenaeum

Biblical Repertory and Princeton Review

British and Foreign Medico-Chirurgical Review

Brownson's Quarterly Review

Builder

Calcutta Review

Catholic World

Chicago Tribune
Christian Examiner
Christian Union
Contemporary Review
Continental Monthly
Cornell Daily Sun
Cornell Era
Cornhill Magazine
Daily Graphic
Dial
Dublin University Magazine
Economist
Edinburgh Review
Examiner
Fortnightly Review
Forum
Fraser's Magazine
Freethought Magazine
Harper's New Monthly Magazine
Homiletic Review
Independent
Index
Isis
Knowledge: An Illustrated Magazine of Science Plainly Worded—Exactly Described
The Leader
London Quarterly Review
Macmillan's Magazine
Magazine of American History with Notes and Queries
Magazine of Natural History
Manchester Guardian
Methodist Quarterly Review
Monthly Record of the Transactions of the Evangelical Alliance
Nation
National Era
National Review
Nature
New Englander
New Orleans Bulletin
New Orleans Monthly Review

New Orleans Monthly Review
New York Daily Tribune
New York Evangelist
New York Herald
New York Times
New York Tribune
New Yorker: The New York University Weekly
Nineteenth Century
North American Review
Osiris
Outlook
Pall Mall Gazette
Popular Science Monthly
Popular Science Review
Presbyterian and Reformed Review
Princeton Review
Putnam's Magazine
Quarterly Review
Radical
Record of the Transactions of the Evangelical Alliance
Reader
Religious Magazine and Monthly Review
Report of the Meetings of the British Association for the Advancement of Science
Saturday Review
Science
Scientific American
Scientific Monthly
Scribner's Monthly
Southern Literary Messenger
Southern Review
Speaker
Spectator
Times
Unitarian Review and Religious Magazine
Universalist Quarterly and General Review
University Quarterly
Watts's Literary Guide
Westminster Review

Books and Pamphlets

Abbott, Lyman. *The Evolution of Christianity*. Boston: Houghton Mifflin, 1892.

Abbott, Lyman. *Henry Ward Beecher. A Sketch of His Career: With Analyses of His Power as a Preacher, Lecturer, Orator, and Journalist, and Incidents and Reminiscences of His Life*. New York: Funk & Wagnalls, 1883.

Abbott, Lyman. *The Theology of an Evolutionist*. Boston: Houghton Mifflin, 1897.

Account of the Proceedings at the Inauguration October 7th 1868. Ithaca, NY: University Press, 1868.

Addresses at the Thirtieth Meeting of the Unitarian Club of California, Held at San Francisco, Cal. April 26, 1897. San Francisco: C. A. Murdock & Co, 1897.

Alger, William R. *Studies of Christianity: or, Timely Thoughts for Religious Thinkers. A Series of Papers by James Martineau*. Boston: American Unitarian Association, 1882.

Allen, Ethan. *Reason the Only Oracle of Man*. Bennington, VT: Haswell and Russell, 1784.

Annual Report of the American Institute of the City of New York, 1869–70. Albany, NY: Argus, 1870.

Arms, William Dallam, ed. *The Autobiography of Joseph Le Conte*. New York: D. Appleton and Co., 1903.

Arnold, Matthew. *God and the Bible: A Review of Objections to "Literature and Dogma."* New York: Macmillan, 1875.

Arnold, Matthew. *Literature and Dogma: An Essay Towards a Better Apprehension of the Bible*. New York: Macmillan, 1874.

Arnold, Matthew. *St. Paul and Protestantism: With an Essay on Puritanism and the Church of England*. London: Smith, Elder, 1870.

Babbage, Charles. *The Exposition of 1851; or, Views of the Industry, the Science, and the Government of England*. London: John Murray, 1851.

Babbage, Charles. *The Ninth Bridgewater Treatise: A Fragment*. London: John Murray, 1837.

Babbage, Charles. *Reflections on the Decline of Science in England*. London: B. Fellowes, 1830.

Barker, George F. *Memoir of John William Draper, 1811–1882: Read Before the National Academy, April 21, 1886*. Washington, DC: National Academy of Sciences, 1886.

Beecher, Henry Ward. *Evolution and Religion*, Parts I and II. Boston: Pilgrim Press, 1885.

Beecher, Lyman. *A Plea for the West*. New York: Leavitt, Lord, 1835.

Birch, Thomas. *The Life of the Most Reverend Dr. John Tillotson*. London: J. and R. Tonson, 1753.

Birch, Thomas. *The Works of Dr. John Tillotson*. 10 vols. London: J. F. Dove, 1820.

Birch, Thomas. *The Works of the Honourable Robert Boyle*. 5 vols. London: A. Miller, 1744.

Briggs, Charles Augustus. *The Authority of Holy Scripture: An Inaugural Address*. New York: Charles Scribner's Sons, 1891.

Brougham, Henry. *A Discourse on the Objects, Advantages, and Pleasures of Science, a New Edition, Illustrated with Engravings*. London: Baldwin and Cradock, 1828.

Burnet, Gilbert. *History of My Own Time*. 2 vols. Oxford: Clarendon Press, 1897.

Burnet, Gilbert. *The History of the Reformation of the Church of England*. 3 vols. Oxford: Oxford University Press, 1829.

Burnet, Gilbert. *A sermon preached at the funeral of the most reverend Father in God, John Tillotson, Lord Archbishop of Canterbury*. London: Richard Chiswell, 1694.

Bushnell, Horace. *Discourses on Christian Nurture*. Boston: Massachusetts Sabbath School Society, 1847.

Bushnell, Horace. *God in Christ: Three Discourses, Delivered at New Haven, Cambridge, and Andover, with a Preliminary Dissertation on Language*. Hartford: Brown and Parsons, 1849.

Bushnell, Horace. *Nature and the Supernatural, as Together Constituting the One System of God*. New York: Charles Scribner, 1858.

Catalogue of the Historical Library of Andrew Dickson White: The Protestant Reformation and its Forerunners. Ithaca, NY: Cornell University Press, 1889.

Chambers, Robert. *Explanations: A Sequel to Vestiges of the Natural History of Creation*. New York: Wiley and Putman, 1846

Chambers, Robert. *Vestiges of the Natural History of Creation*. London: John Churchill, 1844.

Channing, George G., ed. *The Works of William E. Channing*. 6 vols. Boston: James Munroe, 1846.

Chillingworth, William. *The Religion of Protestants, a Safe Way to Salvation*. London: Henry G. Bohn, 1846.

Christmas, Rev. Henry, ed. *Select Works of John Bale*. Cambridge. The University Press, 1849.

Clark, John Spencer, ed. *The Life and Letters of John Fiske*. 2 vols. Boston: Houghton Mifflin, 1917.

Clay, W. L., ed. *Essays on Church Policy*. London: Macmillan, 1868.

Colenso, John William. *The Pentateuch and Book of Joshua Critically Examined*. New York: D. Appleton and Co., 1863.

Coleridge, Samuel Taylor. *Aids to Reflection*. Burlington, VT: Chauncey Goodrich, 1840.

Combe, George. *The Constitution of Man considered in Relation to External Objects.* Boston: Carter and Hendee, 1829.

Combe, George. *On the Relation between Science and Religion.* Edinburgh: Maclachlan and Stewart, 1857.

Cornell, Alonzo. *True and Firm: Biography of Ezra Cornell.* New York: A. S. Barnes, 1884.

The Cornell University. First General Announcement. Albany, NY: Weed, Parsons, 1868.

Cudworth, Ralph. *The True Intellectual System of the Universe.* 2 vols. Andover: Gould and Newman, 1837.

Culverwell, Nathaniel. *An Elegant and Learned Discourse of the Light of Nature.* Edinburgh: Thomas Constable, 1857.

Darwin, Charles. *The Descent of Man and Selection in Relation to Sex.* 2 vols. London: John Murray, 1871.

Darwin, Charles. *On the Origins of Species by Means of Natural Selection; or, The Preservation of Favoured Races in the Struggle for Life.* London: John Murray, 1859.

Darwin, Francis. *The Life and Letters of Charles Darwin.* 2 vols. New York: D. Appleton and Co., 1896.

Davy, Humphry. *Consolations in Travel, Or the Last Days of a Philosopher.* London: John Murray, 1830.

Dawson, John William. *The Origin of the World, According to Revelation and Science.* New York: Harper, 1877.

Derby, James Cephas. *Fifty Years Among Authors: Books and Publishers.* New York: C.W. Carleton, 1884.

Dick, Thomas. *The Christian Philosopher; or, the Connection of Science and Philosophy with Religion.* 2nd ed. Brookfield, MA: E. and G. Merriam, 1828.

Draper, John William. *History of the American Civil War.* 3 vols. New York: Harper and Brothers, 1867–70.

Draper, John William. *History of the Conflict between Religion and Science.* New York: D. Appleton and Co., 1874.

Draper, John William. *A History of the Intellectual Development of Europe.* New York: Harper and Brothers, 1863.

Draper, John William. *Human Physiology, Statical and Dynamical; or, The Conditions and Course of the Life of Man.* New York: Harper and Brothers, 1856.

Draper, John William. *The Influences of Physical Agents on Life: Being an Introductory Lecture to the Course on Chemistry and Physiology.* New York: John A. Gray, 1850.

Draper, John William. *Introductory Lecture in the Course of Chemistry.* New York: Hopkins & Jennings, 1841.

Draper, John William. *Introductory Lecture to the Course of Chemistry: Relations of Atmospheric Air to Animals and Plants.* New York: University of New York, 1844–45.

Draper, John William. *Introductory Lecture to the Course of Chemistry: Relations and Nature of Water.* New York: University of New York, 1845–46.

Draper, John William. *Introductory Lecture on Oxygen Gas.* New York: Joseph H. Jennings, 1848.

Draper, John William. *Scientific Memoirs: Being Experimental Contributions to a Knowledge of Radiant Energy.* New York: Harper and Brothers, 1878.

Draper, John William. *A Text-book on Chemistry: For the use of Schools and Colleges.* New York: Harper and Brothers, 1846.

Draper, John William. *A Text-Book on Natural Philosophy: For the use of Schools and Colleges.* New York: Harper and Brothers, 1847.

Draper, John William. *Thoughts on the Future Civil Policy of America.* New York: Harper and Brothers, 1865.

Draper, John William. *Treatise on the Forces which Produce the Organization of Plants.* New York: Harper and Brothers, 1844.

Draper, Thomas Waln-Morgan. *The Drapers in America, Being a History and Genealogy of those of that name and connection.* New York: John Polhemus Printing Co., 1892.

Drummond, Henry. *Natural Law in the Spiritual Universe.* London: Hodder and Stoughton, 1883.

Duncan, David, ed. *The Life and Letters of Herbert Spencer.* London: Methuen, 1908.

Eliot, Charles William. *Educational Reform: Essays and Addresses.* New York: Century, 1898.

Eliot, Charles William. *The Religion of the Future: A lecture delivered at the close of the eleventh session of the Harvard Summer School of Theology, July 22, 1909.* New York: Frederick A. Stokes, 1909.

Emerson, Edward W., ed. *The Works of Ralph Waldo Emerson.* 14 vols. Boston: Houghton Mifflin, 1883.

Essays and Reviews. London: John W. Parker and Son, 1860.

Featherstone, J. S. *A Tribute of Grateful Remembrance to the Memory of the Rev. John Christopher Draper, Late Superintendent of the Wesleyan Methodist Society in the Sheerness Circuit.* Sheerness, 1829.

Fisk, Ethel F., ed. *The Letters of John Fiske.* New York: Macmillan, 1940.

Fiske, John. *A Century of Science and Other Essays.* Boston: Houghton, Mifflin, 1899.

Fiske, John. *The Destiny of Man Viewed in the Life of His Origin.* Boston: Houghton, Mifflin, 1893.

Fiske, John. *Edward Livingston Youmans, Interpreter of Science for the People: A Sketch of His Life with Selections from his Published Writings and Extracts from his Correspondence with Spencer, Huxley, Tyndall, and Others*. New York: D. Appleton and Co., 1894.

Fiske, John. *The Idea of God as Affected by Modern Knowledge*. Boston: Houghton, Mifflin, 1887.

Fiske, John. *Outlines of Cosmic Philosophy, Based on the Doctrine of Evolution, with Criticisms on the Positive Philosophy*. 2 vols. London: Macmillan, 1874.

Fiske, John. *Through Nature to God*. Boston: Houghton, Mifflin, 1899.

Fisher, George P. *The Life of Benjamin Silliman*. 2 vols. New York: Charles Scribner, 1866.

Fisher, James. *A Spring Day; or, Contemplations on Several Occurrences which Naturally Strike the Eye in that Delightful Season*. New York, 1813.

Foxcroft, H. C. *A supplement to Burnet's history of my own time*. Oxford: Clarendon Press, 1897.

Franklin, Benjamin. *The Private Life of the Late Benjamin Franklin*. London, 1793.

Freedom and Fellowship in Religion: A Collection of Essays and Addresses, edited by a committee of the Free Religious Association. Boston: Roberts Brothers, 1875.

Frothingham, Octavius Brooks. *Recollections and Impressions, 1822–1890*. New York: G. P. Putnam's Sons, 1891.

Froude, J. A. *The Nemesis of Faith*. London: John Chapman, 1849.

Froude, J. A. *Short Studies on Great Subjects*. 4 vols. New York: Charles Scribner's Sons, 1888.

Froude, J. A. *Thomas Carlyle*. 2 vols. London: Longmans, Green, 1882.

Gibbon, Charles, ed. *The Life of George Combe*. 2 vols. London: Macmillan, 1878.

Gibbon, Edward. *The History of the Decline and Fall of the Roman Empire*, with Notes by Rev. H. H. Milman. 12 vols. London: John Murray, 1839.

Gray, Asa. *Darwiniana: Essays and Reviews Pertaining to Darwinism*. New York: D. Appleton and Co., 1888.

Gray, Asa. *Natural Science and Religion: Two Lectures Delivered to the Theological School of Yale College*. New York: Charles Scribner's Sons, 1880.

Greg, William Rathbone. *The Creed of Christendom*. London: John Chapman, 1851.

Haldeman-Julius, Emanuel. *The First Hundred Million*. New York: Simon and Schuster, 1928.

Harper, William Rainey. *Religion and the Higher Life*. Chicago: University of Chicago Press, 1904.

Harper, William Rainey. *The Trend in Higher Education*. Chicago: University of Chicago Press, 1905.

Harris, Thaddeus Mason. *The Beauties of Nature Delineated; or, Philosophical and Pious Contemplations on the Works of Nature, and the Seasons of the Year, Selected from Sturm's Reflections*. Charlestown, MA: Samuel Etheridge, 1801.

Harris, Thaddeus Mason. *The Natural History of the Bible: or a Description of All the Beasts, Birds, Fishes, Insects, Reptiles, Trees, Plants, Metals, Precious Stones, Etc. Mentioned in the Sacred Scriptures. Collected by the best Authorities, and Alphabetically Arranged*. Boston, 1793.

Hartley, David. *Observations on Man, His Frame, His Duty, and His Expectations*. 2 vols. London: S. Richardson, 1749.

Herschel, John F. W. *A Preliminary Discourse on the Study of Natural Philosophy*. London: Longman, 1830; new edition, 1851.

Higgins, W. M. *The Mosaical and Mineral Geologies, Illustrated and Compared*. London: John Scoble, 1832.

Hillebrand, Karl. *Six Lectures on the History of German Thought*. London: Longmans, 1880.

Hutton, Richard H. *Aspects of Religious and Scientific Thought, by the Late Richard Holt Hutton: Selected from the* Spectator. London: Macmillan, 1899.

Hutton, Richard H. *Cardinal Newman*. London: Methuen, 1891.

Hutton, Richard H. *Criticism on Contemporary Thought and Thinkers: Selected from the* Spectator. 2 vols. London: Macmillan, 1894.

Hutton, Richard H. *Essays: Theological and Literary*. 2 vols. London: Macmillan, 1880.

Huxley, Leonard. *Life and Letters of Thomas Henry Huxley*. 2 vols. New York: D. Appleton and Co., 1901.

Huxley, Thomas Henry. *Collected Essays*. 9 vols. New York: D. Appleton and Co., 1916.

Huxley, Thomas Henry. *Essays Upon Some Controverted Questions*. New York: D. Appleton and Co., 1892.

Huxley, Thomas Henry. *Evidence as to Man's Place in Nature*. New York: D. Appleton and Co., 1863.

Huxley, Thomas Henry. *Lay Sermons, Addresses, and Reviews*. New York: D. Appleton and Co., 1870.

Jeans, William T. *Lives of the Electricians: Professors Tyndall, Wheatstone, and Morse*. London: Whittaker, 1887.

Jordan, David Starr. *College and the Man: An Address to American Youth*. Boston: American Unitarian Association, 1907.

Jordan, David Starr. *The Days of a Man: Being Memories of a Naturalist, Teacher and Minor Prophet of Democracy*. 2 vols. New York: World Book, 1922.

Jordan, David Starr. *The Foundation Ideals of Stanford University*. Stanford, CA: Stanford University Press, 1915.

Jordan, David Starr. *The Religion of a Sensible American*. Boston: American Unitarian Association, 1909.

Jordan, David Starr, *Standeth God Within the Shadow*. New York: Thomas Y. Crowell, 1901.

Jordan, Davis Starr. *The Voice of the Scholar: With Other Addresses on the Problems of Higher Education*. San Francisco: Paul Elder, 1903.

Kingsley, Mrs. Charles, ed. *Charles Kingsley, His Letters and Memories of his Life*. 2 vols. London: Macmillan, 1894.

Le Conte, Joseph. *Evolution: Its Nature, Its Evidences, and Its Relation to Religious Thought*. New York: D. Appleton and Co., 1896.

Le Conte, Joseph. *Evolution and its Relation to Religious Thought*. New York: D. Appleton and Co., 1888.

Le Conte, Joseph. *Religion and Science: A Series of Sunday Lectures on the Relation of Natural and Revealed Religion, or the Truths Revealed in Nature and Scripture*. New York: D. Appleton and Co., 1877.

Lecky, W. E. H. *History of the Rise and Influence of the Spirit of Rationalism in Europe*. 2 vols. New York: Appleton, 1872.

Lecky, W. E. H. *The Religious Tendencies of the Age*. London: Saunders, Otley, 1860.

Litchtenberger, Frédéric A. *History of German Theology in the Nineteenth Century*. Translated and edited by William Hastie. Edinburgh: T. & T. Clark, 1889.

Lyell, Charles. *Principles of Geology: Being an Inquiry How Far the Former Changes of the Earth's Surface are Referable to Causes Now in Operation*. 5th ed., 4 vols. London: John Murray, 1837.

Mach, Ernst. *The Science of Mechanics, a Critical and Historical Account of its Development*. Translated by Thomas J. McCormack. Chicago: Open Court Publishing, 1919.

Machen, J. Gresham. *Christianity and Liberalism*, new ed. Grand Rapids, MI: Eerdmans, 2009.

Mackay, Robert William. *The Progress of the Intellect, as Exemplified in the Religious Development of the Greeks and Hebrews*. 2 vols. London: John Chapman, 1850.

Maizeaux, P. Des. *An Historical and Critical Account of the Life and Writings of William Chillingworth*. London: T. Woodward, 1725.

Mather, Cotton. *The Christian Philosopher: A Collection of the Best Discoveries in Nature, with Religious Improvements*. London: Eman. Matthews, 1721.

Maurice, Frederick. *The Life of Frederick Denison Maurice, Chiefly Told in His Own Letters*. 2 vols. London: Macmillan, 1884.

McCabe, Joseph. *A Biographical Dictionary of Modern Rationalists*. London: Watts, 1920.

McCabe, Joseph. *The Conflict Between Science and Religion*. Little Blue Book No. 1211, 1927.
McCosh, James. *The Development Hypothesis: Is it Sufficient?* New York: Robert Carter and Brothers, 1876.
Mill, John Stuart. *Autobiography*. London: Longmans, 1873.
Mill, John Stuart. *Three Essays on Religion*. London: Longmans, 1874.
Miller, Samuel. *The Dangers of Education in Roman Catholic Seminaries: A Sermon, Delivered by Request, Before the Synod of Philadelphia*. Baltimore, MD: Matchett and Neilson, 1838.
Miller, Samuel. *A History of Popery, Including its Origin, Progress, Doctrines, Practice, Institutions, and Fruits, To the Commencement of the Nineteenth Century, by A Watchman*. New York: John P. Haven, 1834.
More, Henry. *An Antidote against Atheism*. London, 1655.
More, Henry. *A Collection of Several Philosophical Writings of Dr. Henry More*. London: James Flesher, 1662.
Morse, Samuel F. B. *Foreign Conspiracy Against the Liberties of the United States, The Numbers of Brutus, Originally Published in the New-York Observer*. New York: Leavitt, Lord, 1835.
Mozley, Rev. T. *Reminiscences Chiefly of Oriel College and the Oxford Movement*. Boston: Houghton, Mifflin, 1882.
Munger, Theodore T. *Freedom of Faith*. Boston: Houghton, Mifflin, 1883.
Newman, Francis William. *Phases of Faith; or, Passages from the History of My Creed*. London: John Chapman, 1850.
Newman, Francis William. *The Religious Weakness of Protestantism*. Ramsgate: Thomas Scott, 1866.
Newman, Francis William. *Thoughts on a Free and Comprehensive Christianity*. Ramsgate: Thomas Scott, 1866.
Newman, John Henry. *Apologia Pro Vita Sua: Being a History of his Religious Opinions*. London: Longmans, Green, Reader, and Dyer, 1882.
Newman, John Henry. *The Arians of the Fourth Century*. London: Longmans, Green, and Co., 1891.
Newman, John Henry. *An Essay on the Development of Christian Doctrine*. London: James Toovey, 1845.
Newman, John Henry. *The Idea of a University*. London: Basil Montagu Pickering, 1873.
Newman, John Henry. *Speech of his Eminence, Cardinal Newman: On the Reception of the* "Biglietto" *at Cardinal Howard's Palace in Rome on the 12th of May 1879*. Rome: Libreria Spithöver, 1879.
Ogden, Robert Morris, ed. *Diaries of Andrew D. White*. Ithaca, NY: Cornell University Press, 1959.

Oort, H., I. Hooykaas, and A. Kuenen, *The Bible for Learners*. 3 vols. Translated by Philip Henry Wicksteed. Boston: Roberts Brothers, 1878.

Paine, Thomas. *The Collected Writings*. Edited by Eric Foner. New York: Literary Classics of the United States, 1995.

Paine, Thomas. *The Complete Writings of Thomas Paine*. Edited by Philip S. Foner. 2 vols. New York: Citadel Press, 1945.

Parker, Theodore. *The Collected Works of Theodore Parker*. 14 vols. London: Trübner, 1876.

Parker, Theodore. *The Critical and Miscellaneous Writings of Theodore Parker*. Boston: Horace B. Fuller, 1843.

Pearson, Rev. George, ed. *Writings and Translations of Myles Coverdale*. Cambridge: The University Press, 1844.

Pfleiderer, Otto. *The Development of Theology in Germany since Kant and its Progress in Great Britain since 1825*. London: Sawn Sonneschein, 1890.

Priestley, Joseph. *A Catechism, for Children, and Young Persons*. London: J. Johnson, 1767.

Priestley, Joseph. *A General History of the Christian Church, From the Fall of the Western Empire to the Present Time*. 4 vols. Northumberland: Andrew Kennedy, 1802.

Priestley, Joseph. *Memoirs of the Rev. Dr. Joseph Priestley*. London, 1809.

Priestley, Joseph. *The Theological and Miscellaneous Works of Joseph Priestley*. Edited by John Towill Rutt. 25 vols. London: George Smallfield, 1817–32.

Proceedings at the Farewell Banquet to Professor Tyndall. New York: D. Appleton and Co., 1873.

Proceedings of the First American Congress of Liberal Religious Societies, Held at Chicago, May 22, 23, 24 & 25, 1895. Chicago: Bloch and Newman, 1894.

Proceedings at the First Annual Meeting of the Free Religious Association, Held in Boston, May 28th and 29th, 1868. Boston: Adams, 1868.

Proceedings at the Ninth Annual Meeting of the Free Religious Association, Held in Boston, June 1 and 2, 1876. Boston: Published by the Free Religious Association, 1876.

Prothero, Rowland E., and G. Bradley. *Life and Correspondence of Arthur Penrhyn Stanley*. 2 vols. London: John Murray, 1893.

Putnam, Samuel P. *400 Years of Freethought*. New York: The Truth Seeker Company, 1894.

Putnam, Samuel P. *Religion a Curse, Religion a Disease, Religion a Lie*. New York: The Truth Seeker Company, 1893.

Reiche, Charles Christopher. *Fifteen discourses on the marvelous works in nature, delivered by a father to his children: calculated to make mankind feel, in every thing, the very presence of a Supreme Being, and to influence their minds with a permanent delight in, and firm reliance upon, the directions of an almighty, all-good, and all-wise Creator, and Governor*. Philadelphia, 1791.

Riley, George. *Beauties of the Creation: or, a New Moral System of Natural History; Displayed in the Most Singular, Curious and Beautiful Quadrupeds, Birds, Insects, Trees, and Flowers. Designed to Inspire Youth with Humanity towards the Brute Creation, and bring them early acquainted with the wonderful Works of the Divine Creator*. Philadelphia, 1796.

Renan, Ernest. *Averroes et L'Averroïsme*. Paris: Auguste Durand, 1852.

Renan, Ernest. *The Life of Jesus*. London: Trübner, 1864.

Report of the Committee on Organization, Presented to the Trustees of the Cornell University. October 21st, 1866. Albany, NY: C. Van Benthuysen and Sons, 1867.

Ripley, Dorothy. *The Extraordinary Conversion, and Religious Experience of Dorothy Ripley, with Her First Voyage and Travels in America*. New York: G. and R. Waite, 1810.

Ritschl, Albrecht. *The Christian Doctrine of Justification and Reconciliation*. Translated and edited by H. R. Mackingtosh and A. B. Macaulay. Edinburgh: T. & T. Clark, 1900.

Robertson, J. M. *The Dynamics of Religion: An Essay in English Culture History*. London: Watts, 1897.

Robertson, J. M. *A History of Freethought in the Nineteenth Century*. 2 vols. London: Watts, 1929.

Robertson, J. M. *A Short History of Freethought: Ancient and Modern*. 2 vols. London: Watts, 1906.

Russell, Bertrand. *Religion and Science*. Oxford: Oxford University Press, 1997.

Russell, Bertrand. *Why I Am Not a Christian and Other Essays on Religion and Related Subjects*. London: George Allen and Unwin, 1957.

Savage, M. J. *The Evolution of Christianity*. Boston: Geo. H. Ellis, 1892.

Savage, M. J. *My Creed*. Boston: Geo. H. Ellis, 1887.

Savage, M. J. *The Religion of Evolution*. Boston: Lockwood, Brooks, 1876.

Savage, M. J. *Religious Reconstruction*. Boston: Geo. H. Ellis, 1888.

Savage, M. J. *Signs of the Times*. Boston: Geo. H. Ellis, 1889.

Schaff, Philip, and S. Irenaeus Prime, eds. *History, Essays, Orations and Other Documents of the Sixth General Conference of the Evangelical Alliance*. New York: Harper and Brothers, 1874.

Schaff, Philip. *Theological Propaedeutic: A General Introduction to the Study of Theology*. New York: Charles Scribner's Sons, 1894.

Schaff, Philip. *America: A Sketch of the Political, Social, and Religious Character of the United States of North America*. New York: Charles Scribner, 1855.

Schaff, Philip. *The Principles of Protestantism, as Related to the Present State of the Church*. Chambersburg, PA: "Publication Office" of the German Reformed Church, 1845.

Schleiermacher, Friedrich D. E. *On Religion: Speeches to its Cultured Despisers*. Translated by Richard Crouter. Cambridge: Cambridge University Press, 1988.

Seeley, John Robert. *Natural Religion*. London: Macmillan, 1882.

Sexton, George. *Scientific Materialism Calmly Considered: Being a reply to the Address Delivered before the British Association, at Belfast, on August 19th, 1874, by Professor Tyndall*. London: J. Burns, 1874.

Silliman, Benjamin. *Outline of the Course of Geological Lectures Given in Yale College*. New Haven, CT: Howe, 1829.

Slugg, J. T. *Woodhouse Grove School: Memorials and Reminiscences*. London: A. Ireland, 1885.

Smith, John. *Select Discourses Treating Theological Topics*. Cambridge: Cambridge University Press, 1859.

Smyth, Newman. *Old Faiths in New Light*. New York: Charles Scribner's Sons, 1879.

Smyth, Newman. *The Orthodox Theology of To-day*. New York: Charles Scribner's Sons, 1881.

Smyth, Newman. *The Religious Feeling: A Study for Faith*. New York: Scribner, Armstrong, 1877.

Spedding, James, ed. *The Life and Letters of Francis Bacon*. 7 vols. London: Longmans, Green, Reader, and Dyer, 1868.

Spedding, James, Robert Leslie Ellis, and Douglas Denon Heath, eds. *The Works of Francis Bacon*. 15 vols. Boston: Houghton, Mifflin, 1900.

Spencer, Herbert. *An Autobiography*. 2 vols. New York: D. Appleton and Co., 1904.

Spencer, Herbert. *First Principles of a New System of Philosophy*. New York: D. Appleton and Co., 1862.

Spencer, Herbert. *The Principles of Sociology*. 3 vols. New York: D. Appleton and Co., 1898.

Sprat, Thomas. *The History of the Royal-Society of London, For the Improving of Natural Knowledge*. London: Printed by T. R. for J. Martyn at the Bell, 1667.

Stair Douglas, Mrs. *The life and selections from the correspondence of William Whewell D.D.* London: Kegan Paul, 1881.

Stephen, Leslie. *History of English Thought in the Eighteenth Century*. 1876.

Stillingfleet, Edward. *The Works of that Eminent and most Learned Prelate, Dr. Edw. Stillingfleet*. 6 vols. London: Printed by J. Heptinstall, 1710.

Sydney, William Connor. *England and the English in the Eighteenth Century: Chapters in the Social History of the Times.* 2 vols. New York: Macmillan, 1892.

Temple, Frederick. *The Present Relations of Science to Religion: A Sermon Preached on Act Sunday, July 1, 1860 Before the University of Oxford, during the Meeting of the British Association.* J. H. & Jas. Parker, Oxford, 1860.

Priestley, Joseph. *The Theological and Miscellaneous Works of Joseph Priestley.* Edited by John Towill Rutt. 25 vols. London: George Smallfield, 1817–1832.

Toogood, John. *The Book of Nature: A Discourse on some of those Instances of the Power, Wisdom, and Goodness of God, which are Within the Reach of common Observations.* Boston, 1802.

Tulloch, John. *Movements of Religious Thought in Britain during the Nineteenth Century.* London: Longmans, Green, 1885.

Tyndall, John. *Fragments of Science: A Series of Detached Essays, Addresses, and Reviews.* 2 vols. New York: D. Appleton and Co., 1898.

Tyndall, John. *New Fragments.* New York: D. Appleton and Co., 1896.

Wace, Henry. *Christianity and Agnosticism: Reviews of Some Recent Attacks on the Christian Faith.* London: Society for Promoting Christian Knowledge, 1905.

Ward, James. *Naturalism and Agnosticism.* New York: Macmillan, 1899.

Wheeler, J. M. *A Biographical Dictionary of Freethinkers of All Ages and Nations.* London: Progressive Publishing, 1889.

Wheeler, J. M. *Frauds and Follies of the Fathers.* London: Freethought Publishing, 1882.

Whewell, William. *History of the Inductive Sciences, From the Earliest to the Present Times.* 3 vols. London: John W. Parker, 1847.

Whichcote, Benjamin. *Moral and Religious Aphorisms: Collected from the Manuscript Papers of the Reverend and Learned Doctor Whichcote.* London: J. Payne, 1703.

White, Andrew Dickson. *Autobiography.* 2 vols. New York: Century, 1905.

White, Andrew Dickson. *European Schools of History and Politics.* Baltimore, MD: N. Murray, 1887.

White, Andrew Dickson. *Evolution and Revolution: An Address Delivered at the Annual Commencement of the University of Michigan, June 26, 1890.* Ann Arbor: University of Michigan, 1890.

White, Andrew Dickson. *A History of the Warfare of Science with Theology in Christendom.* 2 vols. New York: D. Appleton and Co., 1896.

White, Andrew Dickson. *The Message of the Nineteenth Century to the Twentieth.* New Haven, CT: Tuttle, Morehouse and Taylor, 1883.

White, Andrew Dickson. *The Most Bitter Foe of Nations, and the Way to its Permanent Overthrow.* New Haven, CT: T. J. Stafford, 1866.

White, Andrew Dickson. *On Studies in General History and the History of Civilization*. New York: G. P. Putnam's Sons, 1885.

White, Andrew Dickson. *Outlines of Lectures on History, Addressed to the Students of the Cornell University*. Ithaca, NY: Cornell University Press, 1883.

White, Andrew Dickson. *Some Practical Influences of German Thought upon the United States*. Ithaca, NY: Andrus and Church, 1884.

White, Andrew Dickson. *The Warfare of Science*. New York: D. Appleton and Co., 1876.

Wilkins, John. *Of the Principles and Duties of Natural Religion*. London: J. Walthoe, 1675.

Winchell, Alexander. *The Doctrine of Evolution: Its Data, Its Principles, Its Speculations, and Its Theistic Bearings*. New York: Harper and Brothers, 1874.

Youmans, Edward L. *A Class-Book of Chemistry, In Which The Principles Of The Science Are Familiarly Explained And Applied To The Arts, Agriculture, Physiology, Dietetics, Ventilation, And The Most Important Phenomena Of Nature: Designed For The Use Of Academies And Schools, And For Popular Reading*. New York: D. Appleton and Co., 1852

Youmans, Edward L. *The Correlation and Conservation of Forces: A Series of Expositions, by Prof. Grove, Prof. Helmholtz, Dr. Mayer, Dr. Faraday, Prof. Liebig, and Dr. Carpenter, with an Introduction and Brief Biographical Notices of the Chief Promoters of the New Views*. New York: D. Appleton and Co., 1865.

Youmans, Edward L., ed. *The Culture Demanded by Modern Life; A Series of Addresses and Arguments on the Claims of Scientific Education*. New York: D. Appleton and Co., 1867.

SELECTED SECONDARY SOURCES

This bibliography does not represent a complete record of all the works and sources consulted throughout the research for and writing of this book. Included here are the sources most relevant to the discussion.

Abrams, M. H. *Natural Supernaturalism: Tradition and Revolution in Romantic Literature*. Oxford: Oxford University Press, 1971.

Adams, James Eli. *A History of Victorian Literature*. West Sussex: Wiley-Blackwell, 2009.

Ahnert, Thomas. *Religions and the Origins of the German Enlightenment: Faith and the Reform of Learning in the Thought of Christian Thomasius*. New York: University of Rochester Press, 2006.

Allison, Henry E. *Lessing and the Enlightenment*. Ann Arbor: University of Michigan Press, 1966.

Altholz, Josef L. *Anatomy of a Controversy: The Debate over "Essays and Reviews," 1860–1864*. Aldershot: Scholar Press, 1994.

Altholz, Josef L. "The Mind of Victorian Orthodoxy: Anglican Responses to 'Essays and Reviews,' 1860–1864." *Church History*, vol. 54, no. 2 (1982), 186–97.

Altick, Richard D. *The English Common Reader: A Social History of the Mass Reading Public 1800–1900*. Chicago: University of Chicago Press, 1957.

Altick, Richard D. *Victorian People and Ideas*. New York: Norton, 1973.

Altschuler, Glenn C. *Andrew D. White—Educator, Historian, Diplomat*. Ithaca, NY: Cornell University Press, 1979.

Altschuler, Glenn C. "From Religion to Ethics: Andrew D. White and the Dilemma of a Christian Rationalist." *Church History*, vol. 47, no. 3 (Sept. 1978), 308–24.

Applegate, Debbie. *The Most Famous Man in America: The Biography of Henry Ward Beecher*. New York: Doubleday, 2006.

apRoberts, Ruth. *Arnold and God*. Berkeley: University of California Press, 1983.

Ariew, Roger, and Peter Barker, trans. and eds. *Pierre Duhem: Essays in the History and Philosophy of Science*. Indianapolis, IN: Hackett Publishing Company, 1996.

Armstrong, Karen. *The Case for God*. London: Vintage, 2010.

Arx, Jeffrey Paul von. *Progress and Pessimism: Religion, Politics, and History in Late Nineteenth Century Britain*. Cambridge, MA: Harvard University Press, 1985.

Ashton, Rosemary. *142 Strand: A Radical Address in Victorian London*. London: Vintage, 2006.

Ashton, Rosemary. *The German Idea: Four English Writers and the Reception of Herman Thought, 1800–1860*. Cambridge: Cambridge University Press, 1980.

Ashton, Rosemary. *Victorian Bloomsbury*. New Haven, CT: Yale University Press, 2012.

Aston, Nigel. *Christianity and Revolutionary Europe, 1750–1830*. New York: Cambridge University Press, 2002.

Astore, William J. *Observing God: Thomas Dick, Evangelicalism, and Popular Science in Victorian Britain and America*. Aldershot: Ashgate, 2001.

Aubert, Annette G. *The German Roots of Nineteenth-Century American Theology*. Oxford: Oxford University Press, 2013.

Averill, Lloyd J. *American Theology in the Liberal Tradition*. Philadelphia: Westminster Press, 1967.

Ayer, Alfred Jules. *Language, Truth and Logic*. New York: Dover, 1952.

Baldwin, Melinda. *Making "Nature": The History of a Scientific Journal*. Chicago: University of Chicago Press, 2015.

Barnett, S. J. *The Enlightenment and Religion: The Myths of Modernity*. New York: Manchester University Press, 2003.

Barnett, S. J. *Idol Temples and Crafty Priests: The Origins of Enlightenment Anti-clericalism*. New York: St Martin's Press, 1999.

Barth, Karl. *Protestant Theology in the Nineteenth Century: Its Background and History*. Grand Rapids, MI: Eerdmans, 2002.

Barton, Ruth. *The X Club: Power and Authority in Victorian Science*. Chicago: University of Chicago Press, 2018.

Bebbington, David. *Victorian Religious Revivals: Culture and Piety in Local and Global Contexts*. Oxford: Oxford University Press, 2012.

Becker, Carl L. *Cornell University: Founders and the Founding*. Ithaca, NY: Cornell University Press, 1943.

Beer, Gillian. *Darwin's Plots: Evolutionary Narrative in Darwin, George Eliot, and Nineteenth-Century Fiction*. Cambridge: Cambridge University Press, 2000.

Beiser, Frederick C. *Enlightenment, Revolution and Romanticism: The Genesis of Modern German Political Thought, 1790–1800*. Cambridge, MA: Harvard University Press, 1992.

Beiser, Frederick C. *The Romantic Imperative: The Concept of Early German Romanticism*. Cambridge, MA: Harvard University Press, 2003.

Bellot, H. Hale. *University College London 1826–1926*. London: University of London Press, 1929.

Bentley, Michael. *Modernizing England's Past: English Historiography in the Age of Modernism, 1870–1970*. Cambridge: Cambridge University Press, 2005.

Benz, Ernst. *The Mystical Sources of German Romantic Philosophy*. Translated by Blair Reynolds and Eunice M. Pauland. Allison Park, PA: Pickwick Publications, 1983.

Berry, R. J., ed. *The Lion Handbook of Science and Christianity*. Oxford: Lion Handbooks, 2012.

Bibby, Cyril. *T. H. Huxley: Scientist, Humanist and Educator*. New York: Horizon Press, 1960.

Billington, Ray Allen. *The Protestant Crusade, 1800–1880: A Study of the Origins of American Nativism*. New York: Macmillan, 1938.

Bishop, Morris. *A History of Cornell*. Ithaca, NY: Cornell University Press, 1962.

Bourdeau, Michel, Mary Pickering, and Warren Schmaus, eds. *Love, Order, and Progress: The Science, Philosophy, and Politics of Auguste Comte*. Pittsburgh, PA: University of Pittsburgh Press, 2018.

Bowers, J. D. *Joseph Priestley and English Unitarianism in America*. University Park: Pennsylvania State University Press, 2007.

Bowler, Peter J. *Evolution: The History of an Idea*. Berkeley: University of California Press, 2009.

Bowler, Peter J. *The Invention of Progress: The Victorians and the Past*. Oxford: Basil Blackwell Ltd., 1989.

Bowler, Peter J. *Reconciling Science and Religion: The Debate in Early-Twentieth-Century Britain*. Chicago: University of Chicago Press, 2001.

Bowler, Peter J., and Iwan Rhys Morus. *Making Modern Science: A Historical Survey*. Chicago and London: University of Chicago Press, 2005.

Boyer, John W. *"Broad and Christian in the Fullest Sense": William Rainey Harper and the University of Chicago*. Chicago: University of Chicago Press, 2006.

Bozeman, Theodore Dwight. *Protestants in an Age of Science: The Baconian Ideal and Antebellum American Religious Thought*. Chapel Hill: University of North Carolina Press, 1977.

Brady, Ciaran. *James Anthony Froude: An Intellectual Biography of a Victorian Prophet*. Oxford: Oxford University Press, 2013.

Brasch, Frederick E. "List of Foundation Members of the History of Science Society." *Isis*, vol. 7, no. 3 (1925), 371–93.

Brasch Frederick E., and Lavada Hudgens. "The History of Science Society and the David Eugene Smith Festschrift." *Science*, New Series, vol. 83, no. 2158 (1936), 424–26.

Bready, J. Wesley. *England before and after Wesley: The Evangelical Revival and Social Reform*. New York: Russell and Russell, 1971.

Breckman, Warren. *European Romanticism: A Brief History with Documents*. Boston: Bedford-St. Martins, 2008.

Breisach, Ernst. *Historiography: Ancient, Medieval, and Modern*. 3rd ed. Chicago: University of Chicago Press, 2007.

Bremner, G. A., and Jonathan Conlin, eds. *Making History: Edward Augustus Freeman and Victorian Cultural Politics*. Oxford: Oxford University Press, 2015.

Brock, W. H., N. D. McMillan, and R. C. Mollan, eds. *John Tyndall: Essays on a Natural Philosopher*. Dublin: Royal Dublin Society, 1981.

Brock, W. H., and R. M. Macleod. "The Scientists' Declaration: Reflexions on Science and Belief in the Wake of *Essays and Reviews*, 1864–5." *British Journal for the History of Science*, vol. 9, no. 1 (1976), 39–66.

Brockman, John, ed. *What is Your Dangerous Idea?* New York: Harper Perennial, 2007.

Brooke, John Hedley. "Natural Theology and the Plurality of Worlds: Observations on the Brewster-Whewell Debate." *Annals of Science*, 34 (1977), 221–86.

Brooke, John Hedley. "Presidential Address: Does the History of Science have a Future?" *British Journal for the History of Science*, vol. 32, no. 1 (1999): 1–20.

Brooke, John Hedley. *Science and Religion: Some Historical Perspectives*. Cambridge: Cambridge University Press, 1991.

Brooke, John Hedley, and Geoffrey Cantor. *Reconstructing Nature: The Engagement of Science and Religion: Glasgow Clifford Lectures*. Edinburgh: T and T Clark, 1998.

Brooke, John Hedley, and Ian Maclean, eds. *Heterodoxy in Early Modern Science and Religion*. New York: Oxford University Press, 2005.

Brooke, John Hedley, and Ronald L. Numbers, eds. *Science and Religion around the World*. Oxford: Oxford University Press, 2011.

Brooke, John Hedley, Margaret J. Osler, and Jitse Van der Meer, eds. *Science in Theistic Contexts: Cognitive Dimensions*. Chicago: University of Chicago Press, 2001.

Brown, Alan Willard. *The Metaphysical Society: Victorian Minds in Crisis, 1869–1880*. New York: Columbia University Press, 1947.

Brown, Jerry Wayne. *The Rise of Biblical Criticism in America, 1800–1870: The New England Scholars*. Middletown, CT: Wesleyan University Press, 1969.

Buckley, Jerome Hamilton. *The Triumph of Time: A Study of the Victorian Concepts of Time, History, Progress, and Decadence*. Cambridge, MA: Belknap Press of Harvard University Press, 1966.

Buckley, Michael J. *At the Origins of Modern Atheism*. New Haven, CT: Yale University Press, 1987.

Budd, Susan. *Varieties of Unbelief: Atheists and Agnostics in English Society 1850–1960*. New York: Holmes and Meier, 1977.

Burns, Edward McNall. *David Starr Jordan: Prophet of Freedom*. Stanford, CA: Stanford University Press, 1953.

Burrow, John W. *The Crisis of Reason: European Thought, 1848–1914*. New Haven, CT: Yale University Press, 2000.

Burrow, John W. *Evolution and Society: A Study of Victorian Social Theory*. Cambridge: Cambridge University Press, 1966.

Burrow, John W. *A Liberal Descent: Victorian Historians and the English Past*. Cambridge: Cambridge University Press, 1981.

Burrow, John W. *Whigs and Liberals: Continuity and Change in English Political Thought*. Oxford: Oxford University Press, 1988.

Burtt, E. A. *Metaphysical Foundations of Modern Physical Science: A Historical and Critical Essay*. New York: Kegan Paul, Trench, Trübner, 1925.

Bury, John B. *The Idea of Progress: An Inquiry into its Origins and Growth*. London: Macmillan, 1920.

Butler, Lance St. John. *Victorian Doubt: Literary and Cultural Discourses*. New York: Harvester and Wheatsheaf, 1990.

Butterfield, Herbert. *The Origins of Modern Science: 1300–1800*. New York: Macmillan, 1958.

Butterfield, Herbert. *The Whig Interpretation of History*. New York: Norton, 1965.

Byrne, Peter. *Natural Religion and the Nature of Religion: The Legacy of Deism*. New York: Routledge, 1989.

Cahan, David, ed. *From Natural Philosophy to the Sciences: Writing the History of Nineteenth-Century Science*. Chicago: University of Chicago Press, 2003.

Calab, Amanda Mordavsky, ed. *(Re)Creating Science in Nineteenth-Century Britain*. Newcastle: Cambridge Scholars Publishing, 2007.

Calloway, Katherine. *Natural Theology in the Scientific Revolution*. London: Pickering and Chatto, 2014.

Cannon, Susan Faye. *Science in Culture: The Early Victorian Period*. New York: Science History Publications, 1978.

Cannon, Walter F. "John Herschel and the Idea of Science." *Journal of the History of Ideas*, vol. 22, no. 2 (1961), 215–39.

Cannon, Walter F. "The Problem of Miracles in the 1830s." *Victorian Studies*, vol. 4, no. 1 (1960), 4–32.

Cannon, Walter F. "Scientists and Broad Churchmen: An Early Victorian Intellectual Network." *Journal of British Studies* (1964), 65–88.

Cantor, Geoffrey. *Religion and the Great Exhibition of 1851*. New York: Oxford University Press, 2011.

Cantor, Geoffrey. "Science, Providence, and Progress at the Great Exhibition." *Isis*, vol. 103, no. 3 (Sept. 2012), 439–59.

Cantor, Geoffrey, Gowan Dawson, Graeme Gooday, Richard Noakes, Sally Shuttleworth, and Jonathan R. Topham, eds. *Science in the Nineteenth-Century Periodical: Reading the Magazine of Nature*. Cambridge: Cambridge University Press, 2004.

Cantor, Geoffrey, and Sally Shuttleworth, eds. *Science Serialized: Representation of the Sciences in Nineteenth-Century Periodicals*. Cambridge, MA: MIT Press, 2004.

Canuel, Mark. *Religion, Toleration, and British Writing, 1790–1830*. Cambridge: Cambridge University Press, 2002.

Carey, Hilary M., and John Gascoigne, eds. *Church and State in Old and New Worlds*. Leiden: Brill, 2011.

Carter, Paul A. *The Spiritual Crisis of the Gilded Age*. DeKalb: Northern Illinois University Press, 1971.

Cashdollar, Charles D. *The Transformation of Theology, 1830–1890: Positivism and Protestant Thought in Britain and America*. Princeton, NJ: Princeton University Press, 2014.

Catto, Jeremy, ed. *Oriel College: A History*. Oxford: Oxford University Press, 2013.

Cauthen, Kenneth. *The Impact of American Religious Liberalism*. New York: Harper and Row, 1962.

Chadwick, Henry. *Lessing's Theological Writings: Selections in Translation with an Introductory Essay*. Stanford, CA: Stanford University Press, 1957.

Chadwick, Owen. "Gibbon and the Church Historians." *Daedalus*, vol. 105, no. 3 (1976), 111–23.

Chadwick, Owen. *The Secularization of the European Mind in the 19th Century*. Cambridge: Cambridge University Press, 1975.

Chadwick, Owen. *The Victorian Church*. 2 vols. London: Adam and Charles Black, 1966–70.

Champion, Justin. *Pillars of Priestcraft Shaken: The Church of England and Its Enemies, 1660–1730*. Cambridge: Cambridge University Press, 1992.

Cherry, Conrad. *Hurrying Toward Zion: Universities, Divinity Schools, and American Protestantism*. Bloomington: Indiana University Press, 1995.

Chew, Samuel C. *Fruit Among the Leaves*. New York: Appleton-Century-Crofts, 1950.

Christie, John, and Sally Shuttleworth, eds. *Nature Transfigured: Science and Literature, 1700–1900*. Manchester: Manchester University Press, 1989.

Clagett, Marshall. *Greek Science in Antiquity*. London: Abelard-Schuman, 1957.

Clagett, Marshall. *The Science of Mechanics in the Middle Ages*. Madison: University of Wisconsin Press, 1959.

Clark, Christopher, and Wolfram Kaiser. *Culture Wars: Secular-Catholic Conflict in Nineteenth-Century Europe*. Cambridge: Cambridge University Press, 2003.

Clark, Elizabeth A. *Founding the Fathers: Early Church History and Protestant Professors in Nineteenth-Century America*. Philadelphia: University of Pennsylvania Press, 2011.

Clark, George Kitson. *Churchmen and the Condition of England*. London: Methuen, 1973.

Clark, J. C. D. *English Society 1660–1832: Religion, Ideology and Politics during the Ancient Regime*. New York: Cambridge University Press, 2000.

Clark, J. C. D. "Predestination and Progress: Or, Did the Enlightenment Fail?" *Albion*, vol. 35, no. 4 (Winter 2003), 559–89.

Clark, J. C. D. *Revolution and Rebellion: State and Society in England in the Seventeenth and Eighteenth Centuries*. Cambridge: Cambridge University Press, 1986.

Coffey, John, and Paul C. H. Lim, eds. *The Cambridge Companion to Puritanism*. Cambridge: Cambridge University Press, 2008.

Cogliano, Francis. *No King, No Popery: Anti-Catholicism in Revolutionary New England*. Westport, CT: Greenwood, 1995.

Cohen, I. Bernard. "The Eighteenth-Century Origins of the Concept of Scientific Revolution." *Journal of the History of Ideas*, vol. 37, no. 2 (1976), 257–88.

Cohen, I. Bernard. "George Sarton." *Isis*, vol. 48, no. 3 (1957), 286–300.

Cohen, I. Bernard. "The Isis Crises and the Coming of Age of the History of Science Society." *Isis*, vol. 90, Suppl. (1999), S28–S42.

Cohen, I. Bernard. *Revolution in Science*. Cambridge, MA: Belknap Press of Harvard University Press, 1985.

Cohn, Norman. *The Pursuit of the Millennium: Revised and Expanded Edition*. Oxford: Oxford University Press, 1970.

Colish, Marcia L. *Medieval Foundations of the Western Intellectual Traditions, 400–1400*. New Haven, CT: Yale University Press, 2002.

Collinson, Patrick. *The Elizabethan Puritan Movement*. Oxford: Clarendon Press, 1967.

Conant, James B. "George Sarton and Harvard University." *Isis*, vol. 48, no. 3 (1957), 301–5.

Condren, Conal, Stephen Gaukroger, and Ian Hunter, eds. *The Philosophers in Early Modern Europe: The Nature of a Contested Identity*. New York: Cambridge University Press, 2006.

Conser, Jr., Walter H. *God and the Natural World: Religion and Science in Antebellum America*. Columbia: University of South Carolina Press, 1993.

Cooke, Bill. *The Gathering of Infidels: A Hundred Years of the Rationalist Press Association*. Amherst, NY: Prometheus Books, 2004.

Cooke, Bill. "Joseph McCabe: A Forgotten Early Populariser of Science and Defender of Evolution." *Science and Education*, 19 (2010), 461–64.

Copleston, Frederick. *A History of Philosophy*. 9 vols. New York: Doubleday, 1994.

Corsi, Pietro. *Science and Religion: Baden Powell and the Anglican Debate, 1800–1860*. Cambridge: Cambridge University Press, 1988.

Cosslett, Tess. *Science and Religion in the Nineteenth Century*. Cambridge: Cambridge University Press, 1984.

Cosslett, Tess. *The 'Scientific Movement' and Victorian Literature*. New York: St. Martin's Press, 1982.

Coulson, John. *Religion and the Imagination*. Oxford: Clarendon Press, 1981.

Cragg, Gerald R. *The Church and the Age of Reason, 1648–1789*. London: Penguin, 1990.

Crane, Ronald S. "Anglican Apologetics and the Idea of Progress: 1699–1745." *Modern Philology*, vol. 31, nos. 3 and 4 (1934), 273–306, 349–82.

Crombie, A. C. "The Appreciation of Ancient and Medieval Science" and "Six Wings." *British Journal of the Philosophy of Science*, vol. 10 (1959), 164–65.

Crowther, M. A. *Church Embattled: Religious Controversy in Mid-Victorian England*. London: David and Charles, 1970.

Dale, Richard, ed. *The Scientific Achievement of the Middle Ages*. Philadelphia: University of Pennsylvania Press, 1973.

Daniels, George H. *American Science in the Age of Jackson*. Tuscaloosa: University of Alabama Press, 1994.

David, Deirdre, ed. *The Cambridge Companion to the Victorian Novel.* New York: Cambridge University Press, 2012.

Davis, Edward B. "Newton's Rejection of the 'Newtonian World View': The Role of Divine Will in Newton's Natural Philosophy." *Science and Christian Belief,* 3, no. 1 (1991), 103–117.

Davis, Garold N. *German Thought and Culture in England, 1700–1770.* Chapel Hill: University of North Carolina Press, 1969.

Davis, John R. *The Victorians and Germany.* Oxford: Peter Lang, 2007.

Davis, Philip. *The Victorians.* Oxford: Oxford University Press, 2002.

Dawson, Christopher. *Progress and Religion: An Historical Inquiry.* Washington, DC: Catholic University of America Press, 2001 [1929].

Dawson, Gowan. *Darwin, Literature and Victorian Respectability.* Cambridge: Cambridge University Press, 2007.

Dawson, Gowan. "Literature and Science under the Microscope." *Journal of Victorian Culture,* vol. 11, no. 2 (Autumn 2006), 301–15.

Dawson, Gowan, and Bernard Lightman, eds. *Victorian Scientific Naturalism: Community, Identity, Continuity.* Chicago: University of Chicago Press, 2014.

Dawson, Gowan, and Jonathan R. Topham. "Science in the Nineteenth-Century Periodical." *Literature Compass,* 1 (2004), 1–11.

Day, Aidan. *Romanticism.* New York: Routledge, 1996.

De Certeau, Michel. *The Writing of History.* Translated by Tom Conley. New York: Columbia University Press, 1988.

Desmond, Adrian. *Archetypes and Ancestors: Paleontology in Victorian London, 1850–1875.* Chicago: The University of Chicago Press, 1984.

Desmond, Adrian. *Huxley: From Devils' Disciple to Evolution's High Priest.* Reading, MA: Addison-Wesley, 1997.

Desmond, Adrian. *The Politics of Evolution: Morphology, Medicine, and Reform in Radical London.* Chicago and London: University of Chicago Press, 1989.

Desmond, Adrian. "Redefining the X Axis: 'Professionals,' 'Amateurs,' and the Making of Mid-Victorian Biology—A Progress Report." *Journal of the History of Biology,* 34, (2001), 3–50.

Desmond, Adrian, and James Moore. *Darwin.* London: Michael Joseph, 1991.

DeWitt, Anne. *Moral Authority, Men of Science, and the Victorian Novel.* Cambridge: Cambridge University Press, 2013.

Dickens, A. G., and John M. Tonkin. *The Reformation in Historical Thought.* Cambridge, MA: Harvard University Press, 1985.

Diehl, Carl. *Americans and German Scholarship 1770–1870.* New Haven, CT: Yale University Press, 1978.

Dillenberger, John. *Protestant Thought and Natural Science: A Historical Interpretation.* London: Collins, 1961.

Dixon, Thomas, Geoffrey Cantor, and Stephen Pumfrey, eds. *Science and Religion: New Historical Perspectives*. Cambridge: Cambridge University Press, 2010.

Dixon, Thomas. *Science and Religion: A Very Short Introduction*. Oxford: Oxford University Press, 2008.

Dodds, Gregory D. "An Accidental Historian: Erasmus and the English History of the Reformation." *Church History*, vol. 82, no. 2 (2013), 273–92.

Dorf, Philip. *The Builder: A Biography of Ezra Cornell*. New York: Macmillan, 1952.

Dorrien, Gary. *Kantian Reason and Hegelian Spirit: The Idealistic Logic of Modern Theology*. West Sussex: Wiley-Blackwell, 2012.

Dorrien, Gary. *The Making of American Liberal Theology*. 3 vols. Louisville, KY: Westminster John Knox Press, 2001–06.

Duhem, Pierre. *The Origins of Statics: The Sources of Physical Theory*, Translated by Grant F. Leneaux et al. Boston: Springer Science, 1991.

Dupree, A. Hunter. *Asa Gray: American Botanist, Friend of Darwin*. Baltimore, MD: Johns Hopkins University Press, 1988.

Dupree, A. Hunter. *Science in the Federal Government: A History of Policies and Activities*. Baltimore, MD: Johns Hopkins University Press, 1986.

Dykema, Peter A., and Heiko A. Oberman, eds. *Anticlericalism in Late Medieval and Early Modern Europe*. Leiden: E.J. Brill, 1994.

Eisely, Loren C. *Darwin's Century: Evolution and the Men who Discovered It*. London: Gollancz, 1958.

Eisen, Sydney. "Frederic Harrison and Herbert Spencer: Embattled Unbelievers." *Victorian Studies*, vol. 12, no. 1 (1968), 33–56.

Eisen, Sydney, and Bernard V. Lightman, eds. *Victorian Science and Religion: A Bibliography with Emphasis on Evolution, Belief, and Unbelief, Comprised of Works Published from c. 1900–1975*. Hamden, CT: Archon Books, 1984.

Elder, Gregory P. *Chronic Vigour: Darwin, Anglicans, Catholics, and the Development of a Doctrine of Providential Evolution*. New York: University Press of America, 1996.

Ellegård, Alvar. *Darwin and the General Reader: The Reception of Darwin's Theory of Evolution in the British Periodical Press, 1859–1872*. Chicago and London: University of Chicago Press, 1990.

Ellegård, Alvar. *The Readership of the Periodical Press in Mid-Victorian Britain*. Göteborg: Almqvist and Wiksell Stockholm, 1957.

Elliott, Paul. "Erasmus Darwin, Herbert Spencer, and the Origins of the Evolutionary Worldview in British Provincial Scientific Culture, 1770–1850," *Isis*, vol. 94, no. 1 (2003), 1–29.

Ellis, Ieuan. *Seven against Christ: A Study of "Essays and Reviews."* Leiden: Brill, 1980.

Ellison, Robert H., ed. *A New History of the Sermon: The Nineteenth Century*. Leiden: Brill, 2010.

Ellison, Robert H. *The Victorian Pulpit: Spoken and Written Sermons in Nineteenth-Century Britain*. Cranbury, NJ: Susquehanna University Press, 1998.

Engell, James. *The Creative Imagination: Enlightenment to Romanticism*. Cambridge, MA: Harvard University Press, 1981.

Erdozain, Dominic. *The Soul of Doubt: The Religious Roots of Unbelief from Luther to Marx*. Oxford: Oxford University Press, 2015.

Evans, Christopher H. *Liberalism without Illusions: Renewing an American Christian Tradition*. Waco, TX: Baylor University Press, 2010.

Evan, John H., and Michael S. Evans. "Religion and Science: Beyond the Epistemological Conflict Narrative." *Annual Review of Sociology* 34, no. 5 (2008): 1–19.

Fea, John. *Was America Founded as a Christian Nation? A Historical Introduction*. Louisville, KY: Westminster John Knox Press, 2016.

Ferber, Michael. *Romanticism: A Very Short Introduction*. Oxford: Oxford University Press, 2010.

Fergusson, David, ed. *The Blackwell Companion to Nineteenth-Century Theology*. Malden, MA: Blackwell, 2010.

Ferngren, Gary B., ed. *The History of Science and Religion in the Western Tradition: An Encyclopedia*. New York: Garland, 2000.

Ferngren, Gary B., ed. *Science and Religion: A Historical Introduction*. Baltimore, MD: Johns Hopkins University Press, 2002.

Fisch, Menachem, and Simon Schaffer, eds. *William Whewell: A Composite Portrait*. Oxford: Clarendon Press, 1991.

Fleming, Donald. *John William Draper and the Religion of Science*. Philadelphia: University of Pennsylvania Press, 1950.

"Focus: 100 Volumes of *Isis:* The Vision of George Sarton." *Isis,* vol. 100, no. 1 (2009), 58–107.

Forbes, Duncan. *The Liberal Anglican Idea of History*. Cambridge: Cambridge University Press, 1952.

Forstman, Jack. *A Romantic Triangle: Schleiermacher and Early German Romanticism*. Missoula, MT: Scholars Press, 1977.

Foster, Michael B. "The Christian Doctrine of Creation and the Rise of Modern Natural Science." *Mind,* vol. 43, no. 172 (1934), 446–68.

Foster, Michael B. "Christian Theology and Modern Science of Nature, Part 1." *Mind,* vol. 44, no. 176 (1935), 439–66.

Foster, Michael B. "Christian Theology and Modern Science of Nature, Part 2," *Mind,* vol. 45, no. 177 (1936): 1–27.

Fought, C. Brad. *The Oxford Movement: A Thematic History of the Tractarians and Their Times*. University Park: Pennsylvania State University Press, 2003.

Francis, Mark. *Herbert Spencer and the Invention of Modern Life*. Ithaca, NY: Cornell University Press, 2007.

Francis, Mark, and Michael W. Taylor, eds. *Herbert Spencer Legacies*. London: Routledge, 2015.

Frängsmyr, Tore. "Science of History: George Sarton and the Positivist Tradition in the History of Science." *Lychnos*, 74 (1973–74), 104–44.

Fraser, Liam Jerrold. *Atheism, Fundamentalism and the Protestant Reformation: Uncovering the Secret Sympathy*. Cambridge: Cambridge University Press, 2018.

Fuller, Robert. *Religious Revolutionaries: The Rebels Who Reshaped American Religion*. New York: Palgrave Macmillan, 2004.

Funkenstein, Amos. *Theology and the Scientific Imagination: From the Middle Ages to the Seventeenth Century*. Princeton, NJ: Princeton University Press, 1986.

Fyfe, Aileen. *Science and Salvation: Evangelical Popular Science Publishing in Victorian Britain*. Chicago: University of Chicago Press, 2004.

Fyfe, Aileen. *Steam-Powered Knowledge: William Chambers and the Business of Publishing, 1820–1860*. Chicago: University of Chicago Press, 2012.

Fyfe, Aileen, and Bernard Lightman, eds. *Science in the Marketplace: Nineteenth-Century Sites and Experiences*. Chicago: University of Chicago Press, 2007.

Gascoigne, John. *Cambridge in the Age of the Enlightenment: Science, Religion, and Politics from the Restoration to the French Revolution*. Cambridge: Cambridge University Press, 1989.

Gascoigne, John. "From Bentley to the Victorians: The Rise and Fall of British Newtonian Natural Theology." *Science in Context*, vol. 2, no. 2 (1988), 219–56.

Gaukroger, Stephen. *The Collapse of Mechanism and the Rise of Sensibility: Science and the Shaping of Modernity, 1680–1760*. Oxford: Clarendon Press, 2011.

Gaukroger, Stephen. *The Emergence of a Scientific Culture: Science and the Shaping of Modernity, 1210–1685*. Oxford: Clarendon Press, 2006.

Gaukroger, Stephen. *Francis Bacon and the Transformation of Early-Modern Philosophy*. Cambridge: Cambridge University Press, 2001.

Gaukroger, Stephen. *The Natural and the Human: Science and the Shaping of Modernity, 1739–1841*. Oxford: Clarendon Press, 2015.

Gay, Peter. *The Enlightenment*. 2 vols. New York and London: Norton, 1995.

"The George Sarton Memorial Issue." *Isis*, vol. 48, no. 3 (1957), 281–350.

Gillispie, Charles Coulston. *Genesis and Geology: A Study in the Relations of Scientific Thought, Natural Theology, and Social Opinion in Great Britain, 1790–1850*. Cambridge, MA: Harvard University Press, 1996.

Gillespie, Michael Allen. *The Theological Origins of Modernity*. Chicago: University of Chicago Press, 2008.

Gillespie, Sarah Kate. "John William Draper and the Reception of Early Scientific Photography." *History of Photography*, vol. 36, no. 3 (2012), 241-254.

Gjerde, Jon. *Catholicism and the Shaping of Nineteenth-Century America*. New York: Cambridge University Press, 2012.

Glendening, John. *The Evolutionary Imagination in Late-Victorian Novels*. Aldershot: Ashgate, 2007.

Gohdes, Clarence L. F. *The Periodicals of American Transcendentalism*. Durham, NC: Duke University Press, 1931.

Golinski, Jan. *The Experimental Self: Humphry Davy and the Making of a Man of Science*. Chicago: University of Chicago Press, 2016.

Gooch, G. P. *History and Historians in the Nineteenth Century*. London: Longmans, Green, 1920.

Good, H.G. "Edward L. Youmans: A National Teacher of Science." *The Scientific Monthly*, vol. 18, no. 3 (March 1924), 306–17.

Goodspeed, Thomas Wakefield. *A History of the University of Chicago*. Chicago: University of Chicago Press, 1916.

Gordon, Bruce, ed. *Protestant History and Identity in Sixteenth-Century Europe*. 2 vols. Aldershot: Ashgate, 1996.

Gould, F. J. *The Pioneers of Johnson's Court: A History of the Rationalist Press Association from 1899 Onwards*. London: Watts, 1929.

Grafton, Anthony. *Worlds Made by Words: Scholarship and Community in the Modern West*. Cambridge, MA: Harvard University Press, 2009.

Graham, Loren, Wolf Lepenies, and P. Weingart, eds. *Functions and Uses of Disciplinary Histories*. Dordrecht: D. Reidel, 1983.

Grant, Edward. *The Foundation of Modern Science in the Middle Ages: Their Religious, Institutional, and Intellectual Contexts*. Cambridge: Cambridge University Press, 1996.

Grant, Edward. *A History of Natural Philosophy: From the Ancient World to the Nineteenth Century*. Cambridge: Cambridge University Press, 2007.

Grant, Edward. *Science & Religion, 400 B.C.–A.D. 1550: From Aristotle to Copernicus*. Baltimore, MD: Johns Hopkins University Press, 2004.

Grasso, Christopher. *Skepticism and American Faith: From the Revolution to the Civil War*. Oxford: Oxford University Press, 2018.

Greeley, Andrew. *Unsecular Man: The Persistence of Religion*. New York: Schocken, 1972.

Greene, John C. *The Death of Adam: Evolution and Its Impact on Western Thought*. Ames: Iowa State University Press, 1959.

Greene, John C. *Science, Ideology, and Worldview: Essays in the History of Evolutionary Ideas*. Berkeley: University of California Press, 1981.

Green, John C. *American Science in the Age of Jefferson*. Ames: Iowa State University Press, 1984.

Greenwood, Andrea, and Mark W. Harris. *An Introduction to the Unitarian and Universalist Traditions*. Cambridge: Cambridge University Press, 2011.

Gregory, Brad S. *The Unintended Reformation: How a Religious Revolution Secularized Society*. Cambridge, MA: Belknap Press of Harvard University Press, 2012.

Gregory, Frederick. *Nature Lost? Natural Science and the German Theological Traditions of the Nineteenth Century*. Cambridge, MA: Harvard University Press, 1992.

Gregory, Frederick. *Scientific Materialism in Nineteenth-Century Germany*. Dordrecht: Reidel, 1977.

Griffin, Martin I. J. *Latitudinarianism in the Seventeenth-Century Church of England*. Leiden: Brill, 1992.

Griffin, Susan M. *Anti-Catholicism and Nineteenth-Century Fiction*. New York: Cambridge University Press, 2004.

Guralnick, Stanley M. *Science and the Ante-Bellum American College*. Philadelphia: American Philosophical Society, 1975.

Haakonssen, Knud, ed. *Enlightenment and Religion: Rational Dissent in Eighteenth-Century Britain*. New York: Cambridge University Press, 1996.

Haar, Charles M. "E.L. Youmans: A Chapter in the Diffusion of Science in America." *Journal of the History of Ideas*, vol. 9, no. 2 (1948), 193–213.

Hadot, Pierre. *Philosophy as a Way of Life*. Translated by Arnold I. Davidson. Malden, MA: Blackwell, 1995.

Hadot, Pierre. *The Veil of Isis: An Essay on the History of the Idea of Nature*. Translated by Michael Chase. Cambridge, MA: Harvard University Press, 2006.

Hadot, Pierre. *What is Ancient Philosophy?* Translated by Michael Chase. Cambridge, MA: Harvard University Press, 2002.

Hale, Piers J. "Darwin's Other Bulldog: Charles Kingsley and the Popularisation of Evolution in Victorian England." *Science & Education*, vol. 21, no. 7 (2012), 977–1013.

Halévy, Élie. *A History of the English People in the Nineteenth Century*. Translated by E. I. Watkin. 6 vols. London: Ernest Benn, Ltd., 1949–52.

Hardin, Jeff, Ronald L. Numbers, and Ronald A. Binzley, eds. *The Warfare between Science & Religion: The Idea that Wouldn't Die*. Baltimore, MD: Johns Hopkins University Press, 2018.

Harp, Gillis J. *Positivist Republic: Auguste Comte and the Reconstruction of Amer-*

ican Liberalism, 1865–1920. University Park: Pennsylvania State University Press, 1995.

Harris, Sam. *The End of Faith: Religion, Terror, and the Future of Reason*. New York: Norton, 2004.

Harris, Sam. *Letter to a Christian Nation*. New York: Knopf, 2006.

Harrison, Peter. *The Bible, Protestantism, and the Rise of Natural Science*. Cambridge: Cambridge University Press, 1998.

Harrison, Peter, ed. *The Cambridge Companion to Science and Religion*. Cambridge: Cambridge University Press, 2010.

Harrison, Peter. *The Fall of Man and the Foundations of Science*. Cambridge: Cambridge University Press, 2007.

Harrison, Peter. *'Religion' and the Religions in the English Enlightenment*. Cambridge: Cambridge University Press, 1990.

Harrison, Peter. "'Science' and 'Religion': Constructing the Boundaries." *Journal of Religion*, 86 (2006), 81–106.

Harrison, Peter. "Sentiments of Devotion and Experimental Philosophy in Seventeenth-Century England." *Journal of Medieval and Early Modern Studies*, vol. 44, no. 1 (2014), 113–33.

Harrison, Peter. *The Territories of Science and Religion*. Chicago: University of Chicago Press, 2015.

Harrison, Peter, and Jon H. Roberts, eds. *Science Without God? Rethinking the History of Scientific Naturalism*. Oxford: Oxford University Press, 2019.

Harrison, Peter, Ronald L. Numbers, and Michael H. Shank, eds. *Wrestling with Nature: From Omens to Science*. Chicago: University of Chicago Press, 2011.

Hart, Darryl G. *Defending the Faith: J. Gresham Machen and the Crisis of Conservative Protestantism in Modern America*. Baltimore, MD: Johns Hopkins University Press, 1994.

Hart, David Bentley. *Atheist Delusions: The Christian Revolution and Its Fashionable Enemies*. New Haven, CT: Yale University Press, 2009.

Harte, Negley. *The University of London, 1836–1986*. London: Athlone Press, 1986.

Haskins, Charles H. *The Renaissance of the Twelfth Century*. Cambridge, MA: Harvard University Press, 1927.

Haskins, Charles H. *Studies in the History of Medieval Science*. Cambridge, MA: Harvard University Press, 1924.

Haught, John F. *God and the New Atheism: A Critical Response to Dawkins, Harris, and Hitchens*. Louisville, KY: Westminster John Knox Press, 2008.

Haydon, Colin. *Anti-Catholicism in Eighteenth-Century England, c. 1714–80: A Political and Social Study*. Manchester: Manchester University Press, 1993.

Headley, John M. *Luther's View of Church History*. New Haven, CT: Yale University Press, 1963.

Hearnshaw, F. J. C. *The Centenary History of King's College London, 1828–1928*. London: George G. Harrap, 1926.

Helmstadter, Richard, ed. *Freedom and Religion in the Nineteenth Century*. Stanford, CA: Stanford University Press, 1997.

Helmstadter, Richard J., and Bernard Lightman, eds. *Victorian Faith in Crisis: Essays on Continuity and Change in Nineteenth-Century Religious Belief*. London: Macmillan, 1990.

Hempton, David. *Evangelical Disenchantment: Nine Portraits of Faith and Doubt*. New Haven, CT: Yale University Press, 2008.

Henry, John. *Knowledge is Power: How Magic, the Government and an Apocalyptic Vision Inspired Francis Bacon to Create Modern Science*. London: Icon Books, 2002.

Henry, John. *The Scientific Revolution and the Origins of Modern Science*. New York: Palgrave Macmillan, 2008.

Henry, John. *A Short History of Scientific Thought*. New York: Palgrave Macmillan, 2012.

Henson, Louise, Geffrey Cantor, Gowan Dawson, Richard Noakes, Sally Shuttleworth, and Jonathan R. Topham, eds. *Culture and Science in the Nineteenth-Century Media*. Aldershot: Ashgate, 2004.

Herbst, Jurgen. *The German Historical School in American Scholarship: A Study in the Transfer of Culture*. Ithaca, NY: Cornell University Press, 1965.

Herder, Dale M. "American Values and Popular Culture in the Twenties: The Little Blue Books." *Historical Papers*, vol. 6, no. 1 (1971), 289–99.

Hesketh, Ian. "Behold the (Anonymous) Man: J. R. Seeley and the Publishing of *Ecce Homo*." *Victorian Review*, vol. 38, no. 1 (2012), 92–112.

Hesketh, Ian. *Of Apes and Ancestors: Evolution, Christianity, and the Oxford Debate*. Toronto and London: University of Toronto Press, 2009.

Hesketh, Ian. *The Science of History in Victorian Britain: Making the Past Speak*. London: Pickering and Chatto, 2011.

Hesketh, Ian. *Victorian Jesus: J. R. Seeley, Religion, and the Cultural Significance of Anonymity*. Toronto: University of Toronto Press, 2017.

Hewitt, Martin, ed. *The Victorian World*. London: Routledge, 2012.

Heyck, Boyd. *The Transformation of Intellectual Life in Victorian England*. New York: St. Martin's Press, 1982.

Higgitt, Rebekah. *Recreating Newton: Newtonian Biography and the Making of Nineteenth-Century History of Science*. London: Pickering and Chatto, 2007.

Higham, John. *Strangers in the Land: Patterns of American Nativism, 1860–1925*. New Brunswick, NJ: Rutgers University Press, 2002.

Hilts, Victor L. "History of Science at the University of Wisconsin." *Isis*, vol. 75, no. 1 (1984), 63–94.

Himmelfarb, Gertrude. *Victorian Minds*. New York: Harper and Row, 1970.

Hindmarsh, Bruce. *The Spirit of Early Evangelicalism: True Religion in a Modern World*. Oxford: Oxford University Press, 2018.

Hitchens, Christopher. *god is not Great: How Religion Poisons Everything*. New York: Hachette Book Group, 2007.

Hodge, Jonathan, and Gregory Radick, eds. *The Cambridge Companion to Darwin*. Cambridge: Cambridge University Press, 2009.

Hofstadter, Richard. *Social Darwinism in American Thought*. Boston: Beacon Press, 1992.

Holified, E. Brooks. *Theology in America: Christian Thought from the Age of the Puritans to the Civil War*. New Haven, CT: Yale University Press, 2003.

Hooykaas, R. *Religion and the Rise of Modern Science*. Edinburgh: Scottish Academic Press, 1972.

Houghton, Walter Edwards. *The Victorian Frame of Mind, 1830–1870*. New Haven, CT: Yale University Press, 1957.

Hovenkamp, Herbert. *Science and Religion in America, 1800–1860*. Philadelphia: University of Pennsylvania Press, 1978.

Howard, Thomas A. "Commentary: A 'Religious Turn' in European Historiography?" *Church History*, 75, no. 1 (2006): 157–162.

Howard, Thomas A. *Protestant Theology and the Making of the Modern German University*. Oxford: Oxford University Press, 2006.

Howard, Thomas A. "Religion and Modern Europe." *Historically Speaking*, vol. 4, no. 5 (June 2003).

Howard, Thomas A. *Religion and the Rise of Historicism: W. M. L. De Wette, Jacob Burckhardt, and the Theological Origins of Nineteenth-Century Historical Consciousness*. Cambridge: Cambridge University Press, 2000.

Howard, Thomas A. *Remembering the Reformation: An Inquiry into the Meanings of Protestantism*. Oxford: Oxford University Press, 2016.

Howarth, O. J. R. *The British Association for the Advancement of Science: A Retrospect 1831–1921*. London: The Association, 1922.

Howell, Nancy R., Niels Henrik Gregersen, Wesley J. Wildman, Ian Barbour, and Ryan Valentine, eds. *Encyclopedia of Science and Religion*. 2 vols. New York: Macmillan Reference, 2003.

Howsam, Leslie. "An Experiment with Science for the Nineteenth-Century Book Trade: The International Scientific Series." *The British Journal for the History of Science*, vol. 33, no. 2 (June 2000), 187–207.

Hudson, Wayne, Deigo Lucci, and Jeffrey R. Wigelsworth, eds. *Atheism and Deism Revalued: Heterodox Religious Identities in Britain, 1650–1800*. Farnham-Burlington, VT: Ashgate, 2014.

Hudson, Wayne. *The English Deists: Studies in Early Enlightenment.* London: Pickering and Chatto, 2009

Hudson, Wayne. *Enlightenment and Modernity: The English Deists and Reform.* London: Pickering and Chatto, 2009.

Hunt, Lynn, Margaret C. Jacob, and Wijnad Mijnhardt. *The Book That Changed Europe: Picart & Bernard's* Religious Ceremonies of the World. Cambridge, MA: Harvard University Press, 2010.

Hunter, Ian. *Rival Enlightenments: Civil and Metaphysical Philosophy in Early Modern Germany.* Cambridge: Cambridge University Press, 2001.

Hutchison, William R. *The Modernist Impulse in American Protestantism.* Durham, NC: Duke University Press, 1992.

Hutchison, William R. *The Transcendentalist Ministers: Church Reform in the New England Renaissance.* New Haven: Yale University Press, 1959.

Hyman, Gavin. *A Short History of Atheism.* London and New York: I.B. Tauris, 2010.

Iggers, Georg. *The German Conception of History: The National Tradition of Historical Thought from Herder to the Present.* Middletown, CT: Wesleyan University Press, 1983.

Irvine, William. *Apes, Angels, and Victorians: The Story of Darwin, Huxley, and Evolution.* New York: McGraw-Hill, 1955.

Jacob, James R., and Margaret C. Jacob. "The Anglican Origins of Modern Science: The Metaphysical Foundations of the Whig Constitution." *Isis,* vol. 71, no. 2 (1980), 251–67.

Jacob, Margaret C. *The Newtonians and the English Revolution 1689–1720.* Hassocks, Sussex: Harvester Press, 1976.

Jacyna, L. S. "Science and Social Order in the Thought of A. J. Balfour." *Isis,* vol. 71, no. 1 (March 1980), 11–34.

Jacyna, L. S. "Scientific Naturalism in Victorian Britain: An Essay in the Social History of Ideas." PhD diss. University of Edinburgh, 1980.

Jaki, Stanley L. *Uneasy Genius: The Life and Work of Pierre Duhem.* Dordrech: Martinus Nijhoff Publishers, 1987.

Jann, Rosemary. *The Art and Science of Victorian History.* Columbus, Ohio State University Press, 1985.

Jardine, Nicholas et al. eds. *Cultures of Natural History.* Cambridge: Cambridge University Press, 1996.

Jay, Elisabeth. *The Religion of the Heart: Anglican Evangelicalism and the Nineteenth-Century Novel.* Oxford: Clarendon Press, 1979.

Jenson, J. Vernon. "The X Club: Fraternity of Victorian Scientists." *British Journal for the History of Science,* vol. 5, no. 1 (1970), 63–72.

Johnson, Richard Colles, and G. Thomas Tansell. "The Haldeman-Julius 'Little Blue Books' as a Bibliographical Problem." *Papers of the Bibliographical Society of America* 64 (1970), 29–78.

Jones, Tod E. *The Broad Church: A Biography of a Movement*. Lanham, MD: Lexington Books, 2003.

Jordan, Philip D. *The Evangelical Alliance for the United States of America, 1847–1900: Ecumenism, Identity and the Religion of the Republic*. New York: Edwin Mellen Press, 1982.

Kelly, Donald R. *Faces of History: Historical Inquiry from Herodotus to Herder*. New Haven, CT: Yale University Press, 1998.

Kemp, Anthony. *The Estrangement of the Past: A Study in the Origins of Modern Historical Consciousness*. New York: Oxford University Press, 1991.

Keymer, Thomas, and Jon Mee, eds. *The Cambridge Companion to English Literature 1740–1830*. New York: Cambridge University Press, 2004.

Kidd, Thomas S. *Benjamin Franklin: The Religious Life of a Founding Father*. New Haven, CT: Yale University Press, 2017.

Kim, Stephen S. *John Tyndall's Transcendental Materialism and the Conflict Between Religion and Science in Victorian England*. Lewiston, NY: Edward Mellen Press, 1996.

Kirby, James. *Historians and the Church of England: Religion and Historical Scholarship*. Oxford: Oxford University Press, 2016.

Kirby, W. J. Torrance, ed. *A Companion to Richard Hooker*. Leiden: Brill, 2008.

Knight, David. *Science and Spirituality: The Volatile Connection*. London and New York: Routledge, 2004.

Knight, David. *Sources for the History of Science 1660–1914*. Cambridge: Cambridge University Press, 1976.

Knight, David M., and Matthew D. Eddy, eds. *Science and Beliefs: From Natural Philosophy to Natural Science, 1700–1900*. Aldershot: Ashgate, 2005.

Knight, Frances. *The Nineteenth-Century Church and English Society*. Cambridge and New York: Cambridge University Press, 1995.

Knight, Mark, and Emma Mason, eds. *Nineteenth-Century Religion and Literature: An Introduction*. Oxford: Oxford University Press, 2006.

Knights, Ben. *The Idea of the Clerisy in the Nineteenth Century*. Cambridge: Cambridge University Press, 1978.

Kocher, Paul H. *Science and Religion in Elizabethan England*. San Marino, CA: Huntington Library, 1953.

Kohn, David, ed. *The Darwinian Heritage*. Princeton, NJ: Princeton University Press, 1985.

Koyré, Alexandre. *From the Closed World to the Infinite Universe*. Baltimore, MD: Johns Hopkins University Press, 1957.

Kragh, Helge. *An Introduction to the Historiography of Science*. Cambridge: Cambridge University Press, 1987.

Kramnick, Isaac. *The Portable Enlightenment Reader*. New York: Penguin Books, 1995.

Kring, Walter David. *Liberals among the Orthodox: Unitarian Beginnings in New York City: 1819–1839*. Boston: Beacon Press, 1974.

Kuhn, Thomas S. *The Essential Tension: Selected Studies in Scientific Tradition and Change*. Chicago: University of Chicago Press, 1977.

Kuntz, Marian Leathers. *Colloquium of the Seven about Secrets of the Sublime*. University Park: Pennsylvania State University Press, 2008.

Lane, Christopher. *The Age of Doubt: Tracing the Roots of our Religious Uncertainty*. New Haven, CT: Yale University Press, 2011.

Larson, Edward. *Summer for the Gods: The Scopes Trial and America's Continuing Debate over Science and Religion*. New York: Basic Books, 1997.

Larsen, Timothy. *Contested Christianity: The Political and Social Contexts of Victorian Theology*. Waco, TX: Baylor University Press, 2004.

Larsen, Timothy. *Crisis of Doubt: Honest Faith in Nineteenth-Century England*. New York: Oxford University Press, 2008.

Larsen, Timothy. *John Stuart Mill: A Secular Life*. Oxford: Oxford University Press, 2018.

Larsen, Timothy. *A People of One Book: The Bible and the Victorians*. New York: Oxford University Press, 2011.

Lash, Nicholas. *The Beginning and the End of "Religion."* New York: Cambridge University Press, 1996.

Lasky, Melvin J. *Utopia and Revolution: On the Origins of a Metaphor, or Some Illustrations of the Problem of Political Temperament and Intellectual Climate and How Ideas, Ideals and Ideologies Have Been Historically Related*. Chicago: University of Chicago Press, 1976.

Lee, R. Alton. *Publisher for the Masses, Emanuel Haldeman-Julius*. Lincoln: University of Nebraska Press, 2017.

Legaspi, Michael C. *The Death of Scripture and the Rise of Biblical Studies*. Oxford: Oxford University Press, 2010.

Lessl, Thomas M. *Rhetorical Darwinism: Religion, Evolution, and the Scientific Identity*. Waco, TX: Baylor University Press, 2012.

Leverette, Jr., William. "E.L. Youmans' Crusade for Scientific Autonomy and Respectability." *American Quarterly*, 1 (1965), 12–32.

Leverette, Jr., William. "Science and Values: A Study of Edward L. Youmans' *Popular Science Monthly*, 1872–1887." PhD diss. Vanderbilt University, 1963.

Levine, George. *Darwin and the Novelists: Patterns of Science in Victorian Fiction*. Cambridge, MA: Harvard University Press, 1988.

Levine, George. *Dying to Know: Scientific Epistemology and Narrative in Victorian England*. Chicago: University of Chicago Press, 2002.

Levine, George. *Realism, Ethics and Secularism: Essays on Victorian Literature and Science*. Cambridge: Cambridge University Press, 2008.

Lewis, Andrew J. *A Democracy of Facts: Natural History in the Early Republic*. Philadelphia: University of Pennsylvania Press, 2011.

Lightman, Bernard. "Christian Evolutionists in the United States, 1860–1900." *Journal of Cambridge Studies*, vol. 4, no. 4 (2009), 14–22.

Lightman, Bernard, ed. *A Companion to the History of Science*. West Sussex: Wiley Blackwell, 2016.

Lightman, Bernard. "Does the History of Science and Religion Change Depending on the Narrator? Some Atheist and Agnostic Perspectives." *Science and Christian Belief*, 24 (2012), 149–68.

Lightman, Bernard. *Evolutionary Naturalism in Victorian Britain: The "Darwinians" and Their Critics*. Aldershot: Ashgate, 2009.

Lightman, Bernard, ed. *Global Spencerism: The Communication and Appropriation of a British Evolutionist*. Leiden: Brill, 2016.

Lightman, Bernard. "Huxley and Scientific Agnosticism: The Strange History of a Failed Rhetorical Strategy." *British Journal for the History of Science*, vol. 35, no. 3 (2002), 271–89.

Lightman, Bernard. "The International Scientific Series and the Communication of Darwinism." *Journal of Cambridge Studies*, vol. 5, no. 4 (2010), 27–38.

Lightman, Bernard. *The Origins of Agnosticism: Victorian Unbelief and the Limits of Knowledge*. Baltimore, MD: Johns Hopkins University Press, 1987.

Lightman, Bernard. *Victorian Popularizers of Science: Designing Nature for New Audiences*. Chicago and London. The University of Chicago Press, 2007.

Lightman, Bernard. *Victorian Science in Context*. Chicago: University of Chicago Press, 1997.

Lightman, Bernard. "Victorian Sciences and Religions: Discordant Harmonies." *Osiris*, 16 (2001), 343–66.

Lightman, Bernard, and Gowan Dawson, eds. *Victorian Scientific Naturalism: Community, Identity, Continuity*. Chicago: University of Chicago Press, 2014.

Lightman, Bernard, and Michael S. Reidy, eds. *The Age of Scientific Naturalism: Tyndall and His Contemporaries*. London: Pickering and Chatto, 2014.

Lightman, Bernard, and Bennett Zon, eds. *Evolution and Victorian Culture*. Cambridge: Cambridge University Press, 2014.

Lindberg, David C. *The Beginnings of Western Science: The European Scientific Tradition in Philosophical, Religious, and Institutional Context, 600 B.C. to A.D. 1450*. Chicago: University of Chicago Press, 1992.

Lindberg, David C., and Ronald L. Numbers. "Beyond War and Peace: A Reappraisal of the Encounter between Christianity and Science." *Church History*, 55 (1986), 338–54.

Lindberg, David C., and Ronald L. Numbers, eds. *God and Nature: Historical Essays on the Encounter between Christianity and Science*. Berkeley: University of California Press, 1986.

Lindberg, David C., and Ronald L. Numbers, eds. *When Science and Christianity Meet*. Chicago and London: University of Chicago Press, 2003.

Lindberg, David C., and Michael H. Shank, eds. *The Cambridge History of Science*, vol. 2: *Medieval Science*. Cambridge: Cambridge University Press, 2013.

Livingston, James C. *Matthew Arnold and Christianity*. Columbia: University of South Carolina Press, 1986.

Livingston, James C. *Religious Thought in the Victorian Age: Challenges and Reconceptions*. New York and London: Continuum, 2007.

Livingstone, David N. *Adam's Ancestors: Race, Religion, and the Politics of Human Origins*. Baltimore, MD: Johns Hopkins University Press, 2008.

Livingstone, David N. *Darwin's Forgotten Defenders: The Encounter between Evangelical Theology and Evolutionary Thought*. Grand Rapids, MI: Eerdmans, 1987.

Livingstone, David N. *Putting Science in its Place: Geographies of Scientific Knowledge*. Chicago: University of Chicago Press, 2003.

Livingstone, David N. "Science and Religion: Towards a New Cartography." *Christian Scholar's Review*, vol. 26, no. 3 (Spring 1997), 270–92.

Livingstone, David N., D. G. Hart, and Mark A. Noll, eds. *Evangelicals and Science in Historical Perspective*. New York: Oxford University Press, 1999.

Livingstone, David N., and Charles W. J. Withers, eds. *Geographies of Nineteenth-Century Science*. Chicago: University of Chicago Press, 2011.

Lovelace, Richard F. *The American Pietism of Cotton Mather: Origins of American Evangelism*. Eugene, OR: Wipf & Stock, 1979.

Löwith, Karl. *Meaning in History: The Theological Implications of the Philosophy of History*. Chicago: University of Chicago Press, 1949.

Maas, Korey D. *The Reformation and Robert Barnes: History, Theology and Polemic in Early Modern England*. Suffolk: Boydell Press, 2010.

Macaulay, John Allen. *Unitarianism in Antebellum South: The Other Invisible Institution*. Tuscaloosa: University of Alabama Press, 2001.

MacDonald, George E. *Fifty Years of Freethought, Being the Story of the Truth Seeker*. 2 vols. New York: Truth Seeker Company, 1929.

MacLeod, Roy M. *The 'Creed of Science' in Victorian England*. Aldershot: Ashgate, 2000.

MacLeod, Roy M. "The X-Club: A Social Network of Science in Late-Victorian England." *Notes and Records of the Royal Society of London*, vol. 24, no. 2 (1970), 305–22.

Madden, Michael. "Curious Paradoxes: James Anthony Froude's View of the Bible." *Journal of Religious History*, vol. 30, no. 2 (2006), 199–216.

Madison, Charles A. *Book Publishing in America*. New York: McGraw-Hill, 1966.

Manuel, Frank E. *The Prophets of Paris*. New York: Harper and Row, 1965.

Manuel, Frank E. *The Religion of Isaac Newton*. New York: Oxford University Press, 1974.

Manuel, Frank E, ed. *Utopias and Utopian Thought*. Boston: Beacon Press, 1967.

Manuel, Frank E., and Fritzie P. Manuel. *Utopian Thought in the Western World*. Cambridge, MA.: Belknap Press of Harvard University Press, 1979.

Marjorie, Wheeler-Barclay. *The Science of Religion in Britain, 1860–1915*. Charlottesville: University of Virginia Press, 2010.

Marotti, Arthur F. *Religious Ideology and Cultural Fantasy: Catholic and Anti-Catholic Discourses in Early Modern England*. Notre Dame, IN: University of Notre Dame Press, 2005.

Marsden, George M. *Fundamentalism and American Culture: The Shaping of Twentieth-Century Evangelicalism, 1870–1925*. New York: Oxford University Press, 1980.

Marsden, George M. *The Soul of the American University: From Protestant Establishment to Established Nonbelief*. New York: Oxford University Press, 1994.

Marshall, John. "The Ecclesiology of the Latitude-men 1660–1689: Stillingfleet, Tillotson and 'Hobbism.'" *Journal of Ecclesiastical History*, vol. 36, no. 3 (1985), 407–427.

Martin, David. "Does the Advance of Science Mean Secularisation?" *Science and Christian Belief* 19 (2007), 3–14.

Martin, David. *On Secularization: Towards a Revised General Theory*. Burlington, VT: Ashgate, 2005.

Martin, David. *The Religious and the Secular*. London: Routledge and Kegan Paul, 1969.

Martin, David. "The Secularization Issue: Prospect and Retrospect." *British Journal of Sociology* 42, no. 3, (1991): 465–74.

Martin, R. N. D. "The Genesis of a Medieval Historian: Pierre Duhem and the Origins of Statics." *Annals of Science*, vol. 33 (1976): 119-129.

Marty, Martin E., and R. Scott Appleby, eds. *Fundamentalism and Society: Reclaiming the Sciences, the Family, and Education*. Chicago: University of Chicago Press, 1993.

Masson, Scott. *Romanticism, Hermeneutics and the Crisis of the Human Sciences*. Aldershot: Ashgate, 2004.

Matthews, Steven. *Theology and Science in the Thought of Francis Bacon*. Aldershot: Ashgate, 2008.

Masuzawa, Tomoko. *The Invention of World Religions: Or, How European Universalism was Preserved in the Language of Pluralism*. Chicago: University of Chicago Press, 2005.

Meadows, A. J., ed. *Development of Science Publishing in Europe*. Amsterdam: Elsevier Science, 1980.

Menand, Louis, Paul Reitter, and Chad Wellmon, eds. *The Rise of the Research University: A Sourcebook*. Chicago: University of Chicago Press, 2017.

Merrigan, Terrance. "Newman and Theological Liberalism." *Theological Studies* 66 (2005), 605–21.

Merton, Robert K. "George Sarton: Episodic Recollection by an Unruly Apprentice." *Isis*, vol. 76, no. 4 (1985), 470–86.

Merton, Robert K. "Puritanism, Pietism, and Science," *The Sociological Review* 28 (1936): 1–30.

Merton, Robert K. "Science, Technology and Society in Seventeenth Century England." *Osiris*, vol. 4 (1938): 360–632.

Merton, Robert K. *Science, Technology and Society in Seventeenth-Century England*. New York: Harper and Row, 1970.

McGrath, Alister. *The Making of Modern German Christology, 1750–1990*. Grand Rapids, MI: Zondervan, 1994.

McGrath, Alister. *Why God Won't Go Away: Engaging with the New Atheism*. London: Society for Promoting Christian Knowledge, 2011.

McGrath, Alister, and Joanna Collicutt McGrath. *The Dawkins Delusion?* London: Society for Promoting Christian Knowledge, 2007.

McIntire, C. T. *Herbert Butterfield: Historian as Dissenter*. New Haven, CT: Yale University Press, 2004.

McLeod, Hugh. *European Religion in the Age of the Great Cities, 1830–1930*. London: Routledge, 1995.

McLeod, Hugh. *Secularisation in Western Europe, 1848–1914*. New York: St. Martin's Press, 2000.

McLeod, Hugh, and Werner Ustorf, eds. *The Decline of Christendom in Western Europe, 1750–2000*. New York: Cambridge University Press, 2003.

McLoughlin, William G. *The Meaning of Henry Ward Beecher: An Essay on the Shifting Values of Mid-Victorian America, 1840–1870*. New York: Knopf, 1970.

McManus, Howard R. "The Most Famous Daguerreian Portrait: Exploring the History of the Dorothy Catherine Draper Daguerreotype." *The Daguerreian Annual* (1995), 148–71.

McManners, John. *Church and Society in Eighteenth Century France*. 2 vols. Oxford and New York: Clarendon Press, 1998.

Miller, Perry. *The New England Mind: The Seventeenth Century*. Cambridge, MA: Harvard University Press, 1954.

Milward, Peter. *Religious Controversies of the Elizabethan Age: A Survey of Printed Sources*. London: Scholar Press, 1997.

Milward, Peter. *Religious Controversies of the Jacobean Age: A Survey of Printed Sources*. London: Scholar Press, 1978.

Mislin, David. *Saving Faith: Making Religious Pluralism an American Value at the Dawn of the Secular Age*. Ithaca, NY: Cornell University Press, 2015.

Moore, James R. "Evangelicals and Evolution: Henry Drummond, Herbert Spencer, and the Naturalisation of the Spiritual World." *Scottish Journal of Theology*, vol. 38, no. 3 (1985), 383–418.

Moore, James R., ed. *History, Humanity and Evolution: Essays for John C. Greene*. Cambridge: Cambridge University Press, 1989.

Moore, James R. *The Post-Darwinian Controversies: A Study of the Protestant Struggle to Come to Terms with Darwin in Great Britain and America 1879–1900*. Cambridge: Cambridge University Press, 1979.

Morgan, Charles. *The House of Macmillan (1843–1943)*. London: Macmillan, 1943.

Morrell, Jack, and Arnold Thackray. *Gentlemen of Science: Early Years of the British Association for the Advancement of Science*. Oxford: Clarendon Press, 1981.

Nagel, Thomas. *Mind and Cosmos: Why the Materialist Neo-Darwinian Conception of Nature Is Almost Certainly False*. Oxford: Oxford University Press, 2012.

Neurath, Otto. *Philosophical Papers, 1913–1946*. Translated by Robert S. Cohen and Marie Neurath. Dordrecht: Reidel, 1983.

Nichols, James Hastings. *Romanticism in American Theology: Nevin and Schaff at Mercersburg*. Chicago: University of Chicago Press, 1961.

Nisbet, H. B. *Gotthold Ephraim Lessing: His Life, Works, and Thought*. Oxford: Oxford University Press, 2013.

Nisbet, Robert. *History of the Idea of Progress*. New Brunswick: Transaction Publishers, 2009.

Nixon, Jude V. *Victorian Religious Discourse: New Directions in Criticism*. New York: Palgrave Macmillan, 2004.

Noll, Mark A. *America's God: From Jonathan Edwards to Abraham Lincoln*. Oxford: Oxford University Press, 2002.

Noll, Mark A. *Between Faith and Criticism: Evangelicals, Scholarship, and the Bible in America*. Vancouver: Regent College Publishing, 2004.

Noll, Mark A. *The Old Religion in the New World: The History of North American Christianity*. Grand Rapids, MI: Eerdmans, 2002.

Nongbri, Brent. *Before Religion: A History of a Modern Concept*. New Haven: Yale University Press, 2013.

Norman, Edward. *Anti-Catholicism in Victorian England*. London: Allen and Unwin, 1968.

Numbers, Ronald L. "The *American* History of the Science Society or the *Inter-*

national History of Science Society? The Fate of Cosmopolitanism since George Sarton." *Isis*, vol. 100, no. 1 (2009), 103–07.

Numbers, Ronald L. *Darwinism Comes to America*. Cambridge, MA: Harvard University Press, 1998.

Numbers, Ronald L., ed. *Galileo Goes to Jail and other Myths about Science and Religion*. Cambridge, MA: Harvard University Press, 2009.

Numbers, Ronald L. "Science and Religion." *Osiris*, 2nd Series, vol. 1 (1985), 59–80.

Numbers, Ronald L., and Kostas Kampourakis, eds. *Newton's Apple and Other Myths about Science*. Cambridge, MA: Harvard University Press, 2015.

O'Connor, Ralph. "Young-earth Creationists in Early Nineteenth-century Britain? Towards a Reassessment of 'Scriptural Geology.'" *History of Science* 45 (2007), 357–403.

O' Gorman, Francis, ed. *The Cambridge Companion to Victorian Culture*. New York: Cambridge University Press, 2010.

Olby, R. C., G. N. Cantor, J. R. R. Christie, and M. J. S. Hodge, eds. *Companion to the History of Modern Science*. London: Routledge, 1990.

Olson, Richard G. *Science and Religion, 1450–1900: From Copernicus to Darwin*. Baltimore, MD: Johns Hopkins University Press, 2004.

Olson, Richard G. *Science and Scientism in Nineteenth-Century Europe*. Urbana: University of Illinois Press, 2007.

Orr, Robert. *Reason and Authority: The Thought of William Chillingworth*. Oxford: Clarendon Press, 1967.

Osler, Margaret J. *Reconfiguring the World: Nature, God, and Human Understanding from the Middle Ages to Early Modern Europe*. Baltimore, MD: Johns Hopkins University Press, 2010.

Osler, Margaret J., ed. *Rethinking the Scientific Revolution*. Cambridge: Cambridge University Press, 2000.

Osler, Margaret J., and Paul Lawrence Farber, eds. *Religion, Science, and Worldview: Essays in Honor of Richard S. Westfall*. New York: Cambridge University Press, 1985.

Overton, Grant. *Portrait of a Publisher, and the First Hundred Years of the House of Appleton 1825–1925*. New York: D. Appleton and Co., 1925.

Pailin, David A. *Attitudes to Other Religions: Comparative Religion in Seventeenth- and Eighteenth-Century Britain*. Manchester: Manchester University Press, 1984.

Painter, Borden. *The New Atheist Denial of History*. New York: Palgrave Macmillan, 2016.

Pals, Daniel L. *The Victorian "Lives" of Jesus*. San Antonio, TX: Trinity University Press, 1982.

Paradis, James G. *T. H. Huxley: Man's Place in Nature*. Lincoln: University of Nebraska Press, 1978.

Patrides, C. A. ed. *The Cambridge Platonists*. Cambridge: Cambridge University Press, 1969.

Paz, D. G., ed. *Nineteenth-Century English Religious Traditions: Retrospect and Prospect*. Westport, CT: Greenwood Press, 1995.

Paz, D. G. *Popular Anti-Catholicism in Mid-Victorian England*. Stanford, CA: Stanford University Press, 1992.

Pearcey, Nancy R., and Charles B. Thaxton. *The Soul of Science: Christian Faith and Natural Philosophy*. Wheaton, IL: Crossway Books, 1994.

Peden, W. Creighton. *Empirical Tradition in American Liberal Religious Thought, 1860–1960*. New York: Peter Lang, 2010.

Peel, J. D. Y. *Herbert Spencer: The Evolution of a Sociologist*. New York: Basic Books, 1971.

Pelikan, Jaroslav. *The Idea of the University: A Reexamination*. New Haven, CT: Yale University Press, 1992.

Persons, Stow. *Free Religion: An American Faith*. New Haven, CT: Yale University Press, 1947.

Peschier, Diana. *Nineteenth-Century Anti-Catholic Discourse: The Case of Charlotte Brontë*. New York: Palgrave Macmillan, 2005.

Peterson, William S. *Victorian Heretic: Mrs Humphry Ward's Robert Elsmere*. Leicester: Leicester University Press, 1976.

Pfau, Thomas. *Minding the Modern: Human Agency, Intellectual Traditions, and Responsible Knowledge*. South Bend, IN: University of Notre Dame Press, 2013.

Placher, William C. *The Domestication of Transcendence: How Modern Thinking About God Went Wrong*. Louisville, KY: Westminster John Knox Press, 1996.

Plantinga, Alvin. *Where the Conflict Really Lies: Science, Religion, and Naturalism*. Oxford: Oxford University Press, 2011.

Pochmann, Henry A. *German Culture in America: Philosophical and Literary Influences, 1600–1900*. Madison: University of Wisconsin Press, 1957.

Pocock, J. G. A. *Barbarism and Religion: The Enlightenments of Edward Gibbon, 1737–1764*. 5 vols. Cambridge: Cambridge University Press, 2011.

Pocock, J. G. A. "Historiography and Enlightenment: A View of Their History." *Modern Intellectual History*, vol. 5, no. 1 (2008), 83–96.

Porter, Theodore M., and Dorothy Ross, eds. *The Cambridge History of Science, Volume 7: The Modern Social Sciences*. Cambridge: Cambridge University Press, 2003.

Porterfield, Amanda. *Conceived in Doubt: Religion and Politics in the New American Nation*. Chicago: University of Chicago Press, 2012.

Preus, J. Samuel. *Explaining Religion: Criticism and Theory from Bodin to Freud.* Atlanta, GA: Scholars Press, 1996.

Prickett, Stephen. *The Origins of Narrative: The Romantic Appropriation of the Bible.* Cambridge: Cambridge University Press, 1996.

Prickett, Stephen. *Romanticism and Religion: The Tradition of Coleridge and Wordsworth in the Victorian Church.* Cambridge: Cambridge University Press, 1976.

Pyenson, Lewis. *The Passion of George Sarton: A Modern Marriage and Its Discipline.* Philadelphia: American Philosophical Society, 2006.

Raeder, Linda C. *John Stuart Mill and the Religion of Humanity.* Columbia: University of Missouri Press, 2002.

Randall, John Hermon. *The Career of Philosophy.* 3 vols. New York: Columbia University Press, 1965.

Raven, Charles E. *Science, Religion, and the Future: A Course of Eight Lectures.* New York: Macmillan, 1944.

Reardon, Bernard M. G. *Liberal Protestantism.* Stanford, CA: Stanford University Press, 1968.

Reardon, Bernard M. G. *Religion in the Age of Romanticism: Studies in Early Nineteenth-Century Thought.* New York: Cambridge University Press, 1985.

Reardon, Bernard M. G. *Religious Thought in the Nineteenth Century: Illustrated from Writers of the Period.* Cambridge: Cambridge University Press, 1966.

Reeves, Marjorie E. *Joachim of Fiore and the Prophetic Future.* New York: Harper Torchbooks, 1977.

Reuben, Julie A. *The Making of the Modern University: Intellectual Transformation and the Marginalization of Morality.* Chicago: University of Chicago Press, 1996.

Riasanovsky, Nicholas V. *The Emergence of Romanticism.* Oxford: Oxford University Press, 1992.

Richardson, W. Mark, and Wesley J. Wildman, eds. *Religion and Science: History, Method, Dialogue.* New York: Routledge, 1996.

Roberts, Jon H., and James Turner. *The Sacred and the Secular University.* Princeton, NJ: Princeton University Press, 2000.

Roberts, Jon H. *Darwinism and the Divine in America: Protestant Intellectuals and Organic Evolution, 1859–1900.* South Bend, IN: University of Notre Dame Press, 1988.

Robinson, David. *The Unitarians and the Universalists.* Westport, CT: Greenwood Press, 1985.

Rogers, Ben. *A. J. Ayer: A Life.* New York: Grove Press, 1999.

Rogers, Walter P. *Andrew D. White and the Modern University.* Ithaca, NY: Cornell University Press, 1942.

Roney, John B. *The Inside of History: Jean Henri Merle d'Aubigné and Romantic Historiography*. Westport, CT: Greenwood, 1996.

Rowe, Henry K. *History of Andover Theological Seminary*. Newton Center, MA: Andover-Newton Theological Seminary, 1933.

Royle, Edward. *Radicals, Secularists, and Republicans: Popular Freethought in Britain, 1866–1915*. Manchester: Manchester University Press, 1980.

Rudolph, Frederick. *The American College and University: A History*. Athens: University of Georgia Press, 1990.

Ruse, Michael. *Monad to Man: The Concept of Progress in Evolutionary Biology*. Cambridge, MA: Harvard University Press, 1996.

Ruse, Michael. "The Relationship between Science and Religion in Britain, 1830–1870." *Church History*, vol. 44, no. 4 (1975), 505–22.

Ruse, Michael, and Robert J. Richards, eds. *The Cambridge Companion to the "Origin of Species."* Cambridge: Cambridge University Press, 2009.

Russell, Colin A. "The Conflict Metaphor and Its Social Origins," 3–26.

Russell, Colin A. *Cross-Currents: Interactions between Science and Faith*. Grand Rapids, MI: Eerdmans, 1985.

Russell, Colin A., R. Hooykaas, and David C. Goodman. *The 'Conflict Thesis' and Cosmology*. Milton Keynes: Open University Press, 1974.

Russell, Jeffrey Burton. *Inventing the Flat Earth: Columbus and Modern Historians*. Westport, CT and London: Praeger, 1997.

Ryan, Robert M. *The Romantic Reformation: Religious Politics in English Literature, 1789–1824*. Cambridge: Cambridge University Press, 1997.

"Sarton, Science, and History." *Isis*, vol. 75, no. 1 (1984), 1–104.

Sarton, George. "Auguste Comte, Historian of Science: With a Short Digression on Clotilde de Vaux and Harriet Taylor." *Osiris*, vol. 10 (1952), 328–57.

Sarton, George. "The Faith of the Humanist." *Isis*, vol. 3, no. 1 (1920), 3–6.

Sarton, George. "L'histoire de la science." *Isis*, vol. 1, no. 1 (1913), 3–46.

Sarton, George. "The History of Science." *Monist*, vol. 26, no. 3 (1916), 321–65.

Sarton, George. *The History of Science and the New Humanism*. New York: George Braziller, 1956.

Sarton, George. *Horus: A Guide to the History of Science: A First Guide for the Study of the History of Science with Introductory Essays on Science and Tradition*. Waltham, MA: Chronica Botanica Company, 1952.

Sarton, George. *Introduction to the History of Science, Volume 1: From Homer to Omar Kahayyam*. Baltimore: The Williams and Wilkins Company, 1927.

Sarton, George. "The New Humanism." *Isis*, vo. 6, no. 1 (1924), 9–42.

Sarton, George. "The Teaching of the History of Science." *Isis*, vol. 4, no. 2 (1921), 225–49.

Sarton, George. "The Teaching of the History of Science." *Isis*, vol. 13, no. 2 (1930), 272–97.

Sarton, George. "The Teaching of the History of Science." *Scientific Monthly*, vol. 7, no. 13 (1918), 193–211.

Sarton, May. *I Knew a Phoenix: Sketches for an Autobiography*. New York: Norton, 1995.

Schaefer, Richard. "Andrew Dickson White and the History of Religious Future." *Zygon*, vol. 50, no. 1 (2015), 7–27.

Schirmer, Walter F. "German Literature, Historiography and Theology in Nineteenth-Century England." *German Life & Letters*, no. 1 (1947–48), 165–74.

Schlereth, Eric R. *An Age of Infidels: The Politics of Religious Controversy in the Early United States*. Philadelphia: University of Pennsylvania Press, 2013.

Schlossberg, Herbert. *Conflict and Crisis in the Religious Life of Late Victorian England*. New Brunswick, NJ: Transaction Publishers, 2009.

Schlossberg, Herbert. *The Silent Revolution and the Making of Victorian England*. Columbus: Ohio State University Press, 2000.

Schmidt, Leigh Eric. *Restless Souls: The Making of American Spirituality*. Berkeley: University of California Press, 2005.

Schocket, Eric. "Proletarian Paperbacks: The Little Blue Books and Working-Class Culture." *College Literature*, vol. 29, no. 4 (2002), 67–78.

Schofield, Robert E. *The Enlightenment of Joseph Priestley*. 2 vols. University Park: Pennsylvania State University Press, 2004–2009.

Schweber, S. S. "John Herschel and Charles Darwin: A Study of Parallel Lives." *Journal of the History of Biology*, vol. 22, no. 1 (1989), 1–71.

Secord, James A. *Victorian Sensation: The Extraordinary Publication, Reception, and Secret Authorship of "Vestiges of the Natural History of Creation."* Chicago: University of Chicago Press, 2000.

Secord, James A. *Visions of Science: Books and Readers at the Dawn of the Victorian Age*. Oxford: Oxford University Press, 2014.

Sell, Alan P. F. *Mill on God: The Pervasiveness and Elusiveness of Mill's Religious Thought*. Aldershot: Ashgate, 2004.

Sera-Shriar, Efam. *The Making of British Anthropology, 1813–1871*. London: Pickering and Chatto, 2013.

Settanni, Harry. *The Probabilist Theism of John Stuart Mill*. New York: Peter Lang, 1991.

Shapiro, B. J. "Latitudinarianism and Science in Seventeenth-Century England." *Past & Present* 40 (1968), 16–41.

Shattock, Joanne, ed. *The Cambridge Companion to English Literature 1830–1914*. Cambridge: Cambridge University Press, 2010.

Shea, Victor, and William Whitla, eds. *Essays and Reviews: The 1860 Text and Its Reading*. Charlottesville: University Press of Virginia, 2000.

Sheets-Pyenson, Susan. *John William Dawson: Faith, Hope, and Science*. Montreal: McGill-Queen's University Press, 1996.

Short, Edward. *Newman and His Contemporaries*. New York: T and T Clark, 2011.

Shortland, Michael, and Richard Yeo, eds. *Telling Lives in Science: Essays on Scientific Biography*. Cambridge: Cambridge University Press, 1985.

Simon, W. M. *European Positivism in the Nineteenth Century: An Essay in Intellectual History*. Ithaca, NY: Cornell University Press, 1963.

Simonson, Harold P. *Radical Discontinuities: American Romanticism and Christian Consciousness*. Teaneck, NJ: Fairleigh Dickinson University Press, 1983.

Simpson, James. *Permanent Revolution: The Reformation and the Illiberal Roots of Liberalism*. Cambridge, MA: Harvard University Press, 2018.

Sleigh, Charlotte. *Literature and Science*. New York: Palgrave Macmillan, 2011.

Smart, Ninian, John Clayton, Steven T. Katz, and Patrick Sherry. *Nineteenth Century Religious Thought in the West*. 3 vols. Cambridge: Cambridge University Press, 1985.

Smith, Crosbie. *The Science of Energy: A Cultural History of Energy Physics in Victorian Britain*. Chicago: University of Chicago Press, 1998.

Smith, Jonathan Z. *Drudgery Divine: On the Comparison of Early Christianities and the Religions of Late Antiquity*. Chicago: University of Chicago Press, 1990.

Snyder, Laura J. *The Philosophical Breakfast Club: Four Remarkable Friends Who Transformed Science and Changed the World*. New York: Broadway Books, 2011.

Spellman, W.M. *The Latitudinarians and the Church of England, 1660–1700*. Athens: University of Georgia Press, 1993.

Spurr, John. "'Rational Religion' in Restoration England." *Journal of the History of Ideas*, vol. 49, no. 4 (1988), 563–85.

Stanley, Matthew. *Huxley's Church and Maxwell's Demon: From Theistic Science to Naturalistic Science*. Chicago: University of Chicago Press, 2015.

Steinbach, Susie. *Understanding the Victorians: Politics, Culture, and Society in Nineteenth-Century Britain*. London: Routledge, 2012.

Stenger, Victor J. *God: The Failed Hypothesis—How Science Shows that God Does Not Exist*. New York: Prometheus Books, 2007.

Stenger, Victor J. *God and the Folly of Faith: The Incompatibility of Science and Religion*. New York: Prometheus Books, 2012.

Stenger, Victor J. *The New Atheism: Taking a Stand for Science and Reason*. New York: Prometheus Books, 2009.

Stephens, Lester D. *Joseph Le Conte: Gentile Prophet of Evolution*. Baton Rouge: Louisiana State University Press, 1982.

Stern, Fritz, ed. *The Varieties of History: From Voltaire to the Present*. London: Macmillan, 1970.

Stewart, Melville Y., ed. *Science and Religion in Dialogue*. 2 vols. West Sussex: Wiley-Blackwell, 2010.

Stimson, Dorothy. "Puritanism and the New Philosophy in 17th Century England." *Bulletin of the Institute of the History of Medicine*, vol. 3, no. 5 (1935), 321–34.

Stroumsa, Guy G. *A New Science: The Discovery of Religion in the Age of Reason*. Cambridge, MA: Harvard University Press, 2010.

Stump, J. B., ed. *Science and Christianity: An Introduction to the Issues*. West Sussex: Wiley-Blackwell, 2017.

Sullivan, Robert E. *John Toland and the Deist Controversy: A Study in Adaptations*. Cambridge, MA: Harvard University Press, 1982.

Sweet, William, and Richard Feist, eds. *Religion and the Challenges of Science*. Aldershot: Ashgate, 2007.

Taylor, Charles. *A Secular Age*. Cambridge, MA: Belknap Press of Harvard University Press, 2007.

Taylor, Michael. *The Philosophy of Herbert Spencer*. London: Continuum, 2007.

Thackray, Arnold, and Robert K. Merton. "On Discipline Building: The Paradoxes of George Sarton." *Isis*, vol. 63, no. 4 (1972), 473–95.

Thomas, Keith. *Religion and the Decline of Magic: Studies in Popular Beliefs in Sixteenth Century England*. London: Weidenfeld and Nicolson, 1971.

Thorndike, Lynn. *History of Magic and Experimental Science*. 8 vols. New York: Columbia University Press, 1923–58.

Topham, Jonathan R. "Beyond the 'Common Context': The Production and Reading of the Bridgewater Treatises." *Isis*, vol. 89, no. 2 (1998), 233–62.

Topham, Jonathan R. "Science and Popular Education in the 1830s: The Role of the 'Bridgewater Treatises.'" *British Journal for the History of Science*, vol. 25, no. 4 (1992), 397–430.

Topham, Jonathan R. "Scientific Publishing and the Reading of Science in Nineteenth-Century Britain: A Historiographical Survey and Guide to Sources." *Studies in the History and Philosophy of Science*, vol. 31, no. 4 (2000), 559–612.

Troeltsch, Ernst. *Protestantism and Progress: The Significance of Protestantism for the Rise of the Modern World*. Philadelphia: Fortress Press, 1986.

Tuckwell, William. *Pre-Tractarian Oxford: A Reminiscence of the Oriel 'Noetics'*. London: Smith, Elder, 1909.

Tumbleson, Raymond D. *Catholicism in the English Protestant Imagination: Nationalism, Religion, and Literature, 1660–1745*. Cambridge: Cambridge University Press, 1998.

Tumbleson, Raymond D. "'Reason and Religion': The Science of Anglicanism." *Journal of the History of Ideas*, vol. 57, no. 1 (1996), 131–56.

Turner, Frank Miller. *Between Science and Religion: The Reaction to Scientific Naturalism in Late Victorian England*. New Haven, CT: Yale University Press, 1974.

Turner, Frank Miller. *Contesting Cultural Authority: Essays in Victorian Intellectual Life*. Cambridge: Cambridge University Press, 1993.

Turner, Frank Miller. "The Victorian Conflict between Science and Religion: A Professional Dimension." *Isis*, vol. 69, no. 3 (1978), 356–76.

Turner, Frank Miller. "Victorian Scientific Naturalism and Thomas Carlyle." *Victorian Studies*, vol. 18, no. 3 (1975), 325–43.

Turner, James. *Without God, Without Creed: The Origins of Unbelief in America*. Baltimore, MD: Johns Hopkins University Press, 1985.

Tuveson, Ernest L. *Millennium and Utopia: A Study in the Background of the Idea of Progress*. Berkeley: University of California Press, 1949.

Ungureanu, James C. "Relocating the Conflict between Science and Religion at the Foundations of the History of Science." *Zygon*, vol. 53, no. 4 (2018), 1106–30.

Ungureanu, James C. "Science, Religion and the 'New Reformation' of the Nineteenth Century." *Science & Christian Belief*, vol. 31, no. 1 (2019), 41–61.

Ungureanu, James C. "Tyndall and Draper." *Notes and Queries*, vol. 64, no. 1 (2017), 125–28.

Ungureanu, James C. "A Yankee at Oxford: John William Draper at the British Association for the Advancement of Science at Oxford, 30 June 1860." *Notes and Records*, vol. 70, no. 2 (2015), 135–50.

Van Wyhe, John. *Phrenology and the Origins of Victorian Scientific Naturalism*. Aldershot: Ashgate, 2004.

Vann, J. Don, and Rosemary T. VanArsdel, eds. *Victorian Periodicals and Victorian Society*. Toronto: Scholar Press, 1994.

Veysey, Laurence R. *The Emergence of the American University*. Chicago: University of Chicago Press, 1965.

Vickers, Brian, ed. *Francis Bacon: The Major Works*. Oxford: Oxford University Press, 2002.

Vidler, Alec R. *The Church in an Age of Revolution, 1749 to the Present Day*. London: Penguin, 1990.

Walsham, Alexandra. "History, Memory, and the English Reformation." *The Historical Journal*, vol. 55, no. 4 (2012), 899–938.

Walsham, Alexandra. "The Reformation and 'The Disenchantment of the World' Reassessed." *Historical Review*, vol. 51, no. 2 (2008), 497–528.

Walsham, Alexandra. "The Reformation of Generations: Youth, Age and Religious Change in England, c. 1500–1700." *Transactions of the Royal Historical Society*, Sixth Series, vol. 21 (2011), 93–121.

Ward, W. R. *Religion and Society in England 1790–1850*. London: Batsford, 1972.

Warner, Laceye C. *Saving Women: Retrieving Evangelistic Theology and Practice*. Waco, TX: Baylor University Press, 2007.

Warren, Sidney. *American Freethought, 1860–1914*. New York: Columbia University Press, 1943.

Webb, George E. *The Evolution Controversy in America*. Lexington: University Press of Kentucky, 1994.

Weber, Max. *The Protestant Ethic and the Spirit of Capitalism*. Translated by Talcott Parsons. London: Routledge, 1930.

Webster, Charles. *The Great Instauration: Science, Medicine and Reform, 1626–1660*. London: Duckworth, 1975.

Welch, Claude. *Protestant Thought in the Nineteenth Century*. 2 vols. New Haven, CT: Yale University Press, 1972.

Westfall, Richard S. *Science and Religion in Seventeenth-Century England*. Hamden, CT: Archon Books, 1970.

Whatmore, Richard, and Brian Young, eds. *A Companion to Intellectual History*. West Sussex: John Wiley, 2016.

Wheeler, Michael. *The Old Enemies: Catholic and Protestant in Nineteenth-Century English Culture*. Cambridge: Cambridge University Press, 2006.

White, Daniel E. *Early Romanticism and Religious Dissent*. Cambridge: Cambridge University Press, 2006.

White, Edward A. *Science and Religion in American Thought: The Impact of Naturalism*. Stanford, CA: Stanford University Press, 1952.

White, Paul. *Thomas Huxley: Making the "Man of Science."* Cambridge: Cambridge University Press, 2003.

Whitehead, Alfred North. *Science and the Modern World*. New York: Macmillan, 1925.

Whyte, A. G. *The Story of the R.P.A., 1899–1949*. London: Watts, 1949.

Wigelsworth, Jeffrey R. *Deism in Enlightenment England: Theology, Politics, and Newtonian Public Science*. Manchester: Manchester University Press, 2009.

Wigmore-Beddoes, Dennis G. *Yesterday's Radicals: A Study of the Affinity between Unitarianism and Broad Church Anglicanism in the Nineteenth Century*. Cambridge: James Clarke, 1971.

Wilbur, Earl Morse. *Three Prophets of American Liberalism: Channing, Emerson, Parker*. Boston: Beacon Press, 1961.

Wiles, Maurice. *Archetypal Heresy: Arianism through the Centuries*. Oxford: Clarendon Press, 1996.

Willey, Basil. *The Eighteenth-Century Background: Studies on the Idea of Nature in the Thought of the Period*. London: Chatto and Windus, 1950.

Willey, Basil. *More Nineteenth Century Studies: A Group of Honest Doubters*. New York: Harper Torchbooks, 1956.

Willey, Basil. *Nineteenth Century Studies*. New York: Columbia University Press, 1949.

Williams, Daniel Day. *The Andover Liberals: A Study in American Theology*. New York: King's Crown Press, 1941.

Williams, Glanmor. *Reformation Views of Church History*. London: Lutterworth Press, 1970.

Williams, L. Pearce. "A History of Science." *British Journal of the Philosophy of Science*, vol. 11 (1960), 159–61.

Williams, Richard N., and Daniel N. Robinson, eds. *Scientism: The New Orthodoxy*. London: Bloomsbury Academic, 2015.

Wilson, Edwin H. *The Genesis of a Humanist Manifesto*. Washington, DC: Humanist Press, 1995.

Wilson, Leonard G., ed. *Benjamin Silliman and His Circle: The Influence of Benjamin Silliman on Science in America*. New York: Science History Publications, 1979.

Wilson, S. Gordon. *The University of London and Its Colleges*. London: University Tutorial Press, 1923.

Wind, James P. *The Bible and the University: The Messianic Vision of William Rainey Harper*. Atlanta: Scholars Press, 1987.

Wolfe, Gerard R. *The House of Appleton: The History of a Publishing House and Its Relationship to the Cultural, Social, and Political Events that Helped Shape the Destiny of New York City*. Metuchen, NJ: Scarecrow Press, 1981.

Wolffe, John. *God and Greater Britain: Religion and National Life in Britain and Ireland, 1843–1945*. London: Routledge, 1994.

Wolffe, John. *The Protestant Crusade in Great Britain, 1829–1860*. New York: Oxford University Press, 1991.

Womersley, David, John Burrow, and John Pocock, eds. *Edward Gibbon: Bicentenary Essays*. Oxford: Voltaire Foundation, 1997.

Wood, Paul., ed. *Science and Dissent in England, 1688–1945*. Aldershot: Ashgate, 2004.

Woodfield, Malcolm. *R. H. Hutton, Critic and Theologian: The Writings of R. H. Hutton on Newman, Arnold, Tennyson, Wordsworth, and George Eliot*. Oxford: Clarendon Press, 1986.

Woodhead, Linda, ed. *Reinventing Christianity: Nineteenth-Century Contexts.* Aldershot: Ashgate, 2001.

Wyhe, John van. *Phrenology and the Origins of Victorian Scientific Naturalism.* Adlershot: Ashgate, 2004.

Yasukata, Toshimasa. *Lessing's Philosophy of Religion and the German Enlightenment.* New York: Oxford University Press, 2002.

Yeo, Richard. *Defining Science: William Whewell, Natural Knowledge, and Public Debate in Early Victorian Britain.* Cambridge: Cambridge University Press, 1993.

Yerxa, Donald A., *Recent Themes in the History of Science and Religion.* Columbia: University of South Carolina Press, 2009.

Yerxa, Donald A., ed. *Religion and Innovation: Antagonists or Partners?* London: Bloomsbury, 2016.

Young, B. W. "John Jortin, Ecclesiastical History, and the Christian Republic of Letters." *Historical Journal*, vol. 55, no. 4 (2012), 961–81.

Young, B. W. *Religion and Enlightenment in Eighteenth-Century England: Theological Debate from Locke to Burke.* Oxford: Clarendon Press, 1998.

Young, B. W. "'Scepticism in Excess': Gibbon and Eighteenth-Century Christianity." *The Historical Journal*, vol. 40, no. 1 (1998), 179–99.

Young, George Malcolm. *Victorian England: Portrait of an Age.* London: Oxford University Press, 1953.

Young, Robert M. *Darwin's Metaphor: Nature's Place in Victorian Culture.* Cambridge: Cambridge University Press, 1985.

Zachhuber, Johannes. *Theology as Science in Nineteenth-Century Germany: F. C. Baur to Ernst Troeltsch.* Oxford: Oxford University Press, 2013.

INDEX